Decomposition of Graphs

Juraj Bosák

Formerly:

*Mathematical Institute of the Slovak Academy of Sciences,
Bratislava, Czechoslovakia*

KLUWER ACADEMIC PUBLISHERS
DORDRECHT / BOSTON / LONDON

ISBN 0-7923-0747-X
ISBN 80-224-0083-1 (Veda)

Scientific Editor
Prof. Ernest Jucovič, DrSc.

Distributors for the U.S.A. and Canada
Kluwer Academic Publishers,
101 Philip Drive, Norwell, MA 02061, U.S.A.

MATH

Distributors for East European countries, Democratic People's Republic of Korea, People's Republic of China, People's Republic of Mongolia, Republic of Cuba, Socialist Republic of Vietnam
Veda, Publishing House of the Slovak Academy of Sciences,
814 30 Bratislava, Klemensova 19, Czechoslovakia

Distributors for all remaining countries
Kluwer Academic Publishers Group,
P.O. Box 322, 3300 AH Dordrecht, The Netherlands

Original title: Rozklady grafov
First edition published in 1986 by Veda, Bratislava
First English edition published in 1990 by Veda, Bratislava
in co-edition with Kluwer Academic Publishers, Dordrecht, The Netherlands.

© 1990 by *Juraj Bosák*
Translation © *Jozef Širáň and Martin Škoviera*

Printed in Czechoslovakia

Recenzný výtlačok

Decompositions of Graphs

Mathematics and Its Applications (*East European Series*)

Managing Editor:

M. HAZEWINKEL
Centre for Mathematics and Computer Science, Amsterdam, The Netherlands

Volume 47

SERIES EDITOR'S PREFACE

'Et moi, ..., si j'avait su comment en re-
venir, je n'y serais point allé.'

Jules Verne

The series is divergent; therefore we may
be able to do something with it.

O. Heaviside

One service mathematics has rendered the
human race. It has put common sense back
where it belongs, on the topmost shelf next
to the dusty canister labelled 'discarded
nonsense'.

Eric T. Bell

Mathematics is a tool for thought. A highly necessary tool in a world where
both feedback and non-linearities abound. Similarly, all kinds of parts of
mathematics serve as tools for other parts and for other sciences.

Applying a simple rewriting rule to the quote on the right above one finds
such statements as: 'One service topology has rendered mathematical
physics ...'; 'One service logic has rendered computer science ...'; 'One
service category theory has rendered mathematics ...'. All arguably true. And
all statements obtainable this way form part of the raison d'être of this series.

This series, *Mathematics and Its Applications*, started in 1977. Now that
over one hundred volumes have appeared it seems opportune to reexamine
its scope. At the time I wrote

"Growing specialization and diversification have brought a host
of monographs and textbooks on increasingly specialized topics.
However, the 'tree' of knowledge of mathematics and related
fields does not grow only by putting forth new branches. It also
happens, quite often in fact, that branches which were thought to
be completely disparate are suddenly seen to be related. Further,
the kind and level of sophistication of mathematics applied in
various sciences has changed drastically in recent years: measure
theory is used (non-trivially) in regional and theoretical econom-
ics; algebraic geometry interacts with physics; the Minkowsky
lemma, coding theory and the structure of water meet one another in packing and covering theory; quantum fields, crystal
defects and mathematical programming profit from homotopy
theory; Lie algebras are relevant to filtering; and prediction and
electrical engineering can use Stein spaces. And in addition to this
there are such new emerging subdisciplines as 'experimental
mathematics', 'CFD', 'completely integrable systems', 'chaos,
synergetics and large-scale order' which are almost impossible to
fit into the existing classification schemes. They draw upon widely
different sections of mathematics."

By and large, all this still applies today. It is still true that at first sight mathematics seems rather fragmented and that to find, see, and exploit the deeper underlying interrelations more effort is needed and so are books that can help mathematicians and scientists do so. Accordingly MIA will continue to try to make such books available.

If anything, the description I gave in 1977 is now an understatement. To the examples of interaction areas one should add string theory where Riemann surfaces, algebraic geometry, modular functions, knots, quantum field theory, Kac-Moody algebras, monstrous moonshine (and more) all come together. And to the examples of things which can be usefully applied let me add the topic 'finite geometry'; a combination of words which sounds like it might not even exist, let alone be applicable. And yet it is being applied: to statistics via designs, to radar/sonar detection arrays (via finite projective planes), and to bus connections of VLSI chips (via difference sets). There seems to be no part of (so-called pure) mathematics that is not in immediate danger of being applied. And, accordingly, the applied mathematician needs to be aware of much more. Besides analysis and numerics, the traditional workhorses, he may need all kinds of combinatorics, algebra, probability, and so on.

In addition, the applied scientist needs to cope increasingly with the non-linear world and the extra mathematical sophistication that this requires. For that is where the rewards are. Linear models are honest and a bit sad and depressing: proportional efforts and results. It is in the non-linear world that infinitesimal inputs may result in macroscopic outputs (or vice versa). To appreciate what I am hinting at: if electronics were linear we would have no fun with transistors and computers; we would have no TV; in fact you would not be reading these lines.

There is also no safety in ignoring such outlandish things as non-standard analysis, superspace and anticommuting integration, p-adic and ultrametrics space. All three have applications in both electrical engineering and physics. Once, complex numbers were equally outlandish, but they frequently proved the shortest path between 'real' results. Similarly, the first two topics named have already provided a number of 'wormhole' paths. There is no telling where all this is leading — fortunately.

Thus the original scope of the series, which for various (sound) reasons now comprises five subseries: white (Japan), yellow (China), red (USSR), blue (Eastern Europe), and green (everything else), still applies. It has been enlarged a bit to include books treating of the tools from one subdiscipline which are used in others. Thus the series still aims at books dealing with:

— a central concept which plays an important role in several different mathematical and/or scientific specialization areas;

— new applications of the results and ideas from one area of scientific endeavour into another;
— influences which the results, problems and concepts of one field of enquiry have, and have had, on the development of another.

Graphs seem at first sight to be such simple objects that one doubts both their applicability and the possibility of there being deep and fundamental theorems about them: Two misconceptions. As is by now extremely well known, the subject is extremely rich in both applications (and fundamental ones at that) and in deep results. In addition, there are mysterious and fascinating conjectures and problems. Also the subject has grown so large that monographs about the subject as a whole are becoming near impossible to write (and to publish). Instead one aims at stimulating or comprehensive treatises on well established subfields of graph theory. Indeed already in the early 1970's the yearly volume of published papers enriching the theory was around 400.

One such subfield is the theory of decompositions of graphs, a topic which in itself, besides it obvious fundamental theoretical interest, boasts a host of applications; e.g., as the author notes, to coding theory, design of experiments, crystallography, all kinds of networks, serology, and a variety of location problems.

This is a first systematic and thorough treatment of the topic by a leader in the field.

The shortest path between two truths in the real domain passes through the complex domain.

J. Hadamard

La physique ne nous donne pas seulement l'occasion de résoudre des problèmes ... elle nous fait pressentir la solution.

H. Poincaré

Never lend books, for no one ever returns them; the only books I have in my library are books that other folk have lent me.

Anatole France

The function of an expert is not to be more right than other people, but to be wrong for more sophisticated reasons.

David Butler

Bussum, June 1989

Michiel Hazewinkel

CONTENTS

THE AUTHOR

RNDr. Juraj Bosák, DrSc., was born on April 6th, 1933 in Bratislava. As a high school student in 1952 he was the outright winner of the First Czechoslovak Mathematical Olympiad. He studied mathematics at Comenius University in Bratislava and graduated in 1957; then for two years he worked at the Department of Mathematics, Faculty of Science. Since 1959 he was employed in the Mathematical Institute of the Slovak Academy of Sciences. In 1965 he obtained the degree CSc. and in 1981 the degree DrSc. — the highest scientific degree in Czechoslovakia.

During his first period of scientific research he worked in the field of semigroups. In 1963, J. Bosák was

one of the organizers of the Symposium on Graph Theory in Smolenice (Czechoslovakia), which represented one of the first important meetings of graph theorists in the world. From this time onwards, the major part of his scientific work was devoted to graph theory. His results in this field can be divided into three areas: metric properties of graphs, chromatical problems and Hamiltonian problems. He constructed a convex cubic polytope without Hamiltonian cycle, which is the "smallest" known example and is commonly called Bosák's graph.

He was the author of about 30 scientific articles and his results are cited in more than 200 articles of other mathematicians. Furthermore, there are six books by Bosák written in Slovak.

The present book contains the results of some of his investigations in this field spanning more than 20 years. He intended to write a second part, which would have contained mainly the study of the decompositions of graphs into factors with given diameters.

Unfortunately, his serious illness of the last two years and his early death prevented him from finishing that work. Juraj Bosák died on the night before his 54th birthday.

Štefan Znám

INTRODUCTION

This book is a monograph on decompositions of a graph into subgraphs, that is, on decompositions of the edge set of a graph.

The origins of the study of graph decompositions can be seen in various combinatorial problems, most of which emerged in the 19th century. Among them the best known are Kirkman's problem of 15 strolling schoolgirls, Dudeney's problem of 9 handcuffed prisoners, Euler's problem of 36 army officers, Kirkman's problem of knights, Lucas' dancing rounds problem, and the four colour problem. However, the earliest works in this direction are not explicitly related to graph decompositions. The first papers dealing directly with decompositions of graphs (due to J. Petersen, A. B. Kempe, P. G. Tait, P. J. Heawood, D. König and others) appeared soonly afterwards at the turn of the 19th century. Since that time the interest in graph decompositions has been on increase, and a real upsurge is witnessed after 1950. Nowadays, graph decompositions rank among the most prominent areas of graph theory and combinatorics.

Many combinatorial, algebraic, and other mathematical structures are linked to decompositions of graph, which gives their study a great theoretical importance. On the other hand, results on graph decompositions can be applied in coding theory, design of experiments, X-ray crystallography, radioastronomy, radiolocation, computer and communication networks, serology, and other fields. In spite of this, a treatment of the subject in its full extent has not so far appeared in book form, although there are books considering particular questions of this topic.

The main text consists of 15 chapters. In the first three chapters, the basic concepts of the book are introduced and relationships between decompositions and edge-colourings of a graph (including generalizations) are considered. The next two chapters deal with relations (above all, of Nordhaus —Gaddum type) between particular factors of decomposition, as well as with relations between decompositions of several graphs. Chapters 6—15 consider various questions concerning decompositions of a graph into isomorphic subgraphs.

The bibliography on graph decompositions is enormous; therefore only a part of it could be included in the book. The reader interested in other questions such as special types of decompositions, the Ramsey theory, problems related to edge-chromatic number, etc. can consult Fiorini and Wilson (1977), Graham et al. (1980), and Chung et al. (1981).

Because of a vast amount of various results it turned out to be necessary to omit some proofs, especially the long ones. As regards the choice of the results presented, the author tried his best to keep to the core of the topic under discussion. The reader can consolidate his understanding of the material by trying the exercises placed at the end of each chapter. Their results are summed up at the end of the book; they have the same numeration as the corresponding exercises, but preceded by the letter R.

The book requires no more than a basic knowledge of combinatorics, algebra, number theory, set theory, calculus, geometry, etc. However, it does require certain training in mathematical reasoning, although the exposition tends to be elementary.

It gives the author a great pleasure to express gratitude to his children — Elena, Juraj and Martin — for their help with checking the manuscript and compiling the indexes. I am grateful to Š. Znám and J. Plesník for careful reading of the manuscript and a number of valuable comments which resulted in improvement of the original draft. I am also thankful to E. Tomová and I. Dedíková for completing the bibliography. Finally, thanks are due to E. Jucovič, scientific editor of the present volume, for his readiness of undertaking this task.

Juraj Bosák

LIST OF SYMBOLS

R	set of real numbers
Z	set of integers
$\mathbf{N} = \{1, 2, 3, \ldots\}$	set of positive integers (natural numbers)
$\mathbf{Z}_n$	set of residue classes modulo n
$[a_1, a_2, \ldots, a_l]$	ordered l-tuple
$(a_1, a_2, \ldots, a_l)$	unordered l-tuple
$[a_i \mid i \in L]$	ordered family
$(a_i \mid i \in L)$	unordered family
$\|M\|$	cardinality of M
$a \mid b$	a is a divisor of b
g.c.d. X	greatest comon divisor of a set X
l.c.m. X	least common multiple of a set X
$\lfloor a \rfloor$	(lower) integer part of a
$\lceil a \rceil$	upper integer part of a
$\binom{n}{k}$	binomial coefficient
$\ln n$	natural logarithm of n
$o(f(n))$	function g such that $\lim_{n \to \infty} (g(n)/f(n)) = 0$
$O(f(n))$	function g such that $\lim_{n \to \infty} \sup (g(n)/f(n)) < \infty$
$\mathbf{A}(i, j)$	(i, j)th entry of a matrix $\mathbf{A}$
$\mathbf{A}^{\mathsf{T}}$	transpose to a matrix $\mathbf{A}$
$\det \mathbf{A}$	determinant of $\mathbf{A}$
$M(n)$	maximum number of mutually orthogonal matrices of order n
$N(n)$	maximum number of mutually orthogonal Latin squares of order n
$V = V(G)$	vertex set of a graph (or hypergraph) G; point set of a block design
$v = v(G)$	order of a graph (hypergraph) G; number of points of a block design ($v = \|V\|$)
$E = E(G)$	edge set of a graph (hypergraph) G

$e = e(G)$	size of a graph (hypergraph) $G(e = \|E\|)$
uU	edge emanating from u and terminating at U
$e^-(u, U) = e^-(u, U, G)$	number of undirected edges uU in G
$e^\to(u, U) = e^\to(u, U, G)$	number of directed edges uU in G
$e(u, U) = e(u, U, G)$	total number of edges uU in G
$I = I(G)$	incidence of a graph (hypergraph) G
$O = O(G)$	orientation of a graph G
$\varrho(u, U) = \varrho_G(u, U)$	distance of vertices u and U in G
$\varepsilon(u) = \varepsilon_G(u)$	eccentricity of vertex u in G
$\operatorname{id} u = \operatorname{id}(u, G)$	indegree of vertex u in G
$\operatorname{od} u = \operatorname{od}(u, G)$	outdegree of vertex u in G
$\deg u = \deg(u, G)$	degree of vertex u in G
$\deg G$	average degree of a (finite) graph G
$\delta(G)$	minimum degree of a graph G
$\Delta(G)$	maximum degree of a graph G
$D(G)$	g.c.d. $\{\deg u \| u \in V(G)\}$
$D^+(G)$	g.c.d. $\{\operatorname{id} u \| u \in V(G)\}$
$D^-(G)$	g.c.d. $\{\operatorname{od} u \| u \in V(G)\}$
$\operatorname{rad} G$	radius of a graph G
$\operatorname{diam} G$	diameter of a graph G
$\operatorname{grac} G$	gracefulness of a (finite simple) graph G
$\alpha_0(G)$	vertex-covering number
$\alpha_1(G)$	edge-covering number
$\beta_0(G)$	vertex-independence number of a graph G
$\beta_1(G)$	edge-independence number of a graph G
$\varkappa(G)$	vertex-connectivity of a graph G
$\varkappa'(G)$	edge-connectivity of a graph G
$\varrho_1(G)$	vertex-arboricity of a graph G
$\varrho_n(G)$	vertex n-partition number of G
$\chi_n(G)$	n-clique chromatic number of a graph G
$\chi(G) = \chi_2(G) = \varrho_0(G)$	chromatic number of a graph G
$\chi'(G)$	chromatic index of a graph G
$\chi''(G)$	total chromatic number of a graph G
$a(G)$	achromatic number of a graph G
$a'(G)$	achromatic index of a graph G
$a''(G)$	total achromatic number of a graph G
$\psi(G)$	pseudochromatic number of a graph G
$\psi'(G)$	pseudoachromatic index of a graph G
$\psi''(G)$	total pseudoachromatic number of a graph G
$\iota(G)$	number of isolated vertices of a graph G
$c(G)$	number of central vertices of a graph G
$p(G)$	number of peripheral vertices of G

$h(G)$	number of components of a graph G
nG	n copies of a graph G
G^*	digraph obtained from G by replacing each edge of G by a pair of oppositely directed edges
$L(G)$	line graph of a graph G
$T(G)$	total graph of a graph G
$\bar{G}$	complement of a graph G
$\cup \; \bigcup$	union of sets or graphs
$\vee \; \bigvee$	disjoint union of sets or graphs
$\cap \; \bigcap$	intersection of sets or graphs
$\times \; \times$	direct product of graphs, Cartesian product of sets
$\square \; \square$	Cartesian product of graphs
$\boxtimes \; \boxtimes$	strong product of graphs
$G[H]$	composition of graphs G and H
$\cong$	isomorphism
$G[X]$	subgraph of G induced by a set X
$G\{X\}$	subgraph of G generated by a set X
$G - X$	subgraph of G obtained by deleting elements of a set X
$G - u$	vertex-deleted subgraph of G
$G - w$	edge-deleted subgraph of G
$C(u) = C(u, G)$	component of G containing u
$Q(u) = Q(u, G)$	quasi-component of G containing u
Q_n	n-cube
W_v	wheel of order $v \geq 4$
P_v	undirected path of order v
P_∞	one-way infinite path (undirected)
C_∞	two-way infinite path (undirected)
C_v	undirected cycle of order v
S_v	star of order v
D_v	double star of order v (v even)
L_v	1-regular graph of order v
K_v	complete graph of order v
$^\lambda K_v$	λ-fold complete graph of order v
$K_m(n_1, n_2, \ldots, n_m) =$ $= K(n_1, n_2, \ldots, n_m)$	complete m-partite graph with parts of cardinalities $n_1, n_2, \ldots, n_m$
$K(m \times n) = K_m(n, n, \ldots, n)$	complete m-partite graph with parts of cardinality n
$K_m(n_1, n_2, \ldots, n_m; \lambda_1, \lambda_2)$	(λ_1, λ_2)-fold complete m-partite graph with parts of cardinalities $n_1, n_2, \ldots, n_m$
C_v°	directed cycle of order v

$C_v^{\to}$	quasi-cycle of order v
$P_v^{\to}$	directed path of order v
$K_v^{\to}$	transitive tournament of order v
K_v^*	complete digraph of order v
K_v^{**}	complete directed graph of order v
$K_m^*(n_1, n_2, \ldots, n_m)$	complete m-partite digraph with parts of cardinalities $n_1, n_2, \ldots, n_m$
g_v	number of isomorphism classes of simple graphs of order v
s_v	number of isomorphism classes of self-complementary graphs of order v
s_v^*	number of isomorphism classes of self-complementary digraphs of order v
s_v^{**}	number of isomorphism classes of self-complementary directed graphs of order v
$A(G)$	automorphism group of a graph G
$A_V(G)$	vertex automorphism group of a graph G
$A_E(G)$	edge automorphism group of a graph G
$A_1(R)$	automorphism group of a decomposition R
$A_2(R)$	symmetry group of a decomposition R
$H \leftarrow R$	graph H has a decomposition R
$G \mid H$	graph H has a G-decomposition
$G \parallel H$	graph H has a balanced G-decomposition
b, l	number of blocks; number of subgraphs in a decomposition (packing, covering) of a graph
r	number of blocks of a design (or of subgraphs in a graph decomposition) containing a fixed point (vertex)
k	size of a block; order of a subgraph
$\bar{h}(w)$	value of an edge w in a vertex labelling h
$R(n_1, n_2)$	Ramsey number
$DN(2, k, v)$	smallest number of acyclic tournaments of order k that cover the graph K_v^*
$DD(2, k, v)$	largest number of acyclic tournaments of order k that form a packing of K_v^*

SUBGRAPHS, DECOMPOSITIONS, AND COLOURINGS OF GRAPHS

•

In the first three chapters we shall introduce the basic concepts which we shall be dealing with throughout. We shall also outline the main topic of the book as well as possible generalizations and modifications, including those not studied in detail.

Chapter 1

Graphs, subgraphs, walks, diagrams and hypergraphs

1.1 There are many ways of introducing the basic notions of graph theory, in particular the notion of a graph itself. As we shall be dealing with a variety of graphical structures in our book, the term "graph" will have a quite general meaning here. The reader interested in other possible approaches, as well as in motivations of graph-theoretical concepts, is referred to a great number of monographs, e.g. Ore (1962), Harary (1969), Berge (1970), Behzad and Chartrand (1971), R. J. Wilson (1972), and many others.

1.2 By a *partially directed graph* (or, briefly, by a *graph*) we shall understand an ordered quadruple $G = [V, E, I, O]$, where V and E are disjoint sets, and I and O are binary relations from E into V (i.e. subsets of the Cartesian product $E \times V$) such that for each $w \in R$ we have: $1 \le |I(w)| \le 2, |O(w)| \le 1$, and $O(w) \subseteq I(w)$. (The symbol $|M|$ will denote the cardinality of the set M. For $w \in E$, $I(w)$ will be the set of all $u \in V$ for which $[w, u] \in I$; the symbol $O(w)$ has a similar meaning.)

The sets V, E, $V \cup E$ will be called *vertex-set*, *edge-set*, and *element set* of the graph G, respectively. The elements of the set V [resp. E, $V \cup E$] will be referred to as *vertices* [*edges*, *elements*] of the graph G.

The binary relations I and O will be called *incidence* and *orientation* of the graph G, respectively.

Example. There are a number of graphs on 4 vertices and 9 edges. One of them may be defined as follows:

$G = [V, E, I, O]$,

$V = \{u_1, u_2, u_3, u_4\}$,

$E = \{w_1, w_2, w_3, w_4, w_5, w_6, w_7, w_8, w_9\}$,

$I = \{[w_1, u_1], [w_2, u_1], [w_2, u_2], [w_3, u_1], [w_3, u_2], [w_4, u_1], [w_4, u_2], [w_5, u_2],$

$\qquad [w_5, u_3], [w_6, u_2], [w_6, u_3], [w_7, u_2], [w_7, u_3], [w_8, u_3], [w_9, u_3]\}$,

$O = \{[w_2, u_2], [w_5, u_3], [w_6, u_3], [w_7, u_2], [w_8, u_3], [w_9, u_3]\}$,

where the symbols $u_1, u_2, u_3, u_4, w_1, w_2, w_3, w_4, w_5, w_6, w_7, w_8, w_9$ are mutually

4

distinct elements of some set. It is a matter of routine to check that, according to our definition, G is a graph. We recommend the reader to illustrate all the subsequent notions on this example.

1.3 If the element set of a graph G is finite [infinite, empty, non-empty], the graph G is said to be *finite* [*infinite, empty, non-empty*]. A graph with an empty [non-empty] edge set is called a *null* graph [*non-null* graph]. Of course, a graph with an empty vertex set is automatically the empty graph.

1.4 The cardinality $v\,[e]$ of the set $V\,[E]$ will be called the *order* [*size*] of the graph $G = [V, E, I, O]$. Thus, a null graph has zero size, and the empty graph has also zero order. Both the order and size of a finite graph are finite.

1.5 As regards the binary relations I and O, we shall adopt the following terminology.

If $[w, u] \in I$ we say that the edge w and the vertex u are *incident* in G.

If $I(w) = \{u, U\}$, the edge w is said to *join* vertices u and U in G; vertices u and U are *endvertices* (briefly, *ends*) of the edge w. In the case when $u \neq U$ (i.e. if $|I(w)| = 2$) the edge w is sometimes called a *link*. If $|I(w)| = 1$, w is a *loop*. Vertices u and U are said to be *adjacent* if there is a link joining them. Two edges w and W are called *adjacent* if $w \neq W$ and there is a vertex incident to both w and W. A vertex is said to be a *null* vertex [*isolated* vertex] if it is incident to no edge [no link] of the graph in question. Clearly, each null vertex is isolated.

Suppose that the edge w joins vertices u and U in G. In the case when $O(w) \subseteq \{U\}$, the edge w is said to *emanate* from the vertex u and *terminate* at the vertex U in G; such an edge will often be denoted by $w = uU$. If $O(w) = \{U\}$ the edge w is said to be *directed* from u to U; vertices u and U are then called *initial* and *terminal* vertices, respectively, and the ordered pair $[u, U]$ is called an *orientation* of the edge w. In particular, if $u = U$, the initial and terminal vertices of a directed loop coincide (and hence a loop has only one possible orientation). Thus, a directed edge w is characterized by the relation $|O(w)| = 1$. If $|O(w)| = 0$ the edge w is *undirected*.

A graph is *undirected* [*directed*] if all its edges are undirected [directed]. A graph having both undirected and directed edges is called *mixed*.

1.6 Two edges w, W of a graph $G = [V, E, I, O]$ are said to be *parallel* if $I(w) = I(W)$. In addition, if w and W are directed parallel edges, we distinguish between *consistently* and *inconsistently parallel* edges according to whether $O(w) = O(W)$ or not. Note that directed parallel loops are always consistently parallel. Inconsistently parallel edges will often be called *oppositely directed*, as the initial point of one of the two edges is the terminal point of the other.

1.7 An edge w of a graph G is called an *n-fold* edge if $n = |\{W \in E;\ I(w) = I(W)\ \text{and}\ O(w) = O(W)\}|$. The number n is then called the *multiplicity* of the edge w; w is a *multiple* edge if $n \geq 2$.

A graph having at least one *n*-fold edge, $n \geq 2$ will be said to be a *multigraph*. Graphs in which the *multiplicity* of each edge does not exceed λ will be referred to as *λ-fold* graphs.

1.8 A *loopless* graph is a graph without loops; graphs with loops are *pseudographs*. Most of the papers in graph theory have been dealing with undirected loopless graphs without multiple edges; these will be referred to as *simple graphs*. Directed loopless graphs without multiple edges will be called *digraphs*.

1.9 The *degree* of a vertex *u* of *G*, denoted by deg *u*, is the number of edges incident to *u* in *G*, whereby each loop is to be counted twice. A vertex *u* with deg *u* = 0 [1] is called a *null vertex* [*end-vertex*, or *pendant vertex*]. An edge incident with a pendant vertex is called a *pendant edge*. An infinite graph in which all vertices are of finite degree is said to be *locally finite*.

By the *outdegree* of a vertex *u* in *G* we understand the number od *u* of edges emanating from *u*; the *indegree* id *u* is defined analogously by means of edges terminating at *u*. In both cases, each undirected loop is to be counted twice.

Some authors prefer the term *valency* to the term degree. In our book, however, the notion of valency will have a different meaning, as we shall see in 1.11.

1.10 The *maximum degree* $\Delta(G)$ of a *graph G* is the *supremum* of the set of degrees of all vertices of *G*; if *G* is a non-empty finite graph then $\Delta(G)$ is simply the largest degree of a vertex in *G*. The *minimum degree* $\delta(G)$ is defined analogously. Morever, for a non-empty finite graph *G* of order *v* whose degrees of vertices are $n_1, n_2, \ldots, n_v$ we define the average degree as

$$\deg G = (n_1 + n_2 + \ldots + n_v)/v \tag{1}$$

and the *greatest common divisor of degrees of vertices,*

$$D(G) = \text{g.c.d.} \{n_1, n_2, \ldots, n_v\}. \tag{2}$$

If *G* is empty we put

$$\Delta(G) = \delta(G) = \deg G = D(G) = 0. \tag{3}$$

Clearly, for each finite graph *G* it holds

$$\delta(G) \leq \deg G \leq \Delta(G) \tag{4}$$

and

$$D(G) \leq \Delta(G). \tag{5}$$

For a graph without null vertices we even have

$$D(G) \leq \delta(G) \tag{6}$$

1.11 If $\deg u = n$ for each vertex u of G, the graph G is said to be *regular* of *degree n*, or shortly *n-regular*. For $n = 0$ we thus get a *null graph* (compare with 1.3). Three-regular graphs are known as *cubic graphs*.

Similarly, if each vertex of G has the same indegree and outdegree, both equal to n (id $u = $ od $u = n$ for each $u \in V$), then G is said to be *diregular of valency n*, or *n-diregular*.

1.12 A *graph* G will be called *balanced* if id $u = $ od u for each vertex u of G. Note that, according to our definition, all undirected graphs are considered to be balanced. The similar is true for diregular graphs.

1.13 A simple graph is *complete* if any two of its vertices are adjacent. The complete graph of order v will be denoted by K_v. Obviously, K_0 is the empty graph (1.3); the graph K_1 will be called *trivial*.

1.14 By a *complement* (or a *complementary graph*) of a simple graph $G = [V, E, I, O]$ we understand the simple graph $\bar{G}$ with the same vertex set V in which two vertices are adjacent if and only if they are not adjacent in G. For example, the complement of the complete graph K_v is the null graph of order v, denoted by $\bar{K}_v$.

1.15 Let m be a positive integer and $n_1, n_2, \ldots, n_m$ be cardinal numbers. The *complete m-partite graph* $K_m(n_1, n_2, \ldots, n_m)$ is defined as a simple graph whose vertex set is the disjoint union of sets (so-called *parts*) $V_1, V_2, \ldots, V_m$ such that $|V_i| = n_i$ for $i = 1, 2, \ldots, m$, two vertices being adjacent if and only if they belong to different parts.

For $m = 1$ we clearly obtain the null graph of order n_1. In the case $m = 2$ we speak about *complete bipartite graphs*; if $m = 3$ we get, by analogy, *complete tripartite graphs*. The complete m-partite graph $K_m(n, m, \ldots, n)$ in which the parts are of the same cardinality n is called *equipartite*, and denoted by $K(m \times n)$. Complete m-partite graphs are sometimes called also *complete multipartite graphs*.

1.16 For each cardinal number λ and each graph G we define the graph $^\lambda G$ by replacing each edge of G by λ edges with the same endvertices (and the same orientation when replacing directed edges).

We shall often be dealing with the graph $^\lambda K_v$, the so-called *λ-fold complete graph* of order v.

1.17 Let m be a positive integer and $n_1, n_2, \ldots, n_m, \lambda_1, \lambda_2$ be cardinal numbers. The *complete m-partite (λ_1, λ_2)-graph* $K_m(n_1, n_2, \ldots, n_m; \lambda_1, \lambda_2)$ (the subscript m will sometimes be omitted) is defined as a graph whose vertex set is the disjoint union of sets $V_1, V_2, \ldots, V_m$ with $|V_i| = n_i$ for $i = 1, 2, \ldots, m$, two distinct vertices being joined with $\lambda_1 [\lambda_2]$ undirected edges if and only if they belong to the same part (to different parts) V_i. In particular, if $\lambda_1 = 0$ and $\lambda_2 = 1$ we obtain a *complete m-partite graph*, and $\lambda_1 = 0$, $\lambda_2 = \lambda$ — a *λ-fold complete m-partite graph*.

1.18 Let G be an arbitrary graph. Denote by G^* the directed graph obtained by replacing each undirected link [undirected loop] by a pair of oppositely directed parallel edges [by a directed loop] with the same ends.

The graph $^\lambda K_v^*$ is called *λ-fold complete digraph* of order v, and the graph $^\lambda K_m^*(n_1, n_2, ..., n_m)$ — the *λ-fold complete m-partite digraph* with *parts* of *cardinalities* $n_1, n_2, ..., n_m$. Attaching λ-directed loops to each vertex of the graph $^\lambda K_v^*$ we obtain the so-called *λ-fold complete directed graph* $^\lambda K_v^{**}$ of order v. In the case $\lambda = 1$ we shortly speak about *complete digraph*, *complete m-partite digraph*, etc.

1.19 Let G be an undirected graph. The *line graph* $L(G)$ of the graph G is defined as follows: the vertex set of $L(G)$ is equal to the edge set of G, and two vertices of $L(G)$ are adjacent if and only if they are adjacent as edges in G. The *total graph* $T(G)$ of G is the simple graph whose vertex set is formed by the set of elements of G; two vertices are ajacent in $T(G)$ if they are adjacent or incident when considered as elements of G.

1.20 For each positive integer n the complete bipartite graph $S_n = K(1, n - 1)$ will be called a *star of order n*. By a *double-star D_{2n}* of order $2n$ we shall understand the simple graph of order $2n$ having two adjacent vertices of degree n such that each is adjacent to further $n - 1$ pendant vertices. The *n-cube Q_n* is the graph of order 2^n whose vertex set consists of all ordered n-tuples of 0's and 1's, two vertices being adjacent if and only if, as n-tuples, they differ in exactly one coordinate.

1.21 Assigning a definite orientation to each edge of a complete graph we obtain a digraph called *tournament*. A tournament $K_v^\rightarrow$ of order v is said to be *transitive* if there exists a linear ordering of its vertex set V such that for each u, $U \in V$, there is a directed edge uU in $K_v^\rightarrow$ if and only if $u < U$.

1.22 For any two vertices u, U of a graph G let us denote by $e^-(u, U)$ $[e^\rightarrow(u, U)]$ the cardinality of the set of all undirected [directed] edges emanating from u and terminating at U. If u and U are adjacent then the above two numbers determine the multiplicity of the corresponding edge uU. A graph will be said to be *symmetric* if $e^\rightarrow(u, U) = e^\rightarrow(U, u)$ for any two its vertices u, U.

1.23 If it is necessary to point out that the symbols V, v, E, e, I, O, $I(w)$, $O(w)$, $\deg u$, $\mathrm{id}\, u$, $\mathrm{od}\, u$, $e^-(u, U)$ and $e^\rightarrow(u, U)$ are related to a particular graph G, we shall employ the extended notation: $V(G)$, $v(G)$, $E(G)$, $e(G)$, $I(G)$, $O(G)$, $I(w, G)$, $O(w, G)$, $\deg(u, G)$, $\mathrm{id}(u, G)$, $\mathrm{od}(u, G)$, $e^-(u, U, G)$, $e^\rightarrow(u, U, G)$.

1.24 Let $G = (V, E, I, O)$, $G' = (V', E', I', O')$ be graphs. A 1-1 mapping f of the element set of G onto the set of elements of G' is said to be an

8

isomorphism of graphs G and G' if both f and f^{-1} preserve the vertex-set, the edge-set, the incidence and the orientation of graphs, i.e.

$$f(V) = V',\tag{1}$$

$$f(E) = E',\tag{2}$$

$$[w, u] \in I \Leftrightarrow [f(w), f(u)] \in I' \qquad \text{for each } w \in E,\ u \in V,\tag{3}$$

$$[w, u] \in O \Leftrightarrow [f(w), f(u)] \in O' \qquad \text{for each } w \in E,\ u \in V.\tag{4}$$

1.25 A bijective mapping g of the vertex set V of a graph G onto the vertex set V' of a graph G' is called a *vertex isomorphism* of G and G', if g preserves the cardinalities e^- and $e^{\to}$, that is, if for any u and U in V it holds:

$$e^-(u,\ U,\ G) = e^-(g(u),\ g(U),\ G'),$$

$$e^{\to}(u,\ U,\ G) = e^{\to}(g(u),\ g(U),\ G').$$

1.26 An isomorphism [vertex isomorphism] of G onto itself is called an *automorphism* [*vertex automorphism*] of G.

1.27 Every isomorphism f of G and G' determines a vertex isomorphism of G and G': it suffices to take the restriction f_V of the mapping f to the vertex-set V of G. Conversely, every vertex isomorphism of G and G' extends to an isomorphism of G and G' (moreover, for simple graphs this extension is unique). Thus we have:

Lemma. There exists an isomorphism of G and G' if and only if there exists a vertex isomorphism of G and G'.

1.28 Our considerations imply that there is no need to distiguish between the existence of an isomorphism and vertex isomorphism. Often it is not necessary to distinguish even between isomorphisms and vertex isomorphisms themselves, and we usually adopt this convention.

Graphs G ad G' are said to be *isomorphic* if there exists an isomorphism (or vertex isomorphism) of G and G'; in this case we write $G \cong G'$. In some cases we adopt another convetion according to which isomorphic graphs are not distinguished. The other possibility is to replace graphs by *isomorphism classes* (*types*), i.e. classes consisting of all graphs isomorphic to a given graph.

1.29 An *edge isomorphism* of G and G' is a bijection h of $E(G)$ onto $E(G')$ preserving loops, links, directed and undirected edges, and for arbitrary two edges the number of their common endvertices, initial vertices and terminal vertices. If $G = G'$ then h is called an *edge automorphism* of G.

Clearly, the restriction f_E of an isomorphism f of G and G' to the edge-set $E = E(G)$ is always an edge isomorphism. However, not every edge isomorphism extends to an isomorphism. For example, there exists an edge isomorphism of K_3 and S_4 but not an isomorphism.

In case that there exists an edge isomorphism of G and G' we say that G and G' are *edge-isomorphic*.

1.30 Every automorphism [vertex isomorphism, resp. edge isomorphism] of a graph G is a permutation of the set $V(G) \cup E(G)$ [$V(G)$, resp. $E(G)$]. (The term *permutation* will be used for a bijection of a set onto itself even if the set is infinite.) It is easily seen that all automorphisms [vertex automorphisms, edge automorphisms] form a group under the composition of mappings (permutations). This permutation group acts on the set of elements [on the *vertex-set*, resp. *edge-set*] of G and is called the *automorphism* [vertex automorphism, edge automorphism] group of G. We denote it by $A(G)$ [$A_V(G)$, resp. $A_E(G)$].

1.31 The graph $G' = [V', E', I', O']$ is a *subgraph* of a graph $G = [V, E, I, O]$ if the following hold:
  A. $V' \subseteq V, E' \subseteq E, I' \subseteq I, O' \subseteq O$.
  B. For each $w \in E'$ and $u \in V$ it holds that

$$[w, u] \in I \Rightarrow [w, u] \in I', \tag{1}$$

$$[w, u] \in O \Rightarrow [w, u] \in O'. \tag{2}$$

Obviously, every subgraph G' is uniquely determined by the graph G and the sets V' and E'.

For the relation "to be a subgraph of" we borrow the notation of the set inclusion. Thus we write $G' \subseteq G$ if G' is a subgraph of G. If, moreover, $G' \neq G$ then we shall say that G' is a *proper subgraph* of G and write $G' \subset G$. We note that the relation $\subseteq$ is a (partial) order on the set of subgraphs of a graph.

1.32 Subgraphs $G_1, G_2, \ldots$ of a graph G are called *disjoint* [*edge-disjoint*] if no two among $G_1, G_2, \ldots$ have an element [edge] in common.

The *union* [*intersection*] of *subgraphs* $G_1, G_2, \ldots$ of G is the subgraph G' of G whose vertex-set is the union [intersection] of the vertex-sets of $G_1, G_2, \ldots$ and the similar holds for the edge set, incidence and orientation. The union of subgraphs is said to be *disjoint* [*edge-disjoint*] if the subgraphs $G_1, G_2, \ldots$ are disjoint [edge-disjoint]. To denote the union [disjoint union, resp. intersection] we shall use the symbol $\cup$ [$\vee$, $\cap$] and, in the case of sets or families of subgraphs, the symbol $\bigcup$ [$\bigvee$, $\bigcap$]. We often extend this notation to graphs isomorphic with the graphs $G_1, G_2, \ldots$. For instance, if for graphs $G, G_1, G_2, G', G_1', G_2'$ it holds that $G \cong G_1 \vee G_2$, $G' \cong G$, $G_1' \cong G_1$, $G_2' \cong G_2$ then G' will also be called the disjoint union of G_1 and G_2 or G_1' and G_2'. In this case the notation $G' = G_1' \vee G_2'$ will be used. If $G_1 = G_2 = \ldots = G_n$, $n \in N$ then the disjoint union of $G_1, G_2, \ldots, G_n$ will be denoted by nG.

1.33 Let a graph $G = [V, E, I, O]$ and a subset $X \subseteq V \cup E$ be given. Then there exists the smallest subgraph $G' = [V', E', I', O']$ of G which contains all elements of X (obviously, G' is just the intersection of all subgraphs $G'' = [V'', E'', I'', O'']$ for which $X \subseteq V'' \cup E''$). This graph G' is called the *subgraph of G generated by the set X* and denoted by $G' = G\{X\}$. Further there exists the largest subgraph $G_0 = [V_0, E_0, I_0, O_0]$ of G with vertex set $V_0 = V'$. It is called the *subgraph of G induced by the set X* and denoted by $G_0 = G[X]$. The vertices of G_0 are all the vertices in X plus all endvertices of the edges in X. The edges of G_0 are all edges of X whose endpoints are in G_0. The term *induced subgraph of a graph* is often used without explicitly mentioning the set X; it is easily seen that the set X can always be chosen so as to contain vertices only.

If $X \subseteq V$ [resp. $X \subseteq E$] we set $G - X = G[V - X]$ [resp. $G - X = G\{V \cup (E - X)\}$] and say that $G - X$ is obtained by *deleting* or *omitting the elements of X*. In particular, if X consists of a single vertex u [edge w] then we write $G - u [G - w]$ instead of $G - X$.

1.34 By a *factor of a graph G* we mean any subgraph G' of G with the same vertex-set as G, i.e., $V(G') = V(G)$. A factor F of G is uniquely determined by G and the set $E(F)$. For a fixed graph G it is sometimes convenient to make no distinction between F and $E(F)$. We adopt this convention occasionally by describing factors in the form of their edge sets.

A regular factor of degree n (n-regular factor) is also called an *n-factor*. For $n = 0$ and $n = 1$ the terms *null factor* and *linear factor* are sometimes used.

1.35 A *semiwalk from a vertex u to a vertex U* (or just *uU-semiwalk*) is defined as a finite sequence $[u_0, w_1, u_1, w_2, u_2, ..., u_{n-1}, w_n, u_n]$, where $n \geq 0$, $u_0 = u, u_1, u_2, ..., u_n = U$ are vertices of G, w_i (for $i = 1, 2, ..., n$) edges joining u_{i-1} to u_i. The vertex u [resp. U] is called the *initial vertex* [*terminal vertex*] of the *semiwalk*; the number n is the *length of the semiwalk*. A common term for both initial and terminal vertex of a semiwalk is *end-vertex* or simply *end*. If $u = U$ then the semiwalk is *closed*, otherwise it is *open*. We also say that an uU-semiwalk *connects* u and U. A semiwalk is *undirected* [*directed*] if all its edges are undirected [directed] and is *mixed* if it is neither directed nor undirected.

1.36 The notation $[u_0, w_1, u_1, w_2, u_2, ..., u_{n-1}, w_n, u_n]$ is quite awkward so it is often simplified, for instance, $u_0 w_1 u_1 w_2 u_2 ... u_{n-1} w_n u_n$ or $u_0 u_1 u_2 ... u_{n-1} u_n$ and $w_1 w_2 ... w_n$.

Given two semiwalks $S = u_0 w_1 u_1 ... u_{n-1} w_n u_n$ and $S' = u'_0 w'_1 u'_1 ... u'_{m-1} \cdot w'_m u'_m$ with $u_n = u'_0$ we can form their composition $SS' = u_0 w_1 u_1 ... u_n w'_1 u'_1 ... ... u'_{m-1} w'_m u'_m$ which is a $u_0 u'_m$-semiwalk of length $n + m$.

1.37 Sometimes also *infinite semiwalks* are considered. There are two kinds of them. A *one-way infinite semiwalk with initial vertex* u_0 is an infinite sequence $[u_0, w_1, u_1, ..., u_{n-1}, w_n, u_n, ...]$, where w_i $(i = 1, 2, ...)$ is an edge joining u_{i-1} to u_i. A *two-way infinite semiwalk* is a two-way infinite sequence $[..., u_{-1}, w_0, u_0, w_1, u_1, w_2, u_2, ...]$ in which the condition on w_i is satisfied for all $i \in Z$. The simplified notation of 1.36 can be used for infinite semiwalks as well. The composition can be defined for a finite and one-way infinite semipath, or for two one-way infinite semipaths with the same initial vertex (after reversing the order of the first of them).

1.38 The *subgraph of a graph G generated by a semiwalk S* is defined to be the subgraph generated by its elements, that is, formed by the vertices and edges of S.

1.39 *Two vertices u and U of a graph G are said* to be *connected in G if G* contains a *uU-semiwalk*. The subgraph of G generated by all semiwalks with initial vertex u is called the *component of G containing u* and denoted by $C(u) = C(u, G)$. A graph G is said to be *connected* if any two vertices of G are connected, that is, if it has exactly one or no component (in case it is empty). A graph consisting of at least two components is called *disconnected*.

1.40 A set X of elements of a graph $G = [V, E, I, O]$ (that is, $X \subseteq V \cup E$) is called a *cut-set* (or simply a *cut*) *separating in G two elements x and y of G* if the following hold:

A. Every semiwalk containing both x and y has at least one element of X in between.

B. Neither x nor y belongs to X.

C. No proper subset of X satisfies both A and B.

A cut-set composed exclusively of vertices [edges] is called a *vertex cut-set* [*edge cut-set*]. A vertex [edge] which alone forms a cut-set is called a *cut-vertex* [*cut-edge* or *bridge*] of G. A connected subgraph with at least one edge but no cut-vertex is called a *block*. Every maximal (with respect to $\subseteq$) subgraph which is a block is a *block of the graph* in question.

1.41 A semiwalk $u_0 w_1 u_1 w_2 u_2 ... u_{n-1} w_n u_n$ is called a *walk from u to U*, or briefly a *uU-walk*, if $u_0 = u$, $U = u_n$ and for every $i = 1, 2, ..., n$ the edge w_i is either directed and emanates from u_{i-1} and terminates at u_i or it is undirected and joins u_{i-1} to u_i. A semiwalk of length 0 is also considered as a walk.

1.42 We say that two vertices u and U are *strongly connected* in a graph G if there exist in G both a uU-walk and Uu-walk. The *quasi-component of G containing a vertex u* is a subgraph $Q(u) = Q(u, G)$ of G induced by the set of all vertices of G which are strongly connected with u. A graph G is said to be *strongly connected* if any two its vertices are strongly connected, that is, if G has either one or no component (in case G is empty). If in the definition of cut (1.40) the semiwalk in the condition A is replaced by a walk then a *quasi-cut* is obtained.

1.43 If a non-empty set of semiwalks is given (for instance, the set of all

uU-semiwalks for fixed u and U) then a *shortest* [*longest*] *semiwalk of the set* is one with the minimum [maximum] length. Note that a longest path need not exist even if the set of all semiwalks is non-empty (and even in finite graphs).

The *distance* of two vertices u and U in a graph G is the length $\varrho(u, U) =$ $= \varrho_G(u, U)$ of a shortest uU-walk in G. If there is no uU-walk then we set $\varrho(u, U) = \infty$.

In a non-empty undirected connected graph $G = [V, E, I, O]$ the following holds for arbitrary $u, u', u'' \in V$:

A. $\varrho(u, u') \geq 0$.

B. $\varrho(u, u') = 0 \Leftrightarrow u = u'$.

C. $\varrho(u, u') = (u', u)$.

D. $\varrho(u, u') \leq \varrho(u, u') + \varrho(u', u'')$.

Thus $[V, \varrho]$ is a metric space and the function ϱ is called the *metric of G*.

1.44 A semiwalk [walk] is called a *semitrail* [*trail*] if no edge occurs more than once in it. A semiwalk [walk] is called a *semipath* [*path*] if no vertex (and hence no edge) occurs more than once in it.

A closed semitrail [trail] of non-zero length in which no vertex, except the initial and terminal vertex, occurs more than once is called a *semicycle* [*cycle*].

The terms *semipath*, *path*, *semicycle* and *cycle* will usually be applied also to the subgraph which they generate (1.38). The exact sense will always be clear from the context.

The undirected path [undirected cycle] of order v will be denoted by P_v [resp. C_v]. One-way [two-way] infinite undirected path will be denoted by P_∞ [resp. C_∞].

A semicycle (in particular, cycle) is said to be *even* [*odd*] if its length is even [odd].

1.45 A graph without cycles is *acyclic*. A graph is said to be *transitive* if with any two edges uu' and $u'u''$ it contains the edge uu'' (cf. 1.5). In the case of tournaments both notions coincide, which is in agreement with the terminology introduced in 1.21. The orientation of an acyclic [transitive] graph is called *acyclic* [*transitive*].

1.46 If G is an undirected graph then $G^\rightarrow$ will stand for a directed graph obtained from G by means of a certain acyclic orientation. In 1.21 we introduced the notation $K_v^\rightarrow$ for an acyclic tournament, and this complies with our present notation. By $P_n^\rightarrow$ we shall denote the directed path of order v and by $C_v^\rightarrow$ we shall denote the quasi-cycle of order v, which is the graph obtained from the directed cycle of order v by reversing the orientation of one edge. The directed cycle of order v will be denoted by C_n°.

1.47 A semicycle of order v will usually be called *v-gon* and, if v is not explicit, a *polygon*. For small v the terms *bigon*, *triangle*, *quadrilateral*, *pentagon*, *hexagon*, ... can be used.

1.48 A *semiwalk* is called *Hamiltonian* [*Eulerian*] (in honour of W. R. Hamilton [L. Euler]) if it contains all vertices [elements] of the graph in question. A *graph* which has a Hamiltonian cycle is called *Hamiltonian*. A graph which has a closer Eulerian trail is called *Eulerian*.

1.49 A graph containing no odd semicycles is said to be *bipartite*. A special case of a bipartite graph is the complete bipartite graph (1.15). Another special case is a graph which has no semicycles at all; such a graph is called a *forest*. A connected non-empty forest is a *tree*. Thus every component of a forest is a tree. A graph of a finite order $v \geq 3$ is said to be *pancyclic* if it has a cycle of each length 3, 4, ..., v. Obviously, every pancyclic graph is Hamiltonian.

1.50 A *quasi-plane graph* is one whose vertex set is a certain subset of the plane and whose edges are either oriented or non-oriented curves connecting the corresponding vertices and avoiding any other vertex; links are simple arcs and loops are simple closed curves. Incidence and orientation are defined in the natural way.

A quasi-plane graph is called a *plane graph* if its edges do not intersect in inner points and if its vertex set has no accumulation point.

A graph isomorphic to a plane graph is called *planar*. If a graph G is isomorphic to a quasiplanar graph H then H is called a *diagram of the graph* G (in fact, by a diagram we undestand a figure which corresponds to H, the orientation being represented by an arrow).

1.51 A *hypergraph* is an ordered triple $[V, E, I]$, where V and E are disjoint sets and $I \subseteq E \times V$. Hypergraphs can be considered as a generalization of graphs. The concept of a vertex, edge, order, size, incidence, walk, distance of vertices, connected hypergraph, isomorphism, multiple edges, λ-fold hypergraph and many other concepts can be defined analogously as we did for (undirected) graphs (1.2—1.43).

Let t be a cardinal number. A hypergraph is said to be *t-uniform* if each edge of H is incident with exactly t vertices. A 2-uniform hypergraph is essentially the same as a loopless undirected graph.

1.52 By the *complete hypergraph* we understand the hypergraph $[V, E, I]$ where $E = 2^V$ (the set of all subsets of V) and I consists of all pairs w, u, where $w \in E$, $u \in V$ and $u \in w$.

If in the above definition the set 2^V is replaced by the set of all subsets of V of cardinality t then we get the *complete t-uniform hypergraph*.

Finally, given a positive integer m and cardinal numbers t, n_1, n_2, ..., n_m, the complete *m-partite t-uniform hypergraph with parts of cardinalities* n_1, n_2, ..., n_m is the t-uniform hypergraph whose vertex-set is the disjoint union of sets V_1, V_2, ..., V_m (the *parts of the hypergraph*), where $|V_i| = n_i$ ($i = 1, 2, ..., m$) and the edge-set consists of all subsets of V which have cardinality t and contain at most one vertex from each part (cf. 1.15); the incidence is defined again by the condition $u \in w$.

The terms complete hypergraph, complete t-uniform hypergraph and complete m-partite t-uniform hypergraph are extended also to all hypergraphs isomorphic with them (cf. the convention in 1.28).

Exercises

1.53 Can the condition 1.31.B.2 in the definition of a subgraph be omitted?

1.54 Can the condition 1.31.B.1 in the definition of a subgraph be omitted?

Chapter 2

Decompositions and colourings of a graph

2.1 Let M be a set. By a family $[a_i | i \in L]$ of elements of the set M we understand an arbitrary mapping defined on a set L which assigns to each $i \in L$ an element $a_i \in M$. The set L is then usually called the *index-set* and its elements *indices*. Roughly speaking, to form a family we choose elements of M, some of which may repeat, and we provide them with pairwise distinct indices using the whole set L. If L is finite we shall usually employ the set $\{1, 2, ..., l\}$ as L. In this case we write $[a_1, a_2, ..., a_l]$ instead of $[a_i; i \in \{1, 2, ..., l\}]$.

Note that the definition of a family comprises, in some sense, an order of the elements in the family. Therefore we call $[a_1, a_2, ..., a_l]$ an *ordered l-tuple*; in particular, if $l = 2, 3$ and 4 we call it an *ordered pair, triple* and *quadruple*, respectively. Sometimes, however, the order in a family is immaterial, and in this case we use the notation $(a_i | i \in L)$ and forget about the order. A more precise definition can be given by choosing $(a_i; i \in L)$ to be an equivalence class of families, two families being *equivalent* if $b_i = a_{f(i)}$ for some bijection $f: L \to L$. In particular, if $L = \{1, 2, ..., l\}$ we speak of an *l-tuple* and if $l = 2$, 3 and 4 we use the term *pair, triple* and *quadruple*, respectively.

The elements a_i, $i \in L$, are called *members of a family*.

We can pass on to the definition of a graph decomposition: A decomposition of a graph G is a family $(H_i | i \in L)$ of subgraphs of G such that each edge of G is contained in exactly one member of $(H_i | i \in L)$. In this case we write $G \leftarrow (H_i | i \in L)$ to indicate that $(H_i | i \in L)$ is a decomposition of G. A less

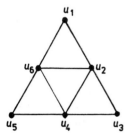
Fig. 2.1.1

16

precise notation $G = \cup (H_i | i \in L)$ can be encountered frequently; it emphasizes the fact that, if G has no null vertices, then G is the edge-disjoint union of its subgraphs H_i.

For example the graph G depicted in Fig. 2.1.1 has a decomposition (H_1, H_2, H_3) into three triangles: $H_1 = u_1 u_2 u_6$ (the triangle with vertices u_1, u_2 and u_6), $H_2 = u_2 u_3 u_4$ and $H_3 = u_4 u_5 u_6$. Obviously, each edge is in exactly one subgraph H_i, $i \in \{1, 2, 3\}$. Hence, $G \leftarrow (H_1, H_2, H_3)$.

2.2 Especially important are two types of decompositions $(H_i | i \in L)$ of a graph G into its subgraphs:

(A) If all subgraphs H_i are non-null (1.11) then a *decomposition into positive* (1.11) *subgraphs* is obtained (as, for instance, the one mentioned above).

(B) If all subgraphs H_i are factors (1.34) then a *decomposition into factors* is obtained.

2.3 There is a close connection between graph decompositions and edge-colourings of graphs. For the sake of completeness we introduce vertex-colourings and total colourings as well.

Given a graph G and a set L, an *edge-colouring* [*vertex-colouring*, resp. *total colouring*] is a mapping of the edge-set [vertex-set, resp. union of the vertex- and edge-set] of G to the set L. In this context the set L is called the *colour-set* and the value of this mapping on an element of G is its *colour*.

2.4 If $l = |L|$ is the cardinality (number of elements) of the set L then a family with index-set L will be occasionally called an *l-family*; similarly, a colouring with colour-set L will be called an *l-colouring*. In the case when $l = 0$ we speak of an *empty family* or *colouring*.

A decomposition of G into factors [*n*-factors (1.34)] will be called a *factorization* [*n-factorization*] of G.

2.5 We establish correspondences between decompositions into positive subgraphs, decompositions into factors and edge-colourings. Decomposition of G given into positive subgraphs, we can construct a decomposition of G into factors by completing every subgraph of the original decomposition into a factor by adding some null vertices. From a decomposition of G into factors we can construct an edge-colouring of G such that an edge is coloured $i \in L$ exactly when it belongs to the factor assigned to i. Finally, from an edge-colouring of G we can construct a decomposition of G into positive subgraphs if to each colour $i \in L$ we assign the subgraph of G generated by edges coloured i.

We shall often make use of the above correspondences and speak of an *edge-colouring corresponding to a decomposition*, etc.

2.6 In the definition of a colouring we did not assume that every element of the colour-set is indeed used as a colour of some edge. If this condition is satisfied, that is, if an edge-colouring of G is a subjective mapping of the set

$E(G)$ onto the colour-set then it is called a *proper colouring*. Similarly, a decomposition of G no member of which is a null subgraph of G is called a *proper decomposition*.

Obviously, proper decompositions into positive subgraphs, proper decompositions into factors and proper edge-colourings of a graph G correspond to each other.

2.7 Employing the correspondences in 2.5 and 2.6 we could restrict ourselves to using only decompositions, either to positive subgraphs or factors, or to using only edge-colourings. Instead, we shall always choose our terminology to be most appropriate to the situation that occurs.

For example, Fig. 2.7.1 illustrates a 3-colouring of K_5 with colour-set $\{1, 2, 3\}$. Each edge is labelled with its colour. The corresponding decomposition (H_1, H_2, H_3) of K_5 into positive subgraphs is depicted in Fig. 2.7.2, and the corresponding decomposition (F_1, F_2, F_3) into factors is in Fig. 2.7.3. In the present case the most natural terminology seems to be the second one (decomposition into positive subgraphs).

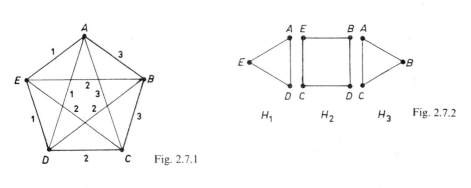

Fig. 2.7.1

Fig. 2.7.2

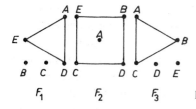

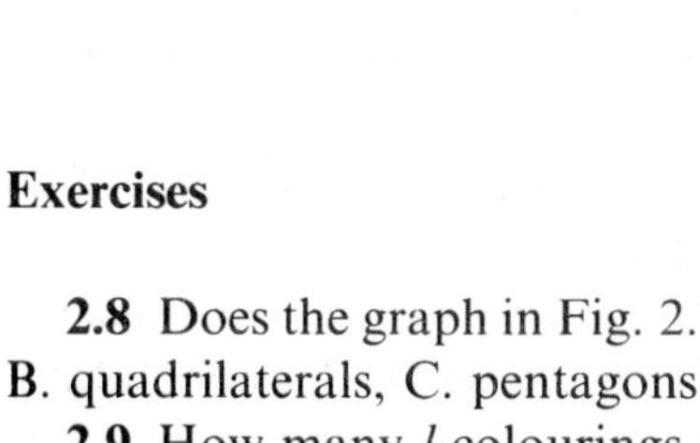

Fig. 2.7.3

Exercises

2.8 Does the graph in Fig. 2.7.1 admit a decomposition into: A. triangles, B. quadrilaterals, C. pentagons?

2.9 How many l-colourings does a graph of size e have?

2.10 Two decompositions $(F_i|i \in L)$ and $(K_i|i \in L)$ of a graph G will be said to be *different* if there exist two edges of G which belong to the same factor in one of the decompositions and to different factors in the other decomposition. If $|L| = l$ then every graph of size e has $l^e/l!$ decompositions into l factors. (We set $0^0 = 1$.)

Chapter 3

Generalizations of graph decompositions

3.1 The concept of a decomposition can be generalized in essentially three ways. In the definition of a decomposition we either replace a graph by a hypergraph (and subgraphs by subhypergraphs) or we replace the word "exactly" by "at most" or "at least". In the first case a *decomposition of a hypergraph* is obtained. To each such a decomposition there corresponds an *edge-colouring of the hypergraph* defined analogously as in 2.5 for graphs. The other two generalizations will be discussed below (3.2—3.3) in more detail.

3.2 A family $(H_i | i \in L)$ of subgraphs of a graph G will be called a *packing of G* (with subgraphs H_i) if the subgraphs H_i are edge-disjoint, that is, if each edge of G is contained in at most one member of $(H_i | i \in L)$. In particular, we can speak of *packing of G with factors* or *positive subgraphs* and of *proper packings* (if each of the subgraphs H_i has at least on edge).

In terms of colourings to each packing of a graph G there corresponds a *partial edge-colouring of G* defined as a mapping of $E(G)$ into the index-set L of the packing. To a proper packing of G there corresponds a *proper partial edge-colouring of G*, that is to say, a mapping of $E(G)$ onto L.

An example of a graph with no decomposition into triangles is exhibited in Fig. 3.2.1. However it admits a packing with (at most) three triangles. We can take, for instance, the triangles T_i, $i \in \{1, 3, 5\}$ whose edges are labelled with the number (colour) i.

3.3 A family $(H_i | i \in L)$ of subgraphs of a graph G will be called a *covering of G* (*with subgraphs H_i*) if each edge of G is contained in at least one member of $(H_i | i \in L)$. In particular, we can again speak of *covering of G with factors* or *positive subgraphs* and of *proper coverings* (*with non-null subgraphs*).

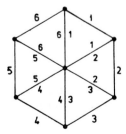

Fig. 3.2.1

The graph in Fig. 3.2.1 has a proper covering with six triangles T_i $(i = 1, 2, 3, 4, 5, 6)$, T_i consisting of edges coloured i.

3.4 In order to find the concept corresponding to a covering in terms of colourings we have to introduce the so-called multicolourings. Multicolourings can also be defined for vertices, edges or all elements of a graph.

Given a graph G and a set L (of colours), an *edge-multicolouring* [*vertex-multicolouring*, resp. *total multicolouring*] is a mapping of the edge-set [vertex-set, resp. union of the vertex- and edge-set] of G to the set 2^L of all subsets of L. Elements of the set assigned to an element of G are called *colours of the element*.

An edge-multicolouring [vertex-multicolouring, resp. total multicolouring] of G is said to be

A. *proper* if every element of L is one of the colours of some edge [vertex, resp. element] of G;

B. *complete* if every edge [vertex, resp. element] of G is assigned some colour in L.

If all values of a multicolouring are *singletons* (1-element sets) then the multicolouring can be identified with a usual colouring.

3.5 The correspondences discussed in 2.5 and 2.6 have their analogues in multicolourings. More specifically, a family of positive subgraphs of a graph G, a family of factors of G and a proper edge-multicolouring of G correspond to each other. Similarly, a proper family of positive subgraphs of G, a proper family of factors of G and a proper multicolouring of G correspond to each other.

3.6 Specializing to coverings we get two further correspondences: coverings with positive subgraphs, coverings with factors and complete edge-multicolourings correspond to each other. Similarly, proper coverings with positive subgraphs, proper coverings with factors and proper complete edge-multicolourings correspond to each other.

Our assertions about the above correspondences can easily be verified and the reader is recommended to do this. The concepts such as the *edge-multicolouring corresponding to a family of subgraphs* of a graph are analogous to similar concepts for colourings and need no further explanation.

3.7 Our convention in 2.4 about l-families and l-colourings (where $l = |L|$ is the cardinality of the index-set or colour-set, respectively) can be extended to l-*packing*, l-*covering* and l-*multicolouring*.

3.8 Different generalizations of graph decompositions mentioned in 3.1—3.3 can be, of course, combined together. Therefore we can speak of *packing* or *covering of a hypergraph* (*by subhypergraphs*). Combining coverings and packings yields nothing new but decompositions of a graph (2.1) or hypergraphs (3.1).

3.9 For the purposes of *enumeration* of decompositions (packings, resp. coverings) it is necessary to state which decompositions (packings, resp. coverings) are considered different. Since these concepts are defined as families of subgraphs it suffices to state this for such families. For the sake of simplicity we confine ourselves with finite families (if necessary, the reader may generalize this to arbitrary families).

A *family* $(G_1, G_2, ..., G_l)$ *of subgraphs of a graph G* is said to be *identical with family* $(H_1, H_2, ..., H_m)$ *of subgraphs of a graph H* if $l = m$ and there exists a permutation α of the set $\{1, 2, ..., l\}$ such that $G_i = H_{\alpha(i)}$ for $i = 1, 2, ..., l$; otherwise they are *different*.

The above two families are said to be *isomorphic* if $l = m$ and there exists an isomorphism f of G and H and a permutation α of the set $\{1, 2, ..., l\}$ such that $fG_i = H_{\alpha(i)}$ for $i = 1, 2, ..., l$; otherwise they are *essentially different*, or *non-isomorphic*. (The symbol fG_i denotes the graph whose vertices and edges are the images in f of vertices and edges of G_i.)

The above definitions of isomorphism and non-isomorphism can naturally be transferred to the corresponding edge-colourings and edge-multicolourings. On the other hand, the equality of colourings or multicolourings is defined as the equality of mappings (see 2.3, 2.9, 2.10, 3.4, 3.12—3.14). The reader is recommended to think over the details.

3.10 This book is devoted to decompositions of graphs. In several places we mention more general results concerning coverings and packings of graphs of hypergraphs or decompositions of hypergraphs. However, these results do not form a systematic theory.

Exercises

3.11 Does the graph in Fig. 3.2.1 admit A. a decomposition into quadrilaterals; B. a 2-packing with quadrilaterals; C. a 4-covering with quadrilaterals; D. a 2-packing with pentagons; E. a 3-covering with pentagons; F. a 2-packing with hexagons; G. a 3-covering with hexagons?

3.12 Every graph of size e has exactly $(l + 1)^e$ partial edge-colourings with colour-set L of cardinality l.

3.13. Every graph of size e has exactly $(l + 1)^e/l!$ different l-packings with factors.

3.14. What is the number of edge-multicolourings of a graph of size e with colour-set L of cardinality l?

RELATIONS BETWEEN FACTORS
OF DECOMPOSITION OF A GRAPH
AND BETWEEN DECOMPOSITIONS
OF GRAPHS

In Chapters 4 and 5 we shall be dealing with relations between separate factors of a decomposition of a graph. The study of sums and products of values of a fixed invariant over the factors of a decomposition, which we undertake in Chapter 4, serves there as the most significant example. Chapter 5 is devoted to other types of relations between factors of decomposition, as well as to relations between decompositions of several graphs. The third topic which could well be included here regards relations between different decompositions of the same graph. However, we did not collect the corresponding results in one place. Instead, we consider this topic in several chapters, together with related questions; see, for example, 2.5, 2.6, 3.9, 11.19, 11.22, 12.33, 12.34 and all the sections dealing with enumeration of decompositions (they are listed in 14.7).

Chapter 4

Nordhaus—Gaddum type theorems

4.1 In 1956, Nordhaus and Gaddum derived the following inequalities for the chromatic number (4.2) of a simple finite graph G of order v and its complement $\bar{G}$ (see Theorem 4.12):

$$\lceil 2\sqrt{v}\rceil \leq \chi(G) + \chi(\bar{G}) \leq v + 1, \tag{1}$$

$$v \leq \chi(G)\chi(\bar{G}) \leq \lfloor ((v + 1)/2)^2 \rfloor . \tag{2}$$

Their paper has had considerable impact, and even today these results are generalized and modified in various directions. To give a brief overview of the topic let us first introduce some basic notions and auxiliary results. The reader is also recommended the survey article of Chartrand and Mitchem (1971).

Note that the right-hand side of (2) can be rewritten in the form

$$\lfloor (v + 1)/2 \rfloor \lceil (v + 1)/2 \rceil , \tag{3}$$

which is sometimes more convenient.

4.2 A *vertex-colouring* of a graph G will be called *regular* if any two adjacent vertices receive different colours. The smallest number of colours for which there exists a regular vertex-colouring of G will be called the *chromatic number of G* and denoted $\chi(G)$. A graph G is *n-chromatic* if $\chi(G) = n$ and *n-colourable* if $\chi(G) \leq n$ (n is a cardinal number).

An *edge-colouring* is said to be *regular* if any two adjacent edges are coloured differently. Similarly, a *total colouring* is called *regular* if any two adjacent or incident elements have different colours. The smallest number of colours for which there exists a regular edge-colouring [total colouring] of G is the *chromatic index* [*total chromatic number*] of G and is denoted $\chi'(G)[\chi''(G)]$.

4.3 It is obvious that for an arbitrary cardinal number n there exists an *n-chromatic* graph G; it is sufficient to take for G the complete graph K_n. Similarly, the complete bipartite graph $K(1, n)$ can serve as an example of a graph of chromatic index n. A graph with total chromatic number n can be constructed as follows. For $n = 0$, G is the empty graph, and for $n \neq 0$ take G of order 1 with the necessary number of loops ($n - 1$ loops for finite n, n loops for n infinite). See also Exercise 4.46.

4.4 A *vertex-colouring* [edge-, or total colouring] is said to be *pseudocomplete* if for any two different colours i, j there exist adjacent vertices [adjacent edges, incident or adjacent elements] coloured i and j. A regular pseudocomplete colouring is called *complete*. The greatest number of colours which can be used in a pseudocomplete vertex-colouring [edge-, or total colouring] of a graph G is the *pseudoachromatic number* [*pseudoachromatic index*, *total pseudoachromatic number*] *of* G, and denoted by $\psi(G)$ [$\psi'(G)$, $\psi''(G)$, respectively]. The greatest number of colours in a complete vertex-colouring [edge-, or total colouring] of G is the *achromatic number* [*achromatic index*, *total achromatic number*] *of* G, denoted by $\alpha(G)$ [$\alpha'(G)$, $\alpha''(G)$].

4.5 In the preceding paragraphs we have introduced nine graph invariants: χ, χ', χ'', ψ, ψ', ψ'', α, α', α''. All are defined for arbitrary graphs. This is clear for finite graphs. For infinite graphs it is a consequence of Bosák and Nešetřil (1976) and relations $\chi'(G) = \chi(L(G))$, $\chi''(G) = \chi(T(G))$ (here $L(G)$, $T(G)$ are the line graph and the total graph of G, respectively); the situation is similar for ψ and α. The following inequalties are obvious for an arbitrary graph G:

$$\chi(G) \leq \alpha(G) \leq \psi(G), \tag{1}$$

$$\chi'(G) \leq \alpha'(G) \leq \psi'(G), \tag{2}$$

$$\chi''(G) \leq \alpha''(G) \leq \psi''(G). \tag{3}$$

4.6 Lemma [Zykov 1949, Chartrand and Polimeni 1974, Plesník 1978]. *Let* $(G_1, G_2, ..., G_l)$ *be a decomposition of a graph* G *into non-empty subgraphs.* *Then*

$$\chi(G) \leq \chi(G_1)\chi(G_2) ... \chi(G_l).$$

Proof. Put $\chi(G_i) = m_i$, $i = 1, ..., l$. Let F_i be the factor of G for which $E(F_i) = E(G_i)$; clearly $\chi(F_i) = \chi(G_i) = m_i$. Assume that a vertex u of G is coloured c_i in an m_i-colouring of F_i. An $m_1 m_2 ... m_l$-colouring of G is then obtained by assigning the ordered l-tuple $[c_1, ..., c_l]$ to each such u. Obviously,

$$\chi(G) \leq m_1 m_2 ... m_l = \chi(G_1)\chi(G_2) ... \chi(G_l).$$

4.7 We shall often deal with the question of finding conditions which would guarantee the existence of a decomposition of the given graph into l factors with prescribed properties. In the case when the factors are assigned upper bounds on their chromatic numbers (the i-th factor is to be m_i-colourable, $i = 1, 2, ..., l$), there is a very simple necessary and sufficient condition.

Theorem [Burr 1980]. *Let G be a graph and $l, m_1, m_2, ..., m_l$ be positive integers. The graph G is decomposable into l factors with chromatic numbers not exceeding $m_1, m_2, ..., m_l$ if and only if $\chi(G) \leq m_1 m_2 \cdot ... \cdot m_l$.*

Proof. Let $\chi(G) \leq m_1 m_2 ... m_l$. Let us colour the vertices of G using (at most) $m_1 m_2 \cdot ... \cdot m_l$ colours, written as l-tuples $[c_1, c_2, ..., c_l]$, where $c_i \leq m_i$, $i = 1, 2, ..., l$. Now let us decompose G into l factors as follows. The i-th factor will be determined by all those edges whose ends receive colours (l-tuples) with the same j-th entries c_j for $1 \leq j \leq i - 1$, but distinct i-th entries c_i. It is easy to see that the chromatic number of the i-th factor does not exceed m_i (since a m_i-colouring of its vertices can be obtained from the i-th entries c_i only).

The reverse implication follows from Lemma 4.6. See also Indzheyan et al. (1979).

4.8 Naturally, Theorem 4.7 (and similarly, Lemma 4.6) could have been stated simply for subgraphs instead of factors. As regards the spirit of both these results, they would fit better into Chapter 5. However, their presentation at this point is justified by their close relationship with subsequent paragraphs.

4.9 Theorem [Plesník 1978]. *Let v and l be positive integers and $G_1, G_2, ..., G_l$ be a decomposition of the complete graph K_v. Then*

$$v \leq \chi(G_1)\chi(G_2)...\chi(G_l),\tag{1}$$

$$\lceil l\sqrt[l]{v}\rceil \leq \chi(G_1) + \chi(G_2) + ... + \chi(G_l).\tag{2}$$

Proof. The first inequality follows from Lemma 4.6 by putting $G = K_v$ (since $\chi(K_v) = v$). The second one is a consequence of (1) and the well known inequality between the arithmetic and geometric mean

$$\sqrt[l]{\chi(G_1)\chi(G_2)...\chi(G_l)} \leq (\chi(G_1) + \chi(G_2) + ... + \chi(G_l))/l.\tag{3}$$

4.10 As is proved in Plesník (1978), the bounds 4.9.1—2 are sharp for infinitely many v; in particular for all those of the form $v = s^l$.

4.11 Lemma [Gupta 1969]. *Let v be a positive integer, G a graph of order v, and $\bar{G}$ its complement. Then*

$$\chi(G) + \psi(\bar{G}) \leq v + 1.$$

For the proof, the reader is referred to Gupta (1969).

4.12 Theorem [Nordhaus and Gaddum 1956]. *Let v be a positive integer, G a graph of order v, and $\bar{G}$ its complement. Then the following holds:*

$$\lceil 2\sqrt{v} \rceil \le \chi(G) + \chi(\bar{G}) \le v + 1, \tag{1}$$

$$v \le \chi(G)\chi(\bar{G}) \le \lfloor ((v+1)/2)^2 \rfloor . \tag{2}$$

Proof. The lower bounds in both (1) and (2) are a consequence of Theorem 4.9. The upper bound in (1) follows from Lemma 4.11 using 4.5.1 (cf. Bollobás 1978, Corollary V.1.4). The upper bound in (2) can be obtained from (1) and 4.9.3:

$$\chi(G)\chi(\bar{G}) = (\sqrt{\chi(G)\chi(\bar{G})})^2 \le ((\chi(G) + \chi(\bar{G}))/2)^2 \le ((v+1)/2)^2 .$$

The rest follows from the fact that $\chi(G)\chi(\bar{G})$ is an integer.

4.13 The bounds given in 4.12 have been subject to thorough investigation from which it is apparent that, in some sense, they cannot be improved. More precisely, the following was proved in Finck (1968) and Stewart (1969): Given a number v, for arbitrary x, y satisfying inequalities

$$\lceil 2\sqrt{v} \rceil \le x + y \le v + 1 \quad \text{and} \quad v \le xy \le \lfloor ((v+1)/2)^2 \rfloor$$

there exists a simple graph G of order v for which $\chi(G) = x$ and $\chi(\bar{G}) = y$. Further results in this direction can be found in Finck (1966), Capobianco and Molluzzo (1978).

4.14 Lemma 4.11 together with 4.5.1 imply the validity of the inequality $\chi(G) + \alpha(\bar{G}) \le v + 1$, proved in Harary and Hedetniemi (1970). This provides a generalization of the upper bound in Theorem 4.12.1. However, even the strongest inequality of that type, namely 4.11, is sharp for all values of v, as it is shown in Gupta (1969).

4.15 In connection with the preceding relations, a natural question of validity of the inequality $\psi(G) + \psi(\bar{G}) \le v + 1$ or at least $\alpha(G) + \alpha(\bar{G}) \le \le v + 1$ arises. But the decomposition of K_4 into 2 paths of order 4 provides an example in which

$$\psi(G) + \psi(\bar{G}) = \psi(P_4) + \psi(\bar{P}_4) = 3 + 3 = 6 = v + 2.$$

A final result can be found in Gupta (1969): For each simple graph G of order v it holds

$$\psi(G) + \psi(\bar{G}) \le \lceil 4v/3 \rceil ,$$

$$\alpha(G) + \psi(\bar{G}) \le \lceil 4v/3 \rceil ,$$

$$\alpha(G) + \alpha(\bar{G}) \le \lceil 4v/3 \rceil .$$

The bounds are exact for every positive integer v.

4.16 In 4.9 we have presented definitive lower bounds on the sum and product of chromatic numbers of the factors of decomposition of a complete graph. Similar definitive upper bounds are not known as yet. In Plesník (1978) the following conjecture was made:

Conjecture. *Let v and l be positive integers and G_1, G_2, ..., G_l the decomposition of the complete graph of order v.* Then

$$\chi(G_1) + \chi(G_2) + \dots + \chi(G_l) \leq v + \binom{l}{2}, \tag{1}$$

$$\chi(G_1)\,\chi(G_2) \dots \chi(G_l) \leq \left\lfloor \left(\frac{v + \binom{l}{2}}{l} \right)^l \right\rfloor. \tag{2}$$

4.17 As the main part of Conjecture 4.16, the inequality (1) can be considered (because (2) can be obtained from (1), 4.9.3 and the fact that the left-hand side of (2) is an integer). The Conjecture 4.16 is true for $l = 2$ (see 4.12) as well as for $l = 3$ (Plesník 1978). Further, it is known (Plesník 1978) that definitive bounds are, for all l, of the form

$$\chi(G_1) + \chi(G_2) + \dots + \chi(G_l) \leq v + t(l),$$

$$\chi(G_1)\,\chi(G_2) \dots \chi(G_l) \leq \left\lfloor \left(\frac{v + t(l)}{l} \right)^l \right\rfloor,$$

where t is a function with values in N_0 and satisfying inequalities

$$\binom{l}{2} \leq t(l) \leq \begin{cases} l - 1 & \text{if } l \leq 2, \\ \sum_{i=2}^{l-1} \binom{l}{i} t(i) & \text{if } l \geq 3. \end{cases}$$

Thus, the conjecture claims that $t(l) = \binom{l}{2}$. Some further results in this direction can be found in Plesník (1978), Erdős (1979), Cook (1984a, b).

4.18 All the generalizations and modifications of the Nordhaus—Gaddum Theorem 4.12 can be derived into four groups: A. the chromatic number is replaced by another graph invariant; B. a greater or an arbitrary number of factors is allowed; C. graphs are replaced by hypergraphs; D. instead of graph decompositions, coverings or packings are considered. Moreover, these groups can be mutually combined. We have already presented examples which belong to the first two groups; the results of Mitchem (1974) belong to the 3rd one. Generalizations to hypergraphs were also investigated in

Jucovič and Olejník (1974) and Olejník (1981) where the authors obtained bounds on $\chi(H) + \chi(\bar{H})$ and $\alpha(H) + \alpha(\bar{H})$ for complementary uniform hypergraphs H and $\bar{H}$. Examples of the 4th group can be found in Dirac (1964), Alavi and Mitchem (1971), Schürger (1974), Cook (1984a, b). In what follows we return to generalizations and modifications of the type A and B. Further results of this kind are in Finck and Sachs (1969), Galvin and Krieger (1971), Arnautov (1972), Jaeger and Payan (1972), Hedayat and Kageyama (1980), Foregger and Foregger (1980), Borowiecki et al. (1982), Cockayne and Thomason (1982).

4.19 An interesting generalization of the concept of chromatic number — the vertex partition number — has been introduced in Lick and White (1970). Let n be a cardinal number. A graph G is said to be *n-degenerated* if the minimum degree of each induced subgraph does not exceed n. The *vertex n-partition number of G* is the smallest number ϱ_n of subsets in a partition of the vertex set of G with the property that each of the subsets induces an n-degenerated subgraph of G. Obviously, ϱ_0 is the same as χ (the chromatic number); ϱ_1 is also called *vertex arboricity of a graph*.

4.20 Theorem. *Let G and $\bar{G}$ be complementary graphs of order v. If $f(n) = \lfloor (1 + 4n + \sqrt{1 + 8n^2})/2 \rfloor$, then*

$$2\sqrt{v/f(n)} \le \varrho_n(G) + \varrho_n(\bar{G}) \le (v + 1 + 2n)/(n + 1), \tag{1}$$

$$v/f(n) \le \varrho_n(G)\,\varrho_n(\bar{G}) \le ((v + 1 + 2n)/2\,(n + 1))^2. \tag{2}$$

The proof is omitted and the reader is referred to Lick and White (1974). (For $n = 0$ see also 4.12, for $n = 1$, Mitchem 1971.)

4.21 A further progress has been made in Cook (1984a) where the following result can be found:

Theorem. *Let $(G_1, G_2, ..., G_l)$ be a decomposition of the complete graph K_v. Then*

$$l\sqrt[l]{\lceil v/(n + 1) \rceil} \le \varrho_n(G_1) + \varrho_n(G_2) + ... + \varrho_n(G_l), \tag{1}$$

$$\lceil v/(n + 1) \rceil \le \varrho_n(G_1)\,\varrho_n(G_2)... \varrho_n(G_l). \tag{2}$$

Proof. Lick and White (1970) proved that $\varrho_n(K_v) = \lceil v/(n + 1) \rceil$.

Using the same method as in the proof of Lemma 4.6 one obtains

$$\varrho_n(K_v) \le \varrho_n(G_1)\,\varrho_n(G_2)... \varrho_n(G_l).$$

Thus we get (2). The inequality (1) follows from (2) using 4.9.3.

4.22 Some of the results concerning upper bounds on the sum and the product of vertex n-partition numbers of factors of a decomposition may be found in Cook (1984a) together with some other related facts. On the whole it can be said that the state of our knowledge here is approximately the same as in the case $n = 0$, i.e. for the chromatic number (4.16, 4.17).

4.23 Let n be a cardinal number. An *n-clique of a graph* is any of its complete subgraphs of order n. In the case when the order is not important we refer simply to a *clique of a graph*.

For $n \geq 2$, the *n-clique chromatic number* $\chi_n(G)$ *of a graph* G was introduced in Sachs and Schäuble (1968) as the smallest number of colours in a vertex-colouring of G in which no n-clique of G is monochromatic. As $\chi_2 = \chi$, we have another generalization of the concept of chromatic number.

4.24 For positive integers n_1, n_2, the *Ramsey number* $R(n_1, n_2)$ is defined as the smallest order of a complete graph such that for every its decomposition (F_1, F_2) either F_1 contains an n_1-clique or F_2 contains an n_2-clique.

Theorem [Achuthan 1980]. *Let G and $\bar{G}$ be complementary graphs of order v and let $n \geq 2$. Then*

$$\lceil \sqrt{4v/(R - 1)} \rceil \leq \chi_n(G) + \chi_n(\bar{G}) \leq \lfloor (v + 2n - 3)/(n - 1) \rfloor , \qquad (1)$$

$$\lceil v/(R - 1) \rceil \leq \chi_n(G) \chi_n(\bar{G}) \leq$$
$$\leq \lfloor \lfloor (v + 2n - 3)/(n - 1) \rfloor /2 \rfloor \lceil \lfloor (v + 2n - 3)/(n - 1) \rfloor /2 \rceil , \qquad (2)$$

where R is the Ramsey number $R(n, n)$.

Proof. See Achuthan (1980). (Note that for $n = 2$ we obtain Theorem 4.12 since $R(2, 2) = 2$.)

4.25 In Achuthan (1980) it is also proved that all the four bounds of Theorem 4.24 are sharp except for the lower bound in (1) in the case when $\lceil \sqrt{4v/(R - 1)} \rceil$ is an odd number and

$$\lceil \sqrt{4v/(R - 1)} \rceil^2 - 1 < 4v/(R - 1).$$

Then, the exact lower bound in (1) is greater by 1.

4.26 Theorem [Vizing 1965, Alavi and Behzad 1971, Capobianco and Molluzzo 1978]. *Let G and $\bar{G}$ be complementary graphs of order v. Then*:

$$2 \lfloor (v + 1)/2 \rfloor - 1 \leq \chi'(G) + \chi'(\bar{G}) \leq v + 2 \lfloor (v - 2)/2 \rfloor ,$$
$$0 \leq \chi'(G) \chi'(\bar{G}) \leq (v - 1)(2 \lfloor v/2 \rfloor - 1).$$

All four bounds are sharp. (For the proof see loc. cit.)

4.27 A result similar to that of Theorem 4.26 for the chromatic index is known also for the total chromatic number [Cook 1974, Nirmala 1979]: If G and $\bar{G}$ are of odd order v then

$$v + 1 \leq \chi''(G) + \chi''(\bar{G}) \leq 2v,$$

$$v \leq \chi''(G)\chi''(\bar{G}) \leq v^2;$$

in the case when v is even, $v \geq 6$ then

$$v + 1 \leq \chi''(G) + \chi''(\bar{G}) \leq 2v - 1,$$

$$v + 1 \leq \chi''(G)\chi''(\bar{G}) \leq v(v-1).$$

These bounds are sharp for all v.

4.28 The *vertex-* [*edge-*] *connectivity of a graph G* is the minimum number $\varkappa(G)[\varkappa'(G)]$ of vertices [edges] the deletion of which results in a disconnected graph or in a graph of order ≤ 1. Our next Lemma completes the inequalities of 1.10.4 as follows:

Lemma. *For every graph G the following holds*:

$$\varkappa(G) \leq \varkappa'(G) \leq \delta(G).$$

Proof. We shall only prove the first inequality, the second one being obvious. Clearly, we may assume that $|V(G)| \geq 2$. Let X be a set of $\varkappa'(G)$ edges, the deletion of which yields a disconnected graph. (Note that the resulting graph cannot have order 1.) Thus $V(G)$ can be partitioned into two sets V_1 and V_2 such that no vertex of V_1 is connected in $G - X$ with a vertex of V_2.

If G contains a pair of non-adjacent vertices $u_1 \in V_1$ and $u_2 \in V_2$ then for each edge in X we can pick up one of its ends which is distinct from both u_1 and u_2. Deleting these vertices we obtain a disconnected graph. Thus (1) holds.

If such vertices u_1 and u_2 do not exist the number of edges in X is at least $|V_1||V_2|$. Deleting all the vertices of $V(G)$ but one results in a graph of order 1. Since the number of deleted vertices does not exceed $|V_1||V_2|$ we get (1) again. This completes the proof.

Let n be a cardinal number. A graph G is said to be *vertex n-connected* (or, simply, *n-connected*) if $\varkappa(G) \geq n$. G will be called *edge n-connected* if $\varkappa'(G) \geq n$.

Theorems of Nordhaus—Gaddum type for the sum and product of vertex-[edge-] connectivities have been proved in Alavi and Mitchem (1971). We now present their generalization.

4.29 **Theorem** [Cook 1984a]. *Let $(G_1, G_2, ..., G_l)$ be a factorization of the complete graph K_v. Then*

$$0 \leq \varkappa(G_1) + \varkappa(G_2) + ... + \varkappa(G_l) \leq \varkappa'(G_1) + \varkappa'(G_2) + ...$$

$$... + \varkappa'(G_l) \leq v - 1, \tag{1}$$

$$0 \leq \varkappa(G_1)\,\varkappa(G_2)\dots\varkappa(G_l) \leq \varkappa'(G_1)\,\varkappa'(G_2)\dots\varkappa'(G_l) \leq ((v-1)/l)^l. \qquad (2)$$

Proof. The lower bounds are trivial. To prove the upper bound in (1) it is sufficient to observe that

$$\varkappa'(G_1) + \varkappa'(G_2) + \dots + \varkappa'(G_l) \leq \delta(G_1) + \delta(G_2) + \dots + \delta(G_l) \leq$$

$$\leq \deg G_1 + \deg G_2 + \dots + \deg G_l = v - 1.$$

The upper bound in (2) follows immediately from 4.9.3.

4.30 As regards the exactness of the bounds in 4.29, the following is easily verified: The upper bound in 4.29.1 is always exact, the lower bound is exact for $v \geq 3$, and $l \geq 3$; the lower bound in 4.29.2 is exact if $v = l = 1$ or if $v \geq 1$ and $l \geq 2$. In case $l = 2$ the lower bound of 4.29.1 can be improved to 1 (since no complete graph can be decomposed into two disconnected factors — see 5.2). This bound is then exact for all $v \in N$.

The situation with the upper bound of 4.29.2 turns out to be more complicated. We shall restrict ourselves to the case $l = 2$. Now the exact upper bounds are $M(v) = 1$ for $v = 4$, $M(v) = \lfloor (v-1)/2 \rfloor \lceil (v-1)/2 \rceil$ for $v \not\equiv 3 \pmod 4$, $v \neq 4$ and $M(v) = (v-3)(v+1)/4$ for $v \equiv 3 \pmod 4$. This was established in Alavi and Mitchem (1971) for the edge-connectivity and in Rao (1975), Li (1977) for the vertex-connectivity, thereby solving implicitly a problem of Chartrand and Mitchem (1971).

4.31 Knowing a Nordhaus—Gaddum type theorem for a certain graph invariant it is sometimes possible to derive a similar theorem for another invariant related to it in an appropriate way. We precede an example of such a relation with several notions followed by one important theorem due to Gallai (in fact we shall prove its generalization).

Let G be a loopless graph. We say that a vertex u of G *covers* and edge w (and conversely, w *covers* u) if u is incident with w. The minimum number of vertices [edges] covering all the edges [non-null vertices] of G is called the *vertex-* [*edge-*] *covering number of* G and denoted by $\alpha_0(G)$ [$\alpha_1(G)$].

A set A of vertices [edges] of G is said to be *independent* if no edge [vertex] of G is incident with more than one vertex [edge] in A. The maximum cardinality of an independent set of vertices [edges] of G is called *vertex-* [*edge-*] *independence number of* G and denoted by $\beta_0(G)$ [$\beta_1(G)$].

The symbol $\iota(G)$ will stand for the number of isolated vertices of G.

4.32 Standard formulations of the following result (see e.g. Gallai 1959, Hanani 1961, p. 95, Lovász 1979, Problem 7.1) require G to be a graph without isolated vertices (or even connected). We shall prove it under more general assumptions, thereby obtaining a generalization of a theorem of Gallai (1959):

Theorem. *Let G be a loopless graph of order v. Then*

$$\alpha_0(G) + \beta_0(G) = v = \alpha_1(G) + \beta_1(G) + \iota(G).$$

Proof. I. Choose a set of $\beta_0(G)$ independent vertices in G. The remaining $v - \beta_0(G)$ vertices cover all edges of G. Hence $\alpha_0(G) \leq v - \beta_0(G)$, that is, $\alpha_0(G) + \beta_0(G) \leq v$.

II. Now choose in G a set of $\alpha_0(G)$ vertices which cover all edges of G. The remaining $v - \alpha_0(G)$ vertices form an independent set. It follows that $\beta_0(G) \geq v - \alpha_0(G)$ and thus $\alpha_0(G) + \beta_0(G) \geq v$.

III. Let F be a factor of G containing exactly $\alpha_1(G)$ edges which cover all vertices of G, except for the isolated ones. Since no edge of F is superfluous, each of them is incident at least with one vertex of degree 1 in F. Induction on the number of edges of F can be used to show that F has $v(F) - e(F) = v - \alpha_1(G)$ components. The number of components of F with at least one edge is $v - \alpha_1(G) - \iota(G)$. Picking up one edge from each component one obtains a set of $v - \alpha_1(G) - \iota(G)$ independent edges of G. Therefore $\beta_1(G) \geq v - \alpha_1(G) - \iota(G)$ and, consequently, $v \leq \alpha_1(G) + \beta_1(G) + \iota(G)$.

IV. Let F' be a factor containing exactly $\beta_1(G)$ edges, all of which are independent. These edges cover $2\beta_1(G)$ vertices of G. Thus there exist $v - 2\beta_1(G) - \iota(G)$ non-isolated vertices of G which are not covered by the edges of F'. Choosing for each of these vertices one edge incident with it and adding the edges of F', we obtain a set of $v - 2\beta_1(G) - \iota(G) + \beta_1(G) = v - \beta_1(G) - \iota(G)$ edges which cover the non-isolated vertices of G. Thus $\alpha_1(G) \leq v - \beta_1(G) - \iota(G)$ whence $v \geq \alpha_1(G) + \beta_1(G) + \iota(G)$.

4.33 In Chartrand and Schuster (1974) it has been shown that for a pair of complementary graphs G and $\bar{G}$ of finite order v one has:

$$\lfloor v/2 \rfloor \leq \beta_1(G) + \beta_1(\bar{G}) \leq 2 \lfloor v/2 \rfloor, \tag{1}$$

$$0 \leq \beta_1(G)\beta_1(\bar{G}) \leq \lfloor v/2 \rfloor^2. \tag{2}$$

The results of Cockayne and Lorimer (1975), Erdős and Schuster (1981) imply, moreover, that

$$\lfloor (v + 1)/3 \rfloor \leq \max\{\beta_1(G), \beta_1(\bar{G})\} \leq \lfloor v/2 \rfloor. \tag{3}$$

The above-mentioned authors also study the problem of the existence of a graph G with prescribed values of $\beta_1(G)$ and $\beta_1(\bar{G})$; the solution given in the latter paper is complete.

If neither G nor $\bar{G}$ have isolated vertices, (1) and (2) above allow some improvement [Laskar and Auerbach, 1978]:

$$\lfloor v/2 \rfloor + 2 \le \beta_1(G) + \beta_1(\bar{G}) \le 2 \lfloor v/2 \rfloor , \tag{4}$$

$$2 \lfloor v/2 \rfloor \le \beta_1(G) \beta_1(\bar{G}) \le \lfloor v/2 \rfloor^2. \tag{5}$$

Employing these inequalities we now derive similar results for the edge-covering number.

4.34 Theorem [see Laskar and Auerbach 1978 for (1) and (2) below]. *If G and $\bar{G}$ are complementary graphs of finite order v without isolated vertices then the following inequalities hold*:

$$2 \lfloor (v + 1)/2 \rfloor \le \alpha_1(G) + \alpha_1(\bar{G}) \le \lceil 3v/2 \rceil - 2 \tag{1}$$

$$\lfloor (v + 1)/2 \rfloor^2 \le \alpha_1(G) \alpha_1(\bar{G}) \le \lfloor (\lceil 3v/2 \rceil - 2)/2 \rfloor .$$
$$. \lceil (\lceil 3v/2 \rceil - 2)/2 \rceil , \tag{2}$$

$$\lfloor (v + 1)/2 \rfloor \le \min\{\alpha_1(G), \alpha_1(\bar{G})\} \le \lfloor (2v + 1)/3 \rfloor . \tag{3}$$

Proof. The lower bounds follow from the following obvious inequalites:

$$\alpha_1(G) \ge \lfloor (v + 1)/2 \rfloor , \qquad \alpha_1(\bar{G}) \ge \lfloor (v + 1)/2 \rfloor .$$

From Theorem 4.32 and from 4.33.4 we deduce that

$$\alpha_1(G) + \alpha_1(\bar{G}) = v - \beta_1(G) + v - \beta_1(\bar{G}) =$$
$$= 2v - (\beta_1(G) + \beta_1(\bar{G})) \le 2v - \lfloor v/2 \rfloor - 2 = \lceil 3v/2 \rceil - 2.$$

This implies that

$$\alpha_1(G) \alpha_1(\bar{G}) = (\sqrt{\alpha_1(G) \alpha_1(\bar{G})})^2 \le ((\alpha_1(G) + \alpha_1(\bar{G}))/2)^2 \le$$
$$\le ((\lceil 3v/2 \rceil - 2)/2)^2.$$

Hence we obtain

$$\alpha_1(G) \alpha_1(\bar{G}) \le \lfloor ((\lceil 3v/2 \rceil - 2)/2)^2 \rfloor =$$
$$= \lfloor (\lceil 3v/2 \rceil - 2)/2 \rfloor \lceil (\lceil 3v/2 \rceil - 2)/2 \rceil .$$

The upper bound in (3) can be obtained as a special case (setting $\iota(G) = \iota(\bar{G}) = 0$) of the following inequality 4.35.3.

4.35 In the general case we have the following result:

Theorem. *Let G and $\bar{G}$ be complementary graphs of order v. Then the following inequalities hold*:

$$\lfloor (v+1)/2 \rfloor \leq \alpha_1(G) + \alpha_1(\bar{G}) \leq \lceil 3v/2 \rceil - 1, \tag{1}$$

$$0 \leq \alpha_1(G)\alpha_1(\bar{G}) \leq \lfloor (\lceil 3v/2 \rceil - 1)/2 \rfloor \lceil (\lceil 3v/2 \rceil - 1)/2 \rceil, \tag{2}$$

$$\lfloor (v+1)/2 \rfloor \leq \min\{\alpha_1(G) + \iota(G), \alpha_1(\bar{G}) + \iota(\bar{G})\} \leq \lfloor (2v+1)/3 \rfloor. \tag{3}$$

Proof. By virtue of Theorem 4.32 and the inequality 4.33.1 we have

$$\alpha_1(G) + \alpha_1(\bar{G}) = 2v - (\beta_1(G) + \beta_1(\bar{G})) - (\iota(G) + \iota(\bar{G})) \geq$$
$$\geq 2v - 2\lfloor v/2 \rfloor - v = \lfloor (v+1)/2 \rfloor.$$

Thus we get the lower bound (1). The lower bound in (2) is obvious. If $\iota(G) + \iota(\bar{G}) = 0$ the upper bound in (1) follows immediately from Theorem 4.34.1 If, on the other hand, $\iota(G) + \iota(\bar{G}) \geq 1$ we get the same bound from Theorem 4.32 and the inequality 4.33.1. Indeed,

$$\alpha_1(G) + \alpha_1(\bar{G}) = 2v - (\beta_1(G) + \beta_1(\bar{G})) - (\iota(G) + \iota(\bar{G})) \leq$$
$$\leq 2v - \lfloor v/2 \rfloor - 1 = \lceil 3v/2 \rceil - 1.$$

The upper bound in (2) can be proved analogously as in 4.34.

Finally, we shall derive (3) from 4.33.3. Using 4.32 the inequality 4.33.3 can be rewritten as follows:

$$\lfloor (v+1)/3 \rfloor \leq \max\{v - \alpha_1(G) - \iota(G), v - \alpha_1(\bar{G}) - \iota(\bar{G})\} \leq \lfloor v/2 \rfloor,$$

that is

$$\lfloor (v+1)/3 \rfloor \leq -\min\{\alpha_1(G) + \iota(G) - v, \alpha_1(\bar{G}) + \iota(\bar{G}) - v\} \leq \lfloor v/2 \rfloor,$$

whence

$$\lfloor (v+1)/2 \rfloor = v - \lfloor v/2 \rfloor \leq \min\{\alpha_1(G) + \iota(G), \alpha_1(\bar{G}) + \iota(\bar{G})\} \leq$$
$$\leq v - \lfloor (v+1)/3 \rfloor = \lfloor (2v+1)/3 \rfloor.$$

4.36 In Chartrand and Schuster (1974), estimate for the sum and product of the vertex independence numbers of two complementary graphs G and $\bar{G}$ of finite order v are also investigated. Among other results it is shown that

$$\beta_0(G) + \beta_0(\bar{G}) \leq v + 1, \tag{1}$$

$$\beta_0(G)\beta_0(\bar{G}) \leq \lfloor (v+1)/2 \rfloor \lceil (v+1)/2 \rceil. \tag{2}$$

The problem of the existence of a graph G with prescribed values of $\beta_0(G)$ and $\beta_0(\bar{G})$ is considered in Erdős and Schuster (1981).

4.37 The same question of the exactness of bounds applies in the case of the results of 4.33—4.36. Since

$$\alpha_1(K_v) = \lceil v/2 \rceil, \; \beta_1(K_v) = \lfloor v/2 \rfloor, \; \beta_0(K_v) = 1,$$

$$\alpha_1(\bar{K}_v) = \beta_1(\bar{K}_v) = 0, \; \beta_0(\bar{K}_v) = v,$$

the lower bounds in 4.33.1—2, 4.35.1—2, and the upper bound in 4.36.1 are exact for each $v \in N$. Further results in this direction can be found in Chartrand and Schuster (1974), Laskar and Auerbach (1978), Gangopadhyay (1981) and in Exercise 4.45. Other modifications of the topic of 4.31—4.37 have been considered in Erdős and Meir (1977), Sampathkumar and Walikar (1979), and for hypergraphs in Olejník (1984).

4.38 Sometimes it may be inconvenient to relate the values of certain invariants of complementary graphs by means of theorems of the Nordhaus —Gaddus type. We shall support this claim by examples of the diameter (4.39) and radius (4.40). First, we introduce some necessary notions.

By the *eccentricity of a vertex u* we shall understand the supremum of distances between u and vertices of G. The maximum [minimum] eccentricity of a vertex of G will be called the *diameter* [*radius*] *of the graph G* and denoted by diam G [rad G].

4.39 When deriving Nordhaus—Gaddum type inequalities for diameters one instantly meets several difficulties:

(1) Factors can have infinite diameter. This can be avoided by adopting appropriate assumptions (e.g., that G and $\bar{G}$ are finite connected graphs).

(2) When the number of vertices is small, exceptional situations may occur. For instance, K_4 cannot be decomposed into two factors of diameter 3; in Cherepanov (1978) it was pointed out that this possibility was passed over in Bosák et al. (1968).

(3) Such inequalities are practically of no use since the paper of Bosák et al. (1968) contains more general information: It is determined for which ordered triples v, d_1, d_2 the complete graph K_v admits a decomposition into two factors with diameters d_1 and d_2, respectively.

(4) Nordhaus—Gaddum type inequalities provide here a less complete picture about the situation that in the case of, say, chromatic numbers. For example, the triple $v = 7$, $d_1 = \text{diam } G = 3$, $d_2 = \text{diam } \bar{G} = 4$ satisfies these inequalities; however, the corresponding decomposition of K_7 does not exist. Therefore we present them just for the sake of completeness:

Theorem. *Let G and $\bar{G}$ be two complementary connected graphs of order* $v \geq 6$. *Then*:

$$4 \leq \operatorname{diam} G + \operatorname{diam} \bar{G} \leq v + 1, \tag{1}$$

$$4 \leq \operatorname{diam} G \operatorname{diam} \bar{G} \leq 2v - 2. \tag{2}$$

Proof follows readily from Theorem 1 and Theorem 5 of Bosák et al. (1968). These results imply that either both diam G and diam $\bar{G}$ are equal to 3 or at least one of them must be 2. (The upper bound of (1) has also been established in Bondy (1968).)

4.40 A similar situation occurs in the case of radii.

Theorem. *Let G and $\bar{G}$ be two connected complementary graphs of order v. Then*:

$$4 \leq \operatorname{rad} G + \operatorname{rad} \bar{G} \leq (v + 4)/2, \tag{1}$$

$$4 \leq \operatorname{rad} G \operatorname{rad} \bar{G} \leq v. \tag{2}$$

Proof follows from Theorem 1 and Theorem 9 of Palumbíny and Znám (1973) since, according to these results, at least one of the graphs G and $\bar{G}$ has radius 2.

Exercises

4.41 Which inequality in 4.39.1—2 does not hold: A. if $v = 4$; B. if $v = 5$?

4.42 [Cook 1984b]. If $P = (G_1, G_2, ..., G_l)$ is a packing of the complete graph K_v and no vertex belongs to more than two graphs in P, then

$$\chi(G_1) + \chi(G_2) + ... + \chi(G_{l-1}) + \psi(G_l) \leq v + \binom{l}{2}.$$

4.43 Prove Conjecture 4.16 under the assumption that no vertex belongs to more than two graphs of the decomposition.

4.44 [Alavi and Mitchem 1971]. Let G and $\bar{G}$ be complementary graphs of order v. Then $\varkappa'(G) \varkappa'(\bar{G}) \leq M(v)$ (see 4.28 and 4.30 for the definition of $M(v)$).

4.45 The upper bound in 4.45.1 is exact for each even v.

4.46 For which cardial numbers n does there exist a loopless graph with total chromatic number n?

4.47 [Entringer et al. 1976]. If G is a finite undirected graph of order v, let $f(G)$ be a sum of all $\binom{v}{2}$ distances of pairs of distinct vertices (for G disconnected we set $f(G) = \infty$). Let G and $\bar{G}$ be complementary graphs of a finite order v. Then

$$f(G) + f(\bar{G}) \geq 3v(v - 1)/2.$$

4.48 [Entriger et al. 1976]. The bound of 4.47 is sharp whenever $v \geq 5$.

4.49 [Erdős and Schuster 1981]. For each $v \in N$ there exists a graph of order v for which $\beta_1(G) = \lfloor (v + 1)/3 \rfloor$ and $\beta_1(\bar{G}) = \lfloor v/3 \rfloor$.

4.50 Is it possible to improve the lower bound in 4.33.3 by more than 1, with the additional assumption that neither G nor $\bar{G}$ have isolated vertices?

Chapter 5

Some relations between complementary graphs and between decompositions of several graphs

5.1 There is a considerable number of papers dealing with relations between complementary factors of graph decompositions. Questions of this kind are considered in many places of this book. Nordhaus—Gaddum type inequalities, which we have studied in the previous chapter, can also be viewed as a particular instance of such a problem. A very strong condition restraining the factors is the requirement of their being isomorphic. We shall be dealing with such decompositions in Chapters 6—15; the special case of two factors in considered in Chapter 14 devoted to self-complementary graphs. In the present chapter we mention just a few relations between complementary factors concerning mostly metric invariants, such as diameter and radius, and connectivity. The final paragraphs of this chapter include a little investigated topic: relations between decompositions of two or more graphs.

5.2 It is easy to decompose a complete graph K_v, with $v \geq 4$, into two connected factors. One simply takes the first factor to be isomorphic with P_v. On the other hand, no complete graph decomposes into two disconnected factors, as can be seen from the following simple result.

Theorem [Bosák et al. 1968]. *If one of two complementary graphs is disconnected then the other is connected and has diameter 1 or 2.*

Proof. Let G and $\bar{G}$ be complementary graphs, where G is disconnected. Then its vertex-set can be partitioned into two non-empty sets V_1 and V_2 in such a way that all edges joining a vertex in V_1 with a vertex in V_2 belong to $\bar{G}$. The result now easily follows.

5.3 Call a *vertex* of a graph *central* [*peripheral*] if its eccentricity is minimal [maximal]. In this case the eccentricity is equal to the radius [diameter] of the graph. Let $c(G)$ and $p(G)$ stand for the number of central and peripheral vertices of a graph G, respectively.

Theorem [Cherepanov 1978]. *Let v, c, p, $\bar{c}$ and $\bar{p}$ be positive integers with $v \geq 7$. A pair of connected complementary graphs G and $\bar{G}$ of order v for which $c(G) = c$, $p(G) = p$, $c(\bar{G}) = \bar{c}$ and $p(\bar{G}) = \bar{p}$ exists if and only if $p \geq 2$, $\bar{p} \geq 2$ and at least one of the following conditions holds:*

A. $v = c = p = \bar{c} = \bar{p}$.
B. $v = c = p = \bar{c} + \bar{p}$.
C. $v = \bar{c} = \bar{p} = c + p$.
D. $v = c = p$ and $v - \bar{c} - \bar{p} \geq 2$.
E. $v = \bar{c} = \bar{p}$ and $v - c - p \geq 2$.
F. $v = c + p = \bar{c} + \bar{p}$, $p \leq \bar{c}$ and $\bar{p} \leq c$.

Proof. See [Cherepanov 1978].

In the above cited paper it is stated without proof that from the previous theorem the following can be obtained: Finite connected complementary graphs G and $\bar{G}$ with $c(G) = c$ and $\bar{c}(G) = \bar{c}$ exist if and only if $(c, \bar{c}) \notin \{(1, 2), (1, 3), (1, 4)\}$

5.4 Theorem [Bosák et al. 1968]. *Let $v \geq 6$ be an integer. For every decomposition of K_v into two connected factors G and $\bar{G}$ there exist at least two vertices in K_v which are cut-vertices neither in G nor in $\bar{G}$.*

Proof. See [Bosák et al. 1968].

Remark. A decomposition of K_5 may happen to have only one vertex u with the above property, as shown in Fig. 5.4.1.

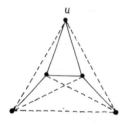

Fig. 5.4.1

5.5 In Akiyama and Harary (1979a) one can find all pairs of complementary graphs both of which have one of the following properties: A. vertex-connectivity equal to 1; B. edge-connectivity equal to 1; C. they are 2-connected; D. bipartite; E. Eulerian; F. have only odd cycles; G. they are forests. A characterization of graphs which have the same girth (that is, the length of the shortest cycle) as their complement is given in Akiyama and Harary (1979c).

5.6 In Nebeský (1973) it is proved that if G and $\bar{G}$ are finite complementary graphs of order ≥ 5 then G is connected and $L(G)$ Hamiltonian or $\bar{G}$ is connected and $L(\bar{G})$ Hamiltonian. This result is improved in Nebeský (1978) as follows: If G and $\bar{G}$ are finite complementary graphs of order ≥ 6 then either G is connected and $L(G)$ is pancyclic or $\bar{G}$ is connected and $L(\bar{G})$ is pancyclic (see 1.49).

5.7 Let $v_1(G)$ be the number of pendant vertices of a graph G. A characterization of graphs G for which $v_1(G) = v_1(\bar{G})$ is given in Akiyama and Harary (1981). It turns out that, in this case, $v_1(G) \in \{0, 1, 2\}$ as follows from the following lemma.

Lemma [Akiyama and Harary 1981]. *Let G and $\bar{G}$ be complementary graphs. If $v_1(G) \geq 2$ then $v_1(\bar{G}) \leq 2$.*

Proof. Let u_1 and u_2 be vertices of degree 1 in G. Then the only vertices which can have degree 1 in $\bar{G}$ are the two vertices adjacent in G with u_1 and u_2.

5.8 If H is a graph, let $H + K_2 \circ K_1$ denote the graph formed from H by adding 4 new vertices u_1, u_2, u_3, and u_4, the edges of the path $u_1 u_2 u_3 u_4$ and the edges joining u_3 and u_4 with all vertices of H.

Theorem [Akiyama and Harary 1981]. *A finite graph G of order $v \geq 4$ has $v_1(G) = v_1(\bar{G}) = 2$ if and only if G is of the form $H + K_2 \circ K_1$, where H is a simple graph of order $v - 4$.*

Proof. It is easily seen that $H + K_2 \circ K_1$ has the required property. Conversely, assume that the graph G is such that $v_1(G) = v_1(\bar{G}) = 2$. Let u_1 and u_2 be the pendant vertices of G and let u_4 and u_3 be the vertices adjacent with u_1 and u_2, respectively. Obviously, u_3 and u_4 are pendant vertices in $\bar{G}$ but other vertices cannot have this property. Hence u_3 and u_4 have degree $v - 2$ in G. Thus, if H is the graph obtained from G by the deletion of the vertices u_1, u_2, u_3, and u_4 then G has the form $H + K_2 \circ K_1$.

Remark. We apply the results of 5.7—5.8 in Chapter 14, namely in 14.8—14.10.

5.9 Let g_v be the number of types (1.28) of simple graphs of order v. In particular we have $g_0 = g_1 = 1$, $g_2 = 2$, $g_3 = 4$, etc.

Corollary [Akiyama and Harary 1981]. *The number of types of simple graphs with finite order $v \geq 4$ having the property $v_1(G) = v_1(\bar{G}) = 2$ is g_{v-4}.*

Proof. Since the mapping $H \to H + K_2 \circ K_1$ is one-to-one the proof immediately follows from Theorem 5.8.

5.10 Further relations between the complementary graphs can be found, for instance, in Kelmans (1967), Nordhaus (1969), Cvetković (1970), Beineke (1971), Ottaviani (1972), Harary and Schwenk (1974), Escalante and Simões-Pereira (1976), Akiyama and Harary (1979a, b, c), Akiyama et al. (1980, 1983), Nebeský (1983), and on several other places of this book. Complementary hypergraphs were studied, for instance, in Guy and Milner (1968) and Duchet (1974); see also 14.19.

5.11 So far we have been dealing with relations between separate factors of a single graph. Interesting possibilities arise when decompositions of two or more graphs are considered instead. Pioneering works in this direction are the papers of Chung et al. (1979, 1981). In these papers the authors investigated decomposition of graphs of the same order and size into the same number of subgraphs in such a way that the subgraphs equally labelled are isomorphic. Below we outline the main results of these papers. Analogous results for uniform hypergraphs are derived in Chung et al. (1982).

5.12 Let v, e, l and n be positive integers with $v \geq 2$. Let $G = (G_1, G_2, ..., G_n)$ be a family of simple graphs of the same order v and the same size e. Assume there exist decompositions

$$G_1 \leftarrow (G_{11}, G_{12}, ..., G_{1l})$$
$$G_2 \leftarrow (G_{21}, G_{22}, ..., G_{2l})$$
$$\cdots\cdots\cdots\cdots\cdots\cdots\cdots\cdots$$
$$G_n \leftarrow (G_{n1}, G_{n2}, ..., G_{nl})$$

such that for each $j \in \{1, 2, ..., l\}$ the graphs $G_{1j}, G_{2j}, ..., G_{nj}$ are mutually isomorphic. Then we call this family of decompositions a U_l-decomposition of the family G.

Let $U(G)$ be the smallest positive integer l for which there exists a U_l-decomposition of the family G. Note that l is always defined since $l \leq e$. Set

$$U_n(v) = \max_e \max_G U(G),$$

where G ranges over all n-families of simple graphs of order v and size e, and e ranges over the set $\left\{1, 2, ..., \binom{v}{2}\right\}$. Clearly we have

$$1 \leq U_n(v) \leq \binom{v}{2}. \tag{1}$$

5.13 Theorem [Chung et al. 1979, 1981]. *Let $n \geq 1$ and $v \geq 2$ be integers. There exist functions $o_n(v)$ such that*

$$U_n(v) = \begin{cases} 1 & \text{if } n = 1, \\ 2v/3 + o_2(v) & \text{if } n = 2, \\ 3v/4 + o_n(v) & \text{if } n \geq 3, \end{cases}$$

and $\lim_{v \to \infty} o_n(v) = 0$ *for* $n = 2, 3, ...$

Proof. I. The equality $U_1(v) = 1$ is obvious.

II. We show that $U_2(v) \geq 2v/3 + o_2(v)$ for $o_2(v) = -1$. To do this we shall construct a 2-family $G = (G_1, G_2)$ as follows: Let G_1 be the star S_v of order v. Further, let G_2 be a graph of order v whose components are the cycles C_3 and one path P_s of order $s = v - 3\lfloor (v-1)/3 \rfloor$. It is easy to see that $s \in \{1, 2, 3\}$ and $G_2 = \lfloor (v-1)/3 \rfloor K_3 \cup P_s$. Note that G_1 and G_2 have the same order v and the same size $v - 1$. Moreover, one easily computes that $U(G) = \lfloor (2v-1)/3 \rfloor \geq 2v/3 - 1$. Hence, $U_2(v) \geq 2v/3 - 1$ also.

III. We show that for each $n \geq 3$ we have

$$U_n(v) \geq 3v/4 + o_n(v), \tag{1}$$

where $o_n(v) = -7\sqrt{2v}/2 - 21/4$.

If $v \leq 8$ then $3v/4 - 7\sqrt{2v}/2 - 21/4 < 3v/4 - 21/4 < 1$ and in view of 5.12.1 the inequality (1) is satisfied. Therefore we may assume that $v > 8$. In this case there exists a unique $m \in N$ such that $8m^2 \leq v < 8(m+1)^2$. Set $v = 8m^2 + q$. Obviously we have

$$0 \leq q \leq 8(m+1)^2 - 8m^2 - 1 = 16m + 7 \leq 4\sqrt{2v} + 7.$$

Now we shall construct an n-family $G = (G_1, G_2, ..., G_n)$ of simple graphs of order v as follows: First let $q = 0$. In this case let G_1 be the star of order $8m^2$. Further, let G_2 be the disjoint union of one copy of P_4 and $2m^2 - 1$ copies of C_4. Finally, for $i = 3, 4, ..., m$, let G_i consist of one copy of $K(2m, 2m)$, $m-1$ copies of C_4, one copy of P_4 and $4m^2 - 4m$ copies of K_2. If $q \neq 0$ we add to each G_i q isolated vertices. Obviously, each of these graphs has order $v = 8m^2 + q$ and size $e = 8m^2 - 1$.

It is not difficult to see (one can use Theorem 9.6, for example) that

$$U(G) = 6m^2 - 2m = 3(v-q)/4 - \sqrt{2(v-q)}/2 \geq$$
$$\geq 3v/4 - 3\sqrt{2v} - 21/4 - \sqrt{2v}/2 = 3v/4 - 7\sqrt{2v}/2 - 21/4,$$

whence

$$U_n(v) \geq 3v/4 - 7\sqrt{2v}/2 - 21/4$$

as required.

IV. To complete the proof it remains to show that

$$U_2(v) \leq 2v/3 + o_2(v), \tag{2}$$
$$U_n(v) \leq 3v/4 + o_n(v) \tag{3}$$

for every $n \geq 3$. We omit the proofs of (2) and (3) and refer the reader to Chung et al. (1979,1981).

5.14 In addition to functions U_n the authors of the above-mentioned papers study the functions $U_n^*(n \in N)$ which are defined similarly as U_n but restricted to bipartite graphs only. Hence, G ranges over all n-families of simple bipartite graphs of order v and size e, and e then, naturally, ranges over the set $\{1, 2, ..., \lfloor v/2 \rfloor \lceil v/2 \rceil \}$. The corresponding results read as follows:

$$U_n^*(v) = \begin{cases} 1 & \text{if } n = 1, \\ v/2 + o_2^*(v) & \text{if } n = 2 \text{ [Chung et al. 1979]}, \\ 3v/4 + o_n^*(v) & \text{if } n = 3 \text{ [Chung et al. 1981]}. \end{cases}$$

where $\lim_{v \to \infty} o^*(v)/v = 0$ for $n \geq 2$ (see also Exercises 5.16—5.17). Comparing the values of U_n and U_n^* we see that, surprisingly, the essential difference encounters only for $n = 2$. Remarkable is also the fact that in both cases the "main term" $3v/4$ for $n \geq 3$ is independent of n.

Exercises

5.15 [Akiyama and Harary 1981]. Every simple graph G with $v_1(G) = v_1(\bar{G})$ (see 5.7) has diameter 3.

5.16 [Chung et al. 1979] (see 5.14). For every positive integer v it holds that $U_2^*(v) \geq v/2 - 1/2$.

5.17 [Chung et al. 1981] (see 5.14). For every integer $n \geq 3$ and arbitrary v it holds that $U_n^*(v) = 3v/4 + o_n^*(v)$, where $\lim_{v \to \infty} o_n^*(v)/v = 0$.

DECOMPOSITIONS OF A GRAPH
INTO ISOMORPHIC SUBGRAPHS

In Chapters 6—15 we shall be dealing with decompositions of complete and other graphs into mutually isomorphic subgraphs. Some of the chapters (6, 7, 12—15) will be mainly devoted to general properties and construction methods of such decompositions while the remaining ones (8—11) to particular types of decompositions.

Chapter 6

Necessary conditions for the existence of a G-decomposition of a graph

6.1 Let a graph G be given. A *G-decomposition of a graph H* is a decomposition of H into subgraphs isomorphic to G. (This definition can be applied to directed, undirected or mixed graphs.) We write $G|H$ whenever a G-decomposition of H exists.

A *decomposition* is said to be *balanced* if every vertex of H belongs to the same number of graphs in the decomposition, more accurately — *r-balanced* if this number is r. We write $G \| H$ if H admits a balanced G-decomposition. The following lemma is obvious:

Lemma. *Let G and H be finite graphs and let both be regular or diregular. Then every G-decomposition of H is balanced.*

6.2 E x a m p l e. The graph in Fig. 6.2.1 clearly has a $P_k^{\rightarrow}$-decomposition (a decomposition into directed paths of order k) for $k = 2, 3, 4$ and 5 but has no $P_k^{\rightarrow}$-decomposition for other k. It is easily verifed that for $k = 2, 3$ and 5 this decomposition cannot be balanced. For $k = 4$ a $P_k^{\rightarrow}$-decomposition can, but need not, be balanced.

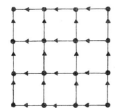

Fig. 6.2.1

6.3 In the subsequent paragraphs we shall often use the symbols $V(G)$, $v(G)$, $E(G)$ and $e(G)$ defined in 1.2, 1.4 and 1.23 as well as od u, id u, deg u, deg G and $D(G)$ introduced in 1.9—1.10.

Theorem. *Let G and H be non-empty graphs with vertex-sets $V(G) = \{u_1, u_2, \ldots, u_k\}$ and $V(H) = \{U_1, U_2, \ldots, U_r\}$, respectively. If there exists a G-decomposition of H into b subgraphs then:*

    A. $k \leq v$ or $e(H) = 0$.

    B. $e(G)b = e(H)$.

C. $e(G)|e(H)$ (*first divisibility condition*).

D. $b|e(H)$ (*second divisibility condition*).

E. *For each* $i \in \{1, 2, ..., v\}$ *there exist non-negative integers* $x_1, x_2, ..., x_k$ *such that*

$$x_1 \operatorname{od} u_1 + x_2 \operatorname{od} u_2 + ... + x_k \operatorname{od} u_k = \operatorname{od} U_i, \tag{1}$$

$$x_1 \operatorname{id} u_1 + x_2 \operatorname{id} u_2 + ... + x_k \operatorname{id} u_k = \operatorname{id} U_i, \tag{2}$$

$$x_1 \deg u_1 + x_2 \deg u_2 + ... + x_k \deg u_k = \deg U_i. \tag{3}$$

Moreover, if the G-decomposition of H is r-balanced, then:

F. $bk = rv$.

G. $ke(H) = rve(G)$.

H. $ve(G)|ke(H)$.

Proof. The statements A and B are obvious; C and D follow from B. To prove E, let x_j ($j = 1, 2, ..., k$) be the number of graphs (isomorphic to G) in the G-decomposition of H in which the copy of the vertex u_j in an isomorphism with G coincides with U_i. The equations (1)—(3) are then clearly satisfied.

If the G-decomposition is r-balanced then F is obvious. Multiplying both sides of B by k and substituting to F we get G, and G implies H.

6.4 Corollary. *If G and H are finite graphs and G|H then D(G)|D(H).*

Proof. We can assume that both G and H are non-empty. By Theorem 6.3 the equation 6.3.3 has an integral solution $[x_1, x_2, ..., x_k]$. Hence

$$D(G) = \text{g.c.d.} \{\deg u_1, \deg u_2, ..., \deg u_k\} | \deg U_i.$$

Since $i \in \{1, 2, ..., v\}$ is arbitrary we have

$$D(G) | \text{g.c.d.} \{\deg U_1, \deg U_2, ..., \deg U_v\} = D(H).$$

6.5 If some special conditions are imposed on G or H further consequences can be derived from the above corollary. We mention one such result (for further results the reader may consult, e.g. Bermond and Sotteau 1976, Wilson 1976, Bermond et al. 1980):

Corollary. *Let G and H be finite graphs and let H be regular. If H admits a G-decomposition then D(G)| deg H (that is, the degree of H is divisible by the greatest common divisor of degrees of the vertices of G).*

Proof follows immediately from Corollary 6.4.

6.6 Another important case of Theorem 6.3 arises when H is an n-*diregular graph* (i.e., for every vertex u of H it holds that $\operatorname{od} u = \operatorname{id} u = n$). We shall need the following characteristic $g(G)$ which can be assigned to an arbitrary finite graph G. If $e(G)$, the size of G, is zero then set $g(G) = 0$. Otherwise let $g(G)$ be the smallest positive integer g for which the following system of two equations

$$x_1 \text{ od } u_1 + x_2 \text{ od } u_2 + \ldots + x_k \text{ od } u_k = g,$$
$$x_1 \text{ id } u_1 + x_2 \text{ id } u_2 + \ldots + x_k \text{ id } u_k = g \tag{1}$$

has an interval solution $[x_1, x_2, \ldots, x_k]$. Here, again, $V(G) = \{u_1, u_2, \ldots, u_k\}$. Since for $g = e(G)$ the system (1) has an obvious solution $x_1 = x_2 = \ldots = = x_k = 1$, the number $g(G)$ is correctly defined.

6.7 Lemma. *If for an integer g the system 6.6.1 has an integral solution then $g(G)|g$.*

Proof. If $g(G) = 0$ then $g = 0$, and the assertion holds. Otherwise let z be the remainder of dividing g by $g(G)$. Clearly, the system 6.6.1 has an integral solution for z in place of g. However, in view of the definition of $g(G)$, this is only possible when $z = 0$. Hence, $g(G)|g$.

6.8 The following theorem in the case when H is a complete graph has been proved in Wilson (1976).

Theorem. *Let G and H be finite directed graphs and let H be n-diregular. If H has a G-decomposition then*

$$e(G)|nv(H), \tag{1}$$
$$g(G)|n. \tag{2}$$

Proof. If $n = 0$ then the assertion is obvious. Therefore, let $n > 0$. Then (1) follows from 6.3.C. Since od $U_i =$ id $U_i = n$ for every i, using 6.3.E we obtain

$$x_1 \text{ od } u_1 + x_2 \text{ od } u_2 + \ldots + x_k \text{ od } u_k = n,$$
$$x_1 \text{ id } u_1 + x_2 \text{ id } u_2 + \ldots + x_k \text{ id } u_k = n.$$

Lemma 6.7 implies $g(G)|n$.

6.9 The necessary conditions (1) and (2) of Theorem 6.8 are in some important cases also asymptotically sufficient, as proved in Wilson (1976):

Theorem. *Let G be a finite digraph with size $e(G)$. Then there exists an integer $n(G)$ such that for all $v \geq n(G)$ it holds: $G|K_v^*$ (where K_v^* is the complete digraph of order v) if and only if*

$$e(G)|(v - 1)v, \tag{1}$$

$$g(G)|v - 1,\tag{2}$$

$g(G)$ *being defined in 6.6.*

Proof. The necessity of the above conditions follows from Theorem 6.8 for $n = v - 1$, where $v(H) = v$. The sufficiency part of the proof is quite involved; it employs designs, finite fields and vector spaces. We refer the reader to Wilson (1976).

6.10 Example. Let G be a digraph depicted in Fig. 6.10.1. In this case the system 6.6.1 has the form

$$x_1 + 3x_2 + 2x_3 + 2x_4 = g,$$
$$3x_1 + x_2 + 2x_3 + 2x_4 = g.$$

It can be easily verified that $g(G) = 2$. By Theorem 6.9, for any sufficiently large v the complete digraph K_v^* has a G-decomposition exactly when $8|v - 1$ and $2|v - 1$, that is, when $v \equiv 1 \pmod 8$.

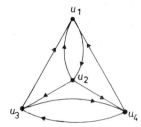

Fig. 6.10.1

6.11 Some special cases of the following corollary can be found in Wilson (1974a), Erdős and Schönheim (1975) and Wilson (1975); see also Hall (1967, p. 248).

Corollary [Wilson 1976]. *Let G be a finite simple graph of size $e(G)$. Then there exists an integer $n(G)$ such that for every $v \geq n(G)$ it holds: The complete graph K_v of order v has a G-decomposition if and only if*

$$2e(G)|(v - 1)v,\tag{1}$$
$$D(G)|v - 1,\tag{2}$$

where $D(G)$ is the g.c.d. of degrees of the vertices of G.

Proof. Let G^* be the digraph obtained from G by replacing each edge of G with a pair of oppositely oriented arcs. Clearly, $e(G^*) = 2e(G)$ and $D(G) = g(G)$ (cf. 6.4—6.7). It is readily seen that K_v has a G-decomposition if and only if K_v^* has a G^*-decomposition. Thus 6.11 now follows from 6.9.

6.12 There are plenty of results concerning necessary or sufficient conditions for the existence of G-decompositions of a given graph H. As a rule, necessary conditions issue from the conditions 6.3.A—C and are often complemented by additional conditions depending on a special choice of G or H. Sufficient conditions are usually proved by construction of corresponding G-decompositions. Most common construction methods include:

A. Cyclic methods (rotation of edges or subgraphs, vertex- or edge-labellings, graceful graphs, Bose's difference method).

B. Application of permutations, finite groups or fields.

C. Application of designs.

D. Composite methods (composition of several G-decompositions).

We remark that in some cases the above methods overlap and the same construction can be viewed as application of different methods. Moreover, some constructions employ several methods in succession. We present a few typical results and outline the methods of their proofs.

6.13 Many papers are devoted to the problem of the existence of G-decompositions of the complete graph K_v or the complete digraph K_v^*. In 6.9—6.11 we have mentioned some asymptotical results of this kind. For the sake of greater generality we shall often consider the λ-fold complete graph $^\lambda K_v$ and digraph $^\lambda K_v^*$ (cf. 1.6). The graphs $^2K_3$ and $^2K_3^*$ are depicted in Fig. 6.13.1.

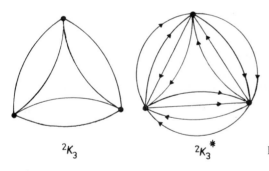

$^2K_3$ $^2K_3^*$ Fig. 6.13.1

6.14 Necessary conditions for the existence of a G-decomposition of the λ-fold complete graph $^\lambda K_v$ can be easily deduced from Theorem 6.3 and its corollaries.

Corollary [Bermond and Sotteau 1976]. *Let G be a finite loopless graph and let $^\lambda K_v$, with $v \geq 2$, have a G-decomposition. Then the following hold:*

$$v(G) \leq v, \tag{1}$$

$$2e(G) | \lambda v(v-1), \tag{2}$$

$$D(G) | \lambda(v-1). \tag{3}$$

If, moreover, $^\lambda K_v$ has a balanced G-decomposition then

$$2e(G)|\lambda(v-1)v(G).\tag{4}$$

Proof. (1) is obvious, (2) follows from 6.3.C, (3) from 6.5 and (4) from 6.3.H.

6.15 Remark. In case G is n-regular we have $D(G) = n$ and $2e(G) = v(G)n$, thus the conditions 6.14.3 and 6.14.4 are (for $v(G) \neq 0$) equivalent.

6.16 A similar result as 6.14 can be derived for G-decompositions of complete λ-fold digraphs (cf. Bermond and Sotteau 1976 and Wilson 1976).

Corollary. *Let G be a finite loopless graph. If $^\lambda K_v^*$, with $v = 2$, has a G-decomposition then the following hold:*

$$v(G) \leq v,\tag{1}$$

$$e(G)|\lambda v(v-1),\tag{2}$$

$$g(G)|\lambda(v-1).\tag{3}$$

If $^\lambda K_v^*$ has a balanced G-decomposition then, moreover, it holds that

$$e(G)|v(G)\lambda(v-1).\tag{4}$$

Proof. It suffices to use Theorems 6.3 and 6.8.

6.17 Remark. It is easy to see that for g.c.d. $D^+(G)[D^-(G)]$ of indegrees [outdegrees] of the vertices of a finite directed loopless graph G we have $D^+(G)|g(G)$ and $D^-(G)|g(G)$. Thus from 6.16.3 one can deduce the following necessary conditions for the existence of a G-decomposition of K_v^* [Bermond and Sotteau 1976]:

$$D^+(G)|\lambda(v-1),\tag{1}$$

$$D^-(G)|\lambda(v-1).\tag{2}$$

If G is a finite loopless n-diregular graph then clearly $g(G) = D^+(G) = = D^-(G)$ and the conditions 6.16.3, 6.17.1 and 6.17.2 coincide.

6.18 Some general results on the existence of G-decompositions including construction methods using basic operations on graphs can be found particularly in Hell and Rosa (1971, 1972) and Bermond and Sotteau (1976); see also 11.23—11.38. For generalizations to hypergraphs see Germa (1978) and references therein. G-decompositions of edge-coloured graphs (above all, modified complete graphs) with applications to whist competitions are investigated in Hanani (1975) and Wilson (1977). *G*-packings, that is, packings of a given graph *H* with graphs isomorphic to *G*, are studied in Hell and Kirkpatrick (1981). In this paper it is required in addition that the vertex-sets of graphs in the packing form a partition of $V(H)$. The computational complexity of the problem whether or not *H* has such a *G*-packing is also considered.

Exercises

6.19 [Bermond and Sotteau 1976]. Let *G* and *H* be finite loopless undirected graphs. Let G_1 be obtained from *G* by assigning an arbitrary orientation to *G*. Assume that G_1 is isomorphic to its converse graph, that is, to the graph obtained from G_1 by reversing the orientation of each edge of G_1. Finally let H^* be the directed graph formed from *H* by replacing each undirected edge by a pair of oppositely directed parallel edges with the same ends (1.18). Prove or disprove the following implications (for notation see 6.1):

A. $G|H \Rightarrow G_1|H^*$.
B. $G|H^* \Rightarrow G|H$.
C. $G_1|H^* \Rightarrow G|^2H$.
D. $G|^2H \Rightarrow G_1|H^*$.

6.20 Let *R* be a decomposition of *H* such that $G|F$ for every $F \in R$. Then $G|R$.

6.21 For arbitrary graphs *F*, *G* and *H* the following holds: If $G|F$ and $F|H$ then $G|H$.

6.22 Let *G* and *K* be arbitrary graphs and $n_1, \lambda_1, \lambda_2, \ldots, \lambda_n$ positive integers. If $G|^{\lambda_1}K, G|^{\lambda_2}K, \ldots, G|^{\lambda_n}K$ then $G|^{\lambda_1 + \lambda_2 + \cdots + \lambda_n}K$.

6.23 Consider the assertions obtained from A. 6.20; B. 6.21; C. 6.22 by assuming that the corresponding decompositions are balanced. Which of them are true?

Chapter 7

Cyclic decompositions, vertex labellings and graceful graphs

7.1 Cyclic decompositions play an important role in the study of graph decompositions, for example in the construction of decompositions of complete graphs into isomorphic subgraphs. A decomposition R of a graph H into subgraphs is said to be *cyclic* if there exists an isomorhism f of H which induces a cyclic permutation f_V of the set $V = V(H)$ and satisfies the following implication: if $G \in R$ then $fG \in R$. Here fG is the subgraph of H with vertex-set $\{fu; u \in V(G)\}$ and edge-set $\{fw; w \in E(G)\}$. The method of cyclic decompositions, which we are going to expound, has been used in many papers, see e.g. Kotzig (1965b), Rosa (1966a, b) and Hartnell (1975). The vertex automorphism f_V will be called a *cyclic (vertex) automorphism of G associated with R*.

7.2 To simplify our investigation of cyclic decompositions of a finite graph H we shall usually assume H to have the vertex-set $V = V(H) = \{0, 1, ..., v-1\}$, $v \geq 3$ and

$$f_V = \begin{pmatrix} 0 & 1 & 2 & ... & v-2 & v-1 \\ 1 & 2 & 3 & ... & v-1 & 0 \end{pmatrix}.$$

(The permutation f_V assigns to an element in the first line the element standing below it.) The mapping f (and sometimes f_V) will be called a *(simple) rotation*, and its i-th iteration an *i-fold rotation* ($i = 0, 1, 2, ..., v$). Note that $f^0 = f^v$ is the identity mapping. The term "rotation" can be applied to subgraphs or their elements; therefore we shall speak of a *rotation of subgraph, vertex* or an *edge*. This terminology is justified by a representation of H in the plane so that the vertices of H coincide with the vertices of a regular n-gon and the edges of H are the sides and diagonals of the n-gon. Rotations of subgraphs, vertices or edges then become rotations around the centre of the n-gon giving this word its natural meaning. Figure 7.2.1 depicts a cyclic decomposition of a graph of order 6 and size 12 (the graph is in fact $K(3 \times 2) = K(2, 2, 2)$) into two isomorphic factors (the first one is represented by solid lines, the second one by dashed lines). Observe that the second factor can be obtained by a simple rotation of the first one. Further rotations yield nothing new since in this case $f^0 = f^2 = f^4$ and $f = f^3 = f^5$.

The following is a useful though simple lemma.

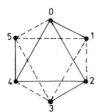

Fig. 7.2.1

Lemma. *Every cyclic decomposition of a finite graph is balanced.*

Proof. Let u and U be two vertices of a finite graph H with a cyclic decomposition R, and let f be the cyclic automorphism of H associated with R. Then, for some $i \in \{0, 1, ..., v - 1\}$, f^i maps u to U. Let $H_1, H_2, ..., H_n$ be the collection of all (mutually distinct) graphs which contain u and belong to R. Since both f^i and its inverse f^{v-i} are automorphisms of h, the graphs $f^i H_1, f^i H_2, ..., f^i H_n$ constitute the collection of all members of R containing U. Hence u and U belong to the same number of graphs of the decomposition R.

7.3 Let R be a cyclic decomposition of a graph H. We can define a decomposition R' of R as follows: two elements of R will fall into the same class if any of them can be obtained from the other by a rotation (possibly applied several times). (For instance, the decomposition R represented in Fig. 7.2.1 has R' consisting of a single class and R' contains two factors.) Choosing one representative from each class we get a *base system of R*. Its elements will be called *base graphs of the decomposition R*. (In the above-mentioned example every base system consists of a single element.) The *length of an edge ij* of a graph with vertex-set $\{0, 1, ..., v - 1\}$ is the number $\min \{|j - i|, v - |j - i|\}$. The length of any edge is thus one of the numbers $0, 1, 2, ..., \lfloor v/2 \rfloor$. Note that an edge with length 0 must be a loop. It is easily seen that a rotation does not change the length of an edge. (The graph in Fig. 7.2.1 has edges of length 1 and 2 but no edges of length 0 and 3.)

We shall make use of cyclic decompositions mostly in the case of complete graphs. However, they are also helpful for construction of decompositions of complete m-partite graphs (7.2, 7.49) and graphs $^\lambda K(m \times n)$ (9.25).

7.4 We now present two more examples by constructing a decomposition of K_{91} into complete subgraphs of order 6 and one into complete subgraphs of order 7. In what follows a decomposition into complete subgraphs of order k will be called a K_k-*decomposition* (this agrees with the general definition in 6.1).

To construct a K_6-decomposition of K_{91} we choose three complete subgraphs of K_{91} with vertex-sets $\{0, 1, 3, 7, 25, 38\}$, $\{0, 5, 20, 32, 46, 75\}$ and $\{0, 8, 17, 47, 57, 80\}$, respectively, as a base system. It is easy to see that for each $i \in \{1, 2, \ldots, 45\}$ there exists exactly one edge of length i in some of the three graphs. Thus applying 91 rotations (including the identity) we obtain a K_6-decomposition of K_{91}. This decomposition was constructed according to Mills (1975b); see also Hanani (1975, p. 304), or Mills (1978a).

Now we shall construct a K_7-decomposition of K_{91}. It is based on the Scheme 111 from the Table in Appendix 1 of Hall (1967). In this case the base system will consist of three complete subgraphs with vertex-sets $\{0, 10, 27, 28, 31, 43, 50\}$, $\{0, 11, 20, 25, 49, 55, 57\}$ and $\{0, 13, 26, 39, 52, 65, 78\}$. These subgraphs contain the edges of length $1, 2, \ldots, 45$ exactly once except for lengths 13, 26 and 39 which occur seven times. It is easy to see that applying 91 rotations of the first two graphs and 13 rotations of the third one yields a K_7-decomposition of K_{91}.

7.5 In general a cyclic decomposition does not necessarily consist of mutually isomorphic graphs, not even in the case when all graphs in the decomposition have the same size. As an example, consider a decomposition of K_9 with vertex-set $\{0, 1, 2, \ldots, 8\}$ into 12 subgraphs of size 3. The corresponding base system is depicted in Fig. 7.5.1. The first base graph has edges of length 3 only, while the second one has edges of length 1, 2 and 4.

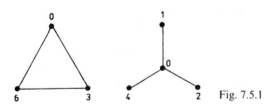

Fig. 7.5.1

It is readily seen that a cyclic decomposition consists of isomorphic graphs exactly when some (equivalently, each) of its base systems has this property. In particular, this is satisfied whenever it has a one-element base system. (A generalization of this situation is a decomposition according to a cyclic group, which will be introduced in 13.4.) The following theorem concerns an important special case.

7.6 Theorem. *Every cyclic decomposition of a finite complete graph K_{2n+1} into $2n + 1$ subgraphs of size n consists of mutually isomorphic graphs, each containing exactly one edge of length i where $i \in \{1, 2, \ldots, n\}$.*

Proof. Let R be a cyclic decomposition of K_{2n+1} into $2n + 1$ subgraphs of size n. If R contains a graph whose edges have mutually distinct lengths then all graphs of R have this property. Hence, all are mutually isomorphic.

Assume that R contains a graph H which has $d \geq 2$ edges of the same length $i \in \{1, 2, ..., n\}$. Choose i in such a way that d will be maximum for H. Then, after several rotations, H transforms into a graph in R which has a common edge with H. Hence these two graphs coincide. Obviously, $d|2n + 1$. Thus rotating H $(2n + 1)/d$ times we get H again. It follows that for each length that occurs in H there are d edges representing this length. Since H has n edges we have $d|n$. Recall that $d|2n + 1$ also, whence $d|2n + 1 - 2n = 1$. This contradiction proves the theorem.

7.7 A decomposition R of a graph H is said to be *symmetric* if $|R| = = |V(H)|$, that is if the number b of subgraphs in R is equal to the number v of vertices of H. This term is related with the so-called symmetric designs which will be discussed in 8.5, 8.7 and 8.20; an analogous condition on such designs can be expressed by the equality $b = v$.

Corollary. *A symmetric decomposition of the complete graph K_v into subgraphs of equal sizes consists of isomorphic graphs.*

Proof. Since the number of elements in the decomposition is v, each of them has size

$$\frac{\binom{v}{2}}{v} = \frac{v - 1}{2}.$$

Hence v is odd. Now, for $v \geq 3$ we can use 7.6 while the case $v = 1$ is obvious.

7.8 Let a *vertex-labelling* h of a graph G (i.e. a mapping of $V(G)$ into a set of numbers) be given. If w is an edge with endvertices u and U we say that the number $\bar{h}(w) = |h(u) - h(U)|$ is the *value of w in the labelling h*.

There is a close connection between cyclic G-decomposition of complete graphs and different kinds of vertex-labellings of G, see Rosa (1967a). Let G be a finite graph of size e. A 1-1 mapping h of $V(G)$ into the set $\{0, 1, 2, ..., 2e\}$ is said to be a ϱ-*labelling* of G if $e = 0$ or else the set of edge-values of h is $\{x_1, x_2, ..., x_e\}$ where $x_i = i$ or $x_i = 2e + 1 - i$, for $i = 1, 2, ..., e$. A special case of a ϱ-labelling is a β-*labelling* [Rosa 1967a] which can be defined as a 1-1 mapping h of $V(G)$ into the set $\{0, 1, 2, ..., e\}$ such that for $e \neq 0$ the edge-values of h constitute the set $\{1, 2, ..., e\}$. β-Labellings are commonly known as *graceful labellings* or *graceful numberings* (the term was introduced in Golomb 1972) and a graph admitting a graceful numbering is called a *graceful graph*. As an example of a graceful graph we can take the *Petersen graph* depicted in Fig. 7.8.1 together with a β-labelling. This example is due to Golomb (1972).

A slightly more general is the concept of a *d-graceful graph*. A graph G of size e is said to be *d-graceful* ($d \in N$) if it admits a vertex labelling h which is a 1-1 mapping of $V(G)$ into the set $\{0, 1, ..., d + e - 1\}$ with the property that for $e \geq 1$ the edge-values of h form the set $\{d, d + 1, ..., d + e - 1\}$. Note that a 1-graceful graph is simply a graceful graph. Basic results and open problems can be found in Slater (1981) and Maheo and Thuillier (1982). In what follows we shall restrict ourselves to the case $d = 1$. The following lemma is obvious.

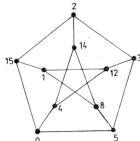

Fig. 7.8.1

Lemma. *In a graceful graph of order v it holds that $e \geq v - 1$.*

Proof. Otherwise there would be no 1-1 mapping $V(G) \rightarrow \{0, 1, ..., e\}$.

The above condition is satisfied in every connected graph. Thus using this condition we can rule out the existence of a graceful labelling just for some disconnected graphs as, for instance, 1-regular graphs of order > 2.

7.9 A further restriction of the concept of a ϱ-labelling leads to α-labellings. A β-*labelling* h of a graph G of size e is said to be an α-*labelling* if there exists a number $\alpha(h) \in \{0, 1, ..., e\}$ such that for every edge uU of G it holds

$$\min \{h(u), h(U)\} \leq \alpha(h) < \max \{h(u), h(U)\}. \tag{1}$$

In Sheppard (1976) a β-labelling (graceful labelling) is called *proper labelling* and α-labelling is called a *balanced proper labelling*.

Lemma. *Let G be a graph of size e and h an α-labelling of G. Then the following hold:*

A. *The number $\alpha(h)$ is uniquely determined by h.*

B. *The mapping $h': V(G) \rightarrow \{0, 1, ..., e\}$, $h'(u) = e - h(u)$ is also an α-labelling of G and $\alpha(h') = e - \alpha(h) - 1$* [Rosa 1965, Sheppard 1976].

C. *G is bipartite* [Rosa 1967a].

Proof. A. Let $\bar{h}(uU) = |h(u) - h(U)| = 1$. Then there is a unique integer $\alpha(h)$ satisfying the inequalities (1), namely $\alpha(h) = \min \{h(u), h(U)\}$.

B. The proof follows from the definition of an α-labelling and the number α.

C. The vertex-set of G can be partitioned into two subsets according to whether $h(u) = \alpha(h)$ or not. Obviously vertices in the same set cannot be adjacent.

7.10 The following theorem explains the importance of the concept of a ϱ-labelling.

Theorem [Rosa 1967a]. *Let G be a finite simple graph of size $e > 0$. A cyclic G-decomposition of K_{2e+1} exists if and only if G has a ϱ-labelling.*

Proof. We can assume K_{2e+1} to have the numbers 0, 1, 2, ..., 2e as its vertices. Then there are $2e + 1$ edges of each length 1, 2, ..., e (7.3). The edge joining the vertices i and j has length $\min\{|i - j|, 2e + 1 - |i - j|\}$.

Now assume that G admits a ϱ-labelling. Construct the subgraphs G_0, G_1, ..., G_{2e} of K_{2e+1} as follows: The edge with end-vertices i and j belongs to G_0 exactly when G contains an edge whose end-vertices are labelled i and j, respectively. A vertex is included in G_0 if there exists some edge incident with that vertex. The graph G_i is then obtained by an i-fold rotation of G_0 (7.1). Each of the subgraphs G_0, G_1, ..., G_{2e} contains exactly one edge of length i for every $i \in \{1, 2, ..., e\}$. Consequently, G_0 is isomorphic with G and the same holds for each G_i, where $i = 1, 2, ..., 2e$. Thus we have constructed a cyclic G-decomposition of K_{2e+1}.

Assume that K_{2e+1} has a cyclic G-decomposition and pick up an arbitrary graph G_0 from this decomposition. By Theorem 7.6, G contains exactly one edge of length i for each $i \in \{1, 2, ..., 2e\}$. Thus, labelling each vertex of G_0 by the number it has as a vertex of K_{2e+1} we obtain a ϱ-labelling of G_0. Since G is isomorphic with G_0 we conclude that G, too, has a ϱ-labelling.

7.11 As we have already mentioned, the concept of a graceful numbering was introduced in Rosa (1967a). Later on, Golomb (1972) has naturally come to the same notion through the following consideration. Let G be a finite graph. Let grac G denote the smallest integer n for which there exists a 1-1 mapping of $V(G)$ into the set $\{0, 1, 2, ..., n\}$ for which the edge-values are mutually distinct. The number grac G is called the *gracefulness* of G. A graceful graph is then defined by the condition grac $G = e(G)$. The gracefulness is always defined, as can be seen from the following result.

Theorem. *Let G be a finite graph of order v and size e. Then the number grac G is defined and the following holds:*

$$e = \text{grac } G \le \text{grac } K_v \le 2^{v-1} - 1.$$

Proof. To show the last inequality it suffices to label the vertices of K_v, denoted by u_i, with $h(u) = 2^i - 1$ for $i = 0, 1, ..., v - 1$. The edge-values then are $\bar{h}(u_0 u_1) = 1$, $\bar{h}(u_2 u_3) = 4$, ..., $\bar{h}(u_{v-2} u_{v-1}) = 2^{v-1}$ or sums of these numbers with distinct summands (e.g., $\bar{h}(u_1 u_4) = \bar{h}(u_1 u_2) + \bar{h}(u_2 u_3) + \bar{h}(u_3 u_4) = 2 + 4 + 8 = 14$). These numbers are clearly mutually distinct, whence grac $K_v \le 2^{v-1} - 1$. The rest of the proof is easy.

7.12 The exact value of the number grac K_v is not known in general. However, for some values of v the upper bound for grac K_v has been improved. For example it is known that grac $K_4 = 6$ (this follows from Theorems 7.11 and 7.12), grac $K_5 = 11$ (see 7.54) and grac $K_6 = 17$, while the upper bound yields just grac $K_4 \le 7$, grac $K_5 \le 15$ and grac $K_6 \le 31$. Along with some related results Erdős has shown (see Golomb 1972) that asymptotically grac $K_v \sim v^2$.

Problem. *Determine the exact value of grac K_v for all v.*

The following result shows that the complete graph K_v is graceful (that is, admits a β-labelling) exactly when $v \le 4$. More general ϱ-labellings and cyclic K_k-decompositions of complete graphs will be considered in 7.55, 7.56 and R 7.56.

Theorem [Golomb 1972]. *The complete graph K_v is graceful if and only if $v \le 4$.*

Proof. If $v \le 4$ it suffices to construct a graceful numbering for K_r. Let the vertex-values form the set $\{0\}$ for K_1, $\{0, 1\}$ for K_2, $\{0, 1, 3\}$ for K_3, and $\{0, 1, 4, 6\}$ for K_4. It is readily verified that these labellings are graceful.

Conversely assume that K_v has a graceful numbering for some $v \ge 5$. Let M be the set of vertex-values of such a labelling. Since K_v necessarily contains an edge with value $e = \binom{v}{2}$ we have $0 \in M$ and $i \in M$. Further there must be an edge with value $e - 1$, thus $1 \in M$ or $e - 1 \in M$. Let $1 \in M$, say (the other case is analogous). Since there exists an edge with value $e - 2$, one of the numbers 2, $e - 2$ or $e - 1$ must be contained in M. However, neither 2 nor $e - 1$ belongs to M for otherwise there would exist two edges of length 1. Hence $\{0, 1, e - 2, e\} \subseteq M$. By assumption, $v \ge 5$ so $e \ge 10$ and hence $e - 4 \ge 6$. Now it is easy to see that to ensure an edge of length $e - 4$ one must allow at least two edges of the same length (1 or 2).

7.13 *Euclidean models* of labelled graphs are also of interest in this context. In this representation to each vertex there corresponds the point of the real line whose value is equal to the label of the vertex. Edge-values are simply the Euclidean distances of the endvertices. Figure 7.13.1 represents the Euclidean model of K_4 with vertex values 0, 1, 4 and 6. It can be imagined as a ruler of length 6 with four slots (0, 1, 4, 6) that can be used to measure any integral distance ≤ 6.

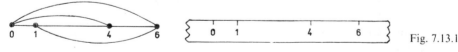

Fig. 7.13.1

7.14 This topic is related to the so-called difference bases studied in additive number theory. A *difference base* with respect to a positive integer e is a set M of integers such that every positive integer $\le e$ can be expressed as a difference of two elements of M. For instance $\{0, 8, 18, 19, 22, 24, 31,$

39} is a difference base with respect to 24. This difference base gives rise to a ϱ-labelling of a graph with 8 vertices and 24 edges. It can also be understood as a ruler enabling to measure any integral distance from 0 up to 24 (see Fig. 7.14.1).

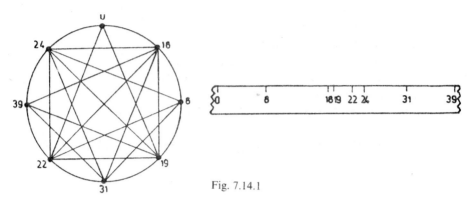

Fig. 7.14.1

A difference base M with respect to a number e is said to be *restricted* if min $M = 0$ and max $M = e$. For instance {0, 1, 4, 7, 9} is a restricted difference base with respect to 9. Using this difference base one can construct a ruler with 5 slots that enables to measure integral distances from 0 up to 9 or a β-labelling of the graph $K_5 - w$ (it is necessary to delete one edge w because two differences are equal: $7 - 4 = 4 - 1$).

A survey of results on difference bases (both restricted and unrestricted) can be found in Miller (1971) and Bermond (1979). For applications (in coding theory, X-ray crystallography, radiolocation, communication networks and radioastronomy) see Bloom and Golomb (1977) and also Bermond et al. (1978d), Bermond (1979, 1982) and Laufer (1982).

7.15 There are a plenty of results stating that a given graph is or is not graceful. We present just a few of them.

Lemma [Rosa 1967a, Golomb 1972, Bodendiek et al. 1976]. *Let G be an Eulerian graph of size e. If $e \equiv 1$ or 2 (mod 4) then G is not graceful.*

Proof. Assume that an Eulerian graph G has a β-labelling. Since G has a closed Eulerian trail, the sum of all edge-values in this trail, and thus in G, is even. On the other hand, the sum of all edge-values is $1 + 2 + ... + e$. Hence, if $e \equiv 1$ or 2 (mod 4), this is an odd number. This proves our lemma.

7.16 Theorem [Rosa 1967a]. A. *The cycle C_e has an α-labelling if and only if $e \equiv 0$ (mod 4).*

B. *The cycle C_e has a β-labelling if and only if $e \equiv 0$ or 3 (mod 4)* (see also Hebbare 1976).

Proof. The necessity follows from Lemmas 7.9 and 7.15. To prove the sufficiency we shall construct the corresponding labellings of $C_e = u_0 u_1 ... u_{e-1}$.

If $e \equiv 0 \pmod 4$, set $h(u_i) = i/2$ for $i = 0, 2, \ldots, e - 2$; $h(u_i) = e - (i - 1)/2$ for $i = 1, 3, \ldots, e/2 - 1$; $h(u_i) = e - (i + 1)/2$ for $i = e/2 + 1, e/2 + 3, \ldots, e - 1$. If $e \equiv 3 \pmod 4$, set $h(u_i) = e - (i - 1)/2$ for $i = 1, 3, \ldots, e - 2$; $h(u_i) = i/2$ for $i = 0, 2, \ldots, (e - 3)/2$; $h(u_i) = i/2 + 1$ for $i = (e + 1)/2$, $(e + 5)/2, \ldots, e - 1$. It is easily verified that in both cases h is a β-labelling and in the first case h is even an α-labelling with $\alpha(h) = e/2 - 1$.

Remark. In Bodendiek et al. (1977a) the authors conjectured that simple graphs which can be obtained by adding an additional edge (chord) to a cycle are always graceful. This conjecture was settled in Delorme et al. (1980). An analogous result for two adjacent chords is proved in Koh et al. (1980), and for three adjacent chords with a common end-vertex see Koh and Punnim (1982).

7.17 Theorem [Rosa 1967a, Golomb 1972]. *Every complete bipartite graph $K(n_1, n_2)$ has an α-labelling.*

Proof. It suffices to label the vertices of the part with cardinality n_1 by the numbers $n_2, 2n_2, \ldots, n_1 n_2$ and the vertices of the part with cardinality n_2 by the numbers $0, 1, 2, \ldots, n_2 - 1$. One easily verifies that this gives rise to an α-labelling h with $\alpha(h) = n_2 - 1$.

7.18 Due to Bermond (1979) and Frucht (1979), Frucht established the gracefulness of *n-prisms*, *wheels* and *crowns* for all $n \geq 3$. An *n-prism* is the graph $C_n \square K_2$ (see 11.23 for the definition of the symbol $\square$), a wheel is constructed from a cycle of length ≥ 3 by adding a vertex joined to any vertex of the cycle, and a crown is formed from a cycle of length ≥ 3 by adding to each vertex a new edge incident with that vertex and with one new end-vertex. The result for 4-prism follows from Theorem 7.18 and for 5-prism from 7.44; wheels are dealt with also in Bange et al. (1979). By Ayel and Favaron (1984), every *helm* (a graph obtained from a crown by adding a new vertex joined to every vertex of the unique cycle which crown contains) is graceful as well. Further, the *n-cube* Q_n is graceful for every n. In fact, a stronger result holds:

Theorem [Maheo 1980, Kotzig 1981]. *The graph Q_n has an α-labelling for each n.*

Proof. It turns out convenient to strengthen even this statement. It is a frequently encountered paradox that a stronger result is easier to prove. In the present case we prove that for every positive integer n there exists an α-labelling h of Q_n and a 1-1 mapping u_n of the set $1, 2, \ldots, 2^n$ onto the vertex-set $V(Q_n)$ (roughly speaking, a numbering u_n of the vertices of Q_n by the numbers $1, 2, \ldots, 2^n$) such that

$$0 = hu_n(1) < hu_n(2) < \ldots < hu_n(2^n) = n2^{n-1}, \tag{1}$$

$$hu_n(2^{n-1}) = (n - 1)2^{n-2}, \tag{2}$$

$$hu_n(2^{n-1} + 1) = (n - 1)2^{n-2} + 1, \tag{3}$$

$$hu_n(i + 2^{n-1} + 1) - hu_n(i + 2^{n-1}) - hu_n(i + 1) + hu_n(i) = 1,$$

$$\text{if } n \geq 2 \text{ and } i \in \{1, 2, \ldots, 2^{n-1} - 1\}, \tag{4}$$

$$\alpha(h) = hu_n(2^{n-1}) = (n - 1)\, 2^{n-2}. \tag{5}$$

We shall prove this by induction on n. The assertion is true for $n = 1$. It is sufficient to designate the two vertices of $Q_1 = K_2$ by $u_1(1)$ and $u_1(2)$ and put $hu_1(2) = 1$. Clearly, h is an α-labelling satisfying the conditions (1)—(5).

Let $k \geq 1$ be an integer and assume that the assertion is true for $n = k$. We shall prove that it is true for $n = k + 1$, too. Since Q_{k+1} can be regarded as the Cartesian product of Q_k and Q_1 ($Q_{k+1} = Q_k \square Q_1$, see 11.23), the vertex-set of Q_{k+1} consists of all pairs $[u_k(i), u_1(j)]$ where $i \in \{1, 2, \ldots, 2^k\}$ and $j \in \{1, 2\}$. By the definition of the Cartesian product, two vertices $[u_k(i), u_1(j)]$ and $[u_k(I), u_1(J)]$ of Q_{k+1} are adjacent in Q_{k+1} whenever $i = I$ and $u_1(j)$ and $u_1(J)$ are adjacent in Q_1 (that is, $j \neq J$) or $j = J$ and the vertices $u_k(i)$ and $u_k(I)$ are adjacent in Q_k.

We define a vertex labelling of Q_{k+1} through the labelling of Q_k as follows (there will be no confusion in using the same symbol h for both):

$$h[u_k(i), u_1(j)] =$$

$$= \begin{cases} hu_k(i) & \text{if } 1 \leq i \leq 2^{k-1}, j = 1, \\ hu_k(i) + (k + 2)\, 2^{k-1} & \text{if } 2^{k-1} + 1 \leq i \leq 2^k, j = 1, \\ hu_k(i + 2^{k-1}) + (k + 1)\, 2^{k-2} & \text{if } 1 \leq i \leq 2^{k-1}, j = 2, \\ hu_k(i - 2^{k-1}) + (k + 1)\, 2^{k-2} & \text{if } 2^{k-1} + 1 \leq i \leq 2^k, j = 2. \end{cases} \tag{6}$$

The numbering u_{k+1} of the vertex-set of Q_{k+1} can be defined in the following way:

$$u_{k+1}(i) = \begin{cases} [u_k(i), u_1(1)] & \text{if } 1 \leq k \leq 2^{k-1}, \\ [u_k(i), u_1(2)] & \text{if } 2^{k-1} + 1 \leq i \leq 2^k, \\ [u_k(i - 2^k), u_1(2)] & \text{if } 2^k + 1 \leq i \leq 2^k + 2^{k-1}, \\ [u_k(i - 2^k), u_1(1)] & \text{if } 2^k + 2^{k-1} + 1 \leq i \leq 2^{k+1}. \end{cases} \tag{7}$$

It is easily seen that u_{k+1} is a 1-1 mapping of the set $\{1, 2, \ldots, 2^{k+1}\}$ onto the vertex-set of Q_{k+1}. Using (7) we can rewrite (6) as follows:

$$hu_{k+1}(i) =$$

$$= \begin{cases} hu_k(i) & \text{if } 1 \leq i \leq 2^{k-1}, \\ hu_k(i - 2^{k-1}) + (k + 1)\, 2^{k-1} & \text{if } 2^{k-1} + 1 \leq i \leq 2^{k-1} + 2^k, \tag{8} \\ hu_k(i - 2^k) + (k + 2)\, 2^{k-1} & \text{if } 2^{k-1} + 2^k + 1 \leq i \leq 2^{k+1}. \end{cases}$$

We shall prove that the mapping h defined by (6) or (8) is an α-labelling satisfying the properties (1)—(5) for $n = k + 1$.

Obviously, h is a mapping of $V(Q_{k+1})$ into the set $\{0, 1, 2, \ldots, e\}$ where $e = (k + 1) 2^k$ is the number of edges of Q_{k+1}. It follows from (7), (8) and the induction hypothesis (including the equalities (1)—(3) for $n = k$) that the mapping h is 1-1. The equalities (2) and (3) can be proved similarly. The proof of (4) can be split into the following cases:

A. $k = 1$.

B. $k \geq 2$, $1 \leq i \leq 2^{k-1} - 1$.

C. $k \geq 2$, $i = 2^{k-1}$.

D. $k \geq 2$, $2^{k-1} + 1 \leq i \leq 2^k - 1$.

The case A is trivial. The cases B and C can be handled using (8) and the induction hypothesis while in the case C one can use, in addition, (1)—(3).

We show that the edge-values in Q_{k+1} form the set $\{1, 2, \ldots, (k + 1) 2^k\}$. It is convenient to distinguish four types of edges $U_1 U_2$ in Q_{k+1}.

E. $U_1 = [u_k(i), u_1(2)]$, $U_2 = [u_k(j), u_1(2)]$, where $2^{k-1} + 1 \leq i \leq 2^k$ and $1 \leq j \leq 2^{k-1}$. The values of these edges form the set $\{1, 2, \ldots, k2^{k-1}\}$.

F. $U_1 = [u_k(i), u_1(1)]$, $U_2 = [u_k(i), u_1(2)]$, where $1 \leq i \leq 2^{k-1}$. The values of these edges form the set $\{k2^{k-1} + 1, k2^{k-1} + 2, \ldots, (k + 1) 2^{k-1}\}$.

G. $U_1 = [u_k(i), u_1(1)]$, $U_2 = [u_k(i), u_1(2)]$, where $2^{k-1} + 1 \leq i \leq 2^k$. The values of these edges form the set $\{(k + 1) 2^{k-1} + 1, (k + 1) 2^{k-1} + 2, \ldots, (k + 2) 2^{k-1}\}$.

H. $U_1 = [u_k(i), u_1(1)]$, $U_2 = [u_k(j), u_1(1)]$, where $1 \leq i \leq 2^{k-1}$ and $2^{k-1} + 1 \leq j \leq 2^k$. The values of these edges form the set $\{(k + 2) 2^{k-1} + 1, (k + 2) 2^{k-1} + 2, \ldots, (k + 1) 2^k\}$.

To establish the above assertions it is sufficient to use (6) and the induction hypothesis and in the cases F and G, in addition, the equalities (1)—(4). The union of the four sets of values is the required set $\{1, 2, \ldots, (k + 1) 2^k\}$.

From (1), (6), (7) and the induction hypothesis for (5) it can easily be seen that each edge $u_{k+1}(i) u_{k+1}(j)$ of Q_{k+1} $(1 \leq i < j \leq 2^{k+1})$ satisfies the inequality $i \leq 2^k < j$. Therefore we can set $a(h) = hu_{k+1}(2^k) = k2^{k-1}$ (see 7.9). This proves (5) and thereby the whole theorem.

7.19 Further results on graceful graphs, α-, β- and ϱ-labellings as well as other labellings can be found in Ringel (1964), Rosa (1967a), Gardner (1972), Guy (1973), Kotzig (1973b, 1975, 1981), Bodendiek et al. (1976, 1977b), Hebbare (1976, 1981), Sheppard (1976), Guy (1977), Bermond et al. (1978d), Chen (1978), Hoede and Kuiper (1978), Kotzig and Turgeon (1978, 1980), Bange et al. (1979), Frucht (1979), Koh et al. (1979a, b, e, 1980a, b), Bange et al. (1980), Gangopadhyay et al. (1980), Maheo (1980), White (1980), Chung and Hwang (1981), Slater (1981) and in the survey article of Bermond (1979). It is remarkable that this topic has applications in radioastronomy, see Bermond et al. (1978c), Bermond (1982) and references therein.

We shall now concentrate on the special case when the graphs in question are trees. Since a finite tree G of size e has $e + 1$ edges, every β-labelling of its vertices is a bijection between $V(G)$ and the set $V = \{0, 1, 2, ..., e\}$. This means that the vertex-values exhaust the set V.

The following famous conjecture of Ringel has become a vital initiation to the research in the field of G-decompositions of graphs.

Conjecture [Ringel 1964]. *Let T be a tree of size e. Then K_{2e+1} has a T-decomposition (that is, a decomposition into $2e + 1$ subgraphs isomorphic to T).*

7.20 Although Ringel's conjecture is still open there have meanwhile been stated even stronger conjectures. For instance, Kotzig (see Rosa 1967a) believes the following to be true:

Conjecture. *For every finite tree T of size e the complete graph K_{2e+1} has a cyclic T-decomposition.*

By virtue of Theorem 7.10 this conjecture is equivalent with the statement that *every finite tree has a ϱ-labelling*.

7.21 A number of partial results suggests that the following even stronger conjecture of Kotzig and Rosa is probably true:

Conjecture [Rosa 1967a, Kotzig 1973b]. *Every finite tree is graceful (that is, admits a β-labelling).*

Much effort of many mathematicians was exerted in proving the conjectures 7.19—7.21. Proofs are known for a number of particular trees (including all trees of order <17 [Rosa 1965, 1967a]). The interested reader is referred to Golomb (1972), Guy (1973, 1977), Cahit and Cahit (1974), Haggard and McWha (1975), Cahit (1977), Koh et al. (1977, 1979c, e, 1980a, 1981), Rosa (1977), Gyárfás and Lehel (1978), Huang and Rosa (1978), Kotzig and Turgeon (1978), Rogers (1978), Bange et al. (1979), Bermond (1979), Beth and Sprague (1979), Chung and Hwang (1981), Morris (1981), Slater (1981), Poljak and Sůra (1982, 1983), to survey papers Guy and Klee (1971), Guy (1973), Bloom (1979), Koh et al. (1979b), or directly to the most important basic sources Rosa (1967a), Kotzig (1973b), Stanton and Zarnke (1973), Huang et al. (1982) (according to Guy, review MR 82g: 05041 of the paper of Bloom 1979). Among many other results it is mentioned that there exist finite trees with no α-labelling (7.51). We shall only consider here the most important results.

7.22 A subgraph Z_G will be called the *foundation of a graph G* if Z_G arises from G by the deletion of vertices with degree zero, pendant vertices and pendant edges. The foundation of a tree with ≤ 2 vertices is empty. Otherwise (if it has at least two edges) its foundation is again a tree.

Since the orientation of edges is immaterial for β-labellings we shall only consider undirected graphs. Let T be a finite undirected tree. If T has no vertices of degree ≥ 3 then it is called a *path* (the term *snake* is also used in this case). If T is not a path but Z_T is, then T is called a *caterpillar*. If T is not a caterpillar but Z_T is, then T is called a *lobster*.

7.23 Theorem [Rosa 1967a]. *If T is a path or caterpillar then T has an α-labelling.*

Proof. Let T be a path or a caterpillar. The theorem is true if $|V(T)| < 3$ (see also Fig. 7.23.1). Otherwise the foundation Z_T of T is a path $u_0 u_1 \ldots u_n$, $n \geq 0$. For $i = 0, 1, 2, \ldots, n$ let A_i denote the set of all pendant vertices adjacent to u_i. Order the vertices of T to a sequence of the form $\{u_0, A_1, u_2, A_3, u_4, A_5, \ldots, u_5, A_4, u_3, A_2, u_1, A_0\}$, the vertices of A_i (if there are any) being ordered arbitrarily. Then assign to members u of this sequence the values $h(u) = 0, 1, 2, \ldots$, following the order. The resulting labelling is clearly an α-labelling of T with $\alpha(h) = h(u_n)$ or $h(u_n) - 1$. (For β-labellings see Cahit and Cahit 1974.)

Examples of α-labellings can be seen in Figs. 7.23.1 and 7.23.2.

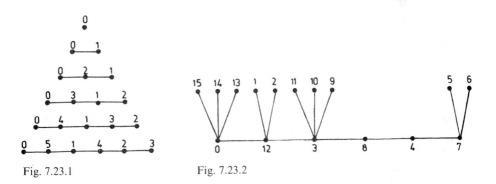

Fig. 7.23.1   Fig. 7.23.2

7.24 Theorem (cf. Huang and Rosa 1978). *Every lobster has a ϱ-labelling.*

Proof. Let T be a lobster of size e whose foundation is a caterpillar $T' = Z_T$ of size e'. By Theorem 7.23, T' has an α-labelling h'. Then h' has the numbers $0, 1, 2, \ldots, e'$ as vertex-values and $1, 2, \ldots, e'$ as edge-values. Define a ϱ-labelling h of T as follows: If $u \in V(T')$ then set $h(u) = h'(u)$. For vertices u and U in $V(T) - V(T')$ let u' and U' be their uniquely determined neighbours in $V(T')$, respectively. Order the vertices in $V(T) - V(T')$ in such a way that $h(u') < h(U')$ implies u precedes U. Now assign to any vertex u in $V(T) - V(T')$ one of the values $e' + 1, e' + 2, \ldots, 2e$ in such a way that edges incident with the vertices in $V(T) - V(T')$ take values $e' + 1, e' + 2, \ldots, e$, following the order. Clearly this gives rise to a ϱ-labelling of T.

An example of a ϱ-labelling of a lobster is given in Fig. 7.24.1. The edges of a caterpillar forming the foundation of the lobster are drawn thick. In this example $e = 22$ and $e' = 10$.

7.25 From Theorems 7.23 and 7.24 it follows that every path and every caterpillar satisfies Conjecture 7.21 and every path, caterpillar or lobster satisfies Conjecture 7.20 and therefore also Ringel's Conjecture 7.19.

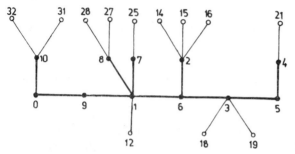

Fig. 7.24.1

7.26 In proving the existence of cyclic G-decompositions of complete graphs the following result is often useful.

Theorem [Rosa 1967a]. *Let G be a graph of size e. If G has an α-labelling then K_v has a balanced cyclic G-decomposition for each positive integer $v \equiv 1$ (mod $2e$).*

Proof. Let the vertex set of K_v consist of integers 0, 1, 2, ..., $v-1$, where $v = 2en + 1$ and $n \geq 1$ is an integer. Let h be an α-labelling of G. Note that G is bipartite by Lemma 7.9. Moreover, if V_1 is the set of all $u \in V(G)$ with $h(u) \leq \alpha(h)$ and $V_2 = V(G) - V_1$ then (V_1, V_2) is a bipartition of G.

We shall define the required cyclic G-decomposition by describing its base system $G^{(1)}, G^{(2)}, ..., G^{(n)}$. Let $(V_1^{(i)}, V_2^{(i)})$ be the bipartition of $G^{(i)}$ corresponding to the bipartition (V_1, V_2) of G in an isomorphism $G \to G^{(i)}$. More exactly, $V_1^{(1)} = V_1^{(2)} = ... = V_1^{(n)}$ will consist of all vertices $h(u)$ where $u \in V_1$, and $V_2^{(i)}$ will consist of all vertices $h(u') + (i-1)e$, where $u' \in V_2(G)$. Vertices $h(u) \in V_1^{(i)}$ and $h(u') + (i-1)e \in V_2^{(i)}$ will be adjacent in $G^{(i)}$ if and only if uu' is an edge in G. It is easy to verify that each $G^{(i)}$ is isomorphic to G. The n graphs $G^{(1)}, G^{(2)}, ..., G^{(n)}$ are readily seen to have ne edges altogether with mutually distinct lengths $\in \{1, 2, ..., ne\}$. Applying i-fold rotations for $i = 0$, 1, ..., $v-1$ one obtains vn graphs forming a cyclic G-decomposition of K_v. By Lemma 7.2, this decomposition is balanced.

7.27 Corollary. *In the four cases listed below there exist a cyclic balanced G-decomposition of K_v:*

A. $G = C_{4n}$ and $v \equiv 1$ (mod $8n$) [Kotzig 1965b].

B. $G = P_{n+1}$ and $v \equiv 1$ (mod $2n$) [Hung 1977, Huang and Rosa 1978].

C. $G = K(n_1, n_2)$ and $v \equiv 1$ (mod $2n_1 n_2$) [Huang and Rosa 1973b].

D. $G = Q_n$ and $v = 1$ (mod $n2^n$) [Kotzig 1981].

Proof. The proof follows from the previous theorem using an appropriate choice of e and the fact that the graph in question has an α-labelling. In case A we choose $e = 4n$ and use Theorem 7.16, in case B $e = n$

and we use Theorem 7.23, in case C $e = n_1 n_2$ and Theorem 7.17 is used, finally in case D $e = n2^{n-1}$ and Theorem 7.18 is applied.

7.28 We now show how vertex-labellings can be employed to construct other decompositions of complete graphs. We shall also outline a connection of these questions with Heffter's difference problems, Skolem pairs and cyclic Steiner triple systems.

7.29 A *Steiner triple system* of order $v \in N$ (named after J. Steiner 1853) is a system of triples, or more accurately, of 3-element sets such that each pair (2-element set) of elements belongs to exactly one triple of the system. A Steiner triple system with elements 0, 1, 2, ..., $v - 1$ is called *cyclic* if with each triple $\{a_1, a_2, a_3\}$ it also contains the triple $\{a_1 + 1, a_2 + 1, a_3 + 1\}$ where addition is taken modulo v. An example of a Steiner triple system of order 9 is the system (012, 036, 137, 238, 345, 047, 148, 246, 678, 058, 156, 257), where $\{a_1, a_2, a_3\}$ is abbreviated simply as $a_1 a_2 a_3$. Obviously this system is not a cyclic one.

It is easy to see that there exists a 1-1 correspondence between K_3-decompositions of K_v and Steiner triple systems of order v. Similarly there exists a 1-1 correspondence between cyclic K_3-decompositions of K_v and cyclic Steiner triple systems of order v.

7.30 For $\lambda = 1$, $G = K_3$, $e(G) = 3$ and $D(G) = 2$ Corollary 6.14.2—3 implies:

Lemma. *If there exists a K_3-decomposition of K_v (or equivalently, a Steiner triple system of order v) then $v \equiv 1$ or 3 (mod 6).*

We shall now pursue the question whether the converse statement is true. First we consider the case $v \equiv 1$ (mod 6). It will be shown that in this case there exists even a cyclic K_3-decomposition of K_v (a cyclic Steiner triple system of order v). We shall need several auxiliary results.

7.31 Lemma (see Heffter 1897 and Colbourn and Mathon 1980). *Let v and n be positive integers with $v = 6n + 1$. The following implication then holds: If the set $\{1, 2, ..., 3n\}$ or the set $\{1, 2, ..., 3n - 1, 3n + 1\}$ can be decomposed into n triples $\{a, b, c\}$ with the property that $a + b = c$ or $a + b = v - c$ then there exists a cyclic K_3-decomposition of K_v (and hence a cyclic Steiner triple system of order v).*

P r o o f. We shall prove the graphical version. In the complete graph K_v with $v = 6n + 1$ with the vertices 0, 1, 2, ..., $6n$ we shall construct triangles of the form $(0, a, c)$ where $\{a, b, c\}$ are the triples of a decomposition of $\{1, 2, ..., 3n\}$ or $\{1, 2, ..., 3n - 1, 3n + 1\}$ with the required properties. The edges of such a triangle will have lengths a, b and c. The lengths of edges of these triangles will be 1, 2, ..., $3n$, therefore they can be considered as members of a base system for a K_3-decomposition of K_v.

7.32 Let n be a positive integer. A set M consisting of $2n$ integers is called (in honour of T. Skolem) a *Skolem set of order n* (see Skolem 1957, 1958,

Rosa 1966c, Colbourn and Mathon 1980, Colbourn et al. 1983b) if it can be written in the form $M = \{a_1, b_1, a_2, b_2, ..., a_n, b_n\}$, where

$$b_1 = a_1 + 1$$
$$b_2 = a_2 + 2$$
$$................$$
$$b_n = a_n + n$$
$$\tag{1}$$

For instance, the set $\{1, 2, 3, 4, 5, 7\} = \{2, 3, 5, 7, 1, 4\}$ is a Skolem set of order 3.

7.33 Lemma [Skolem 1958, Bermond et al. 1978a, Colbourn and Mathon 1980]. *If* $\{1, 2, ..., 2n\}$ *or* $\{1, 2, ..., 2n - 1, 2n + 1\}$ *is a Skolem set then the set* $\{1, 2, ..., 3n\}$ *or* $\{1, 2, ..., 3n - 1, 3n + 1\}$ *can be decomposed into triples* $\{a, b, c\}$ *with the property that* $a + b \in \{c, 6n + 1 - c\}$.

Proof. Write the Skolem set as $\{a_1, b_1, ..., a_n, b_n\}$ so that the equalities 7.32.1 are satisfied. Then the triples $\{i, a_i + n, b_i + n\}$ for $i = 1, 2, ..., n$ form a decomposition of the set $\{1, 2, ..., 3n\}$ or $\{1, 2, ..., 3n - 1, 3n + 1\}$ with the required properties.

7.34 We remark that the problem to find a decomposition of Lemma 7.33 is sometimes called the *First Heffter's difference problem* for the number n (named after L. Heffter: see Heffter 1897 and Colbourn and Mathon 1980). We shall solve this problem in the following lemma.

7.35 Lemma [Skolem 1957, O'Keefe 1961, Rosa 1966c, Hilton 1969, Colbourn and Mathon 1980]. *If* n *is a positive integer such that* $n \equiv 0$ *or* 1 (mod 4) *then* $\{1, 2, ..., 2n\}$ *is a Skolem set. If* $n \equiv 2$ *or* 3 (mod 4) *then* $\{1, 2, ..., 2n - 1, 2n + 1\}$ *is a Skolem set.*

Proof. It is sufficient to express the above-mentioned set in the form $M_n = \{a_1, b_1, ..., a_n, b_n\}$ so that the property 7.32.1 is satisfied. For $n \leq 9$ this can be done as follows. $M_1 = \{1, 2\}$, $M_2 = \{1, 2, 3, 5\}$, $M_3 = \{2, 3, 5, 7, 1, 4\}$, $M_4 = \{1, 2, 5, 7, 3, 6, 4, 8\}$, $M_5 = \{1, 2, 7, 9, 3, 6, 4, 8, 5, 10\}$, $M_6 = \{10, 11, 2, 4, 6, 9, 1, 5, 3, 8, 7, 13\}$, $M_7 = \{2, 3, 10, 12, 4, 7, 9, 13, 1, 6, 5, 11, 8, 15\}$, $M_8 = \{2, 3, 11, 13, 4, 7, 10, 14, 1, 6, 9, 15, 5, 12, 8, 16\}$, $M_9 = \{3, 4, 13, 15, 6, 9, 12, 16, 2, 7, 11, 17, 1, 8, 10, 18, 5, 14\}$. For $n \geq 10$ we distinguish four cases. In each of them we only give a formula for a_i $(i = 1, 2, ..., n)$; b_i is then found as $a_i + i$.

I. $n \equiv 2$ (mod 4), $n = 4s + 2$, $s \geq 2$. We set $a_1 = 7s + 3$; $a_i = 6s + (i - 1)/2$ for $i = 3, 5, ..., 2s - 1$; $a_{2s+1} = 4s + 2$; $a_i = 6s - (i - 7)/2$ for $i = 2s + 3, 2s + 5, ..., 4s - 1$; $a_{4s+1} = 2s + 1$; $a_i = 2s - (i - 2)/2$ for $i = 2, 4, ..., 4s$; $a_{4s+2} = 4s + 3$.

II. $n \equiv 3$ (mod 4), $n = 4s - 1, s \geq 3$. In this case we set $a_1 = s$; $a_i = 2s - (i - 1)/2$ for $i = 3, 5, ..., 2s - 3$; $a_{2s-1} = 2s$; $a_{2s+1} = s - 1$; $a_i = 2s - (i + 1)/2$ for $i = 2s + 3, 2s + 5, ..., 4s - 3$; $a_{4s-1} = 4s$; $a_i = 6s - 1 - i/2$ for $i = 2, 4, ..., 4s - 4$; $a_{4s-2} = 2s + 1$.

III. $n \equiv 0 \pmod 4$, $n = 4s$, $s \geq 3$. In this case we set $a_1 = s$; $a_i = 2s - (i - 1)/2$ for $i = 3, 5, \ldots, 2s - 3$; $a_{2s-1} = 2s$; $a_{2s+1} = s - 1$; $a_i = 2s - (i + 1)/2$ for $i = 2s + 3, 2s + 5, \ldots, 4s - 3$; $a_{4s-1} = 2s + 1$; $a_i = 6s - i/2$ for $i = 2, 4, \ldots, 4s$.

IV. $n \equiv 1 \pmod 4$, $n = 4s + 1$, $s \geq 3$. Then $a_1 = s + 1$; $a_i = 2s - (i - 3)/2$ for $i = 3, 5, \ldots, 2s - 3$; $a_{2s-1} = 2s + 2$; $a_i = 2s - (i - 1)/2$ for $i = 2s + 1$, $2s + 3, \ldots, 4s - 1$; $a_{4s+1} = 2s + 1$; $a_i = 6s + 2 - i/2$ for $i = 2, 4, \ldots, 4s$.

7.36 Theroem [Peltesohn 1939, Skolem 1957, O'Keefe 1961, Rosa 1966c, Hilton 1969, Colbourn and Mathon 1980]. *Let v be an integer with $v \equiv 1$ or 3 (mod 6) and $v \neq 9$. Then there exists a cyclic K_3-decomposition of K_v (or equivalently, a cyclic Steiner triple system of order v).*

Proof. If $v \equiv 1 \pmod 6$ the proof follows from Lemmas 7.31, 7.33 and 7.35. For $v \equiv 3 \pmod 6$ the proof is similar, however, instead of the First Heffter's difference problem we have to solve the *Second Heffter's difference problem*: it is required to find a decomposition of the set $\{1, 2, \ldots, 2n, 2n + 2, 2n + 3, \ldots, 3n + 1\}$ or $\{1, 2, \ldots, 2n, 2n + 2, 2n + 3, \ldots, 3n, 3n + 2\}$ into n triples $\{a, b, c\}$ with the property that $a + b \in \{c, 6n + 1 - c\}$. In order to get such a decomposition it suffices to show that the set $\{1, 2, \ldots, n, n + 2, n + 3, \ldots, 2n, 2n + 2\}$ is a Skolem set for $n \equiv 1$ or 2 (mod 4) and $n \neq 1$ or that the set $\{1, 2, \ldots, n, n + 2, n + 3, \ldots, 2n + 1\}$ is a Skolem set for $n \equiv 0$ or 3 (mod 4), respectively. This can be proved using a similar construction as in Lemma 7.35 (for details see Rosa 1966c, Hilton 1969, Colbourn and Mathon 1980). With the usual order of elements in a Skolem set $\{a_1, b_1, \ldots, a_n, b_n\}$, a base system (7.3) of a K_3-decomposition of K_v ($v = 6n + 3$) consists of triangles with vertices 0, i, and $b_i + n$ ($i = 1, 2, \ldots, n$) and one additional triangle with vertices 0, $2n + 1$, and $4n + 2$.

7.37 Corollary [Peltesohn 1939, Skolem 1957, O'Keefe 1961, Rosa 1966c, Colbourn and Mathon 1980]. *A cyclic K_3-decomposition of K_v (or equivalently, a cyclic Steiner triple system of order v) exists if and only if $v \equiv 1$ or 3 (mod 6) and $v \neq 9$.*

Proof. The sufficiency follows from Theorem 7.36. The necessity of the condition $v \equiv 1$ or 3 (mod 6) follows from Lemma 7.30. We shall prove the necessity of the condition $v \neq 9$. Assume that there exists a cyclic K_3-decomposition of K_9. Then graphs in a base system (7.3) for such a decomposition must contain exactly one edge of each length 1, 2, and 4 and exactly three edges of length 3. Obviously no edge of length 3 can belong to a graph of the decomposition which contains an edge of length 1, 2 or 4. Hence the edges of length 3 form one subgraph. The other graphs of a base system then contain edges of length 1, 2 and 4. But this is impossible since a triangle with one edge of length 1 and another one of length 2 has the third edge of length 3.

7.38 Remarks. A. All solutions of the First [resp. Second] Heffter's difference problem for $u = 1, 2, ..., 8$ can be found in Colbourn (1982a). They are employed to enumerate different or even non-isomorphic (8.34) cyclic Steiner triple systems of order $v = 6n + 1$ [resp. $6n + 3$]. For example, if $n = 7$ then there exist precisely 3091 [resp. 2178] solutions of the First [Second] Heffter's difference problem, 395 648 [278 784] different and 9508 [11 616] non-isomorphic Steiner triple systems on a fixed 43- [45-] element set.

B. In 8.16.C [resp. 8.16.D] we deal with generalizations of Corollary 7.37 to cyclic [resp. 1-rotational] K_3-decompositions of λK_v.

7.39 Corollary [Kirkman 1847, Reiss 1859, Moore 1893]. *A K_3-decomposition of K_v (a Steiner triple system of order v) exists if and only if $v \equiv 1$ or 3 (mod 6).*

Proof. The necessity follows from Lemma 7.30. The sufficiency for $n \neq 9$ follows from Theorem 7.36 and for $n = 9$ from the construction in 7.29.

7.40 Let L_v denote a 1-regular graph of order v. Lemma 7.35 enables to construct, for each positive integer e, a ϱ-labelling of the graph L_{2e} with vertex-values $0, 1, ..., 2e - 1$ or $0, 1, ..., 2e - 2, 2e$ and edge-values $1, 2, ..., e$. It suffices to make use of the Skolem set $M_e = \{a_1, b_1, ..., a_e, b_e\}$ and to label the end-vertices of the i-th edge ($i = 1, 2, ..., e$) with the numbers $a_i - 1$ and $b_i - 1$, respectively (for $e = 9$ see Fig. 7.40.1). If we do not require the edges to have the values $1, 2, ..., e$ (but do require the same just for their lengths), a considerably simpler construction can be used as can be seen from the proof of the following lemma.

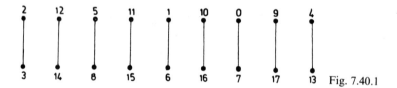

Fig. 7.40.1

7.41 Lemma. *The graph L_{2e} has a ϱ-labelling with vertex-values $0, 1, ..., 2e - 1$ and it holds $L_{2e} | K_{2e+1}$.*

Proof. If $u_i u_i'$ is the i-th edge of L_{2e} ($i = 1, 2, ..., e$) then set $h(u_i) = i - 1$ and $h(u_i') = 2e - i$. It is easy to see that h is a ϱ-labelling with vertex-values $0, 1, ..., 2e - 1$. The second statement follows from Theorem 7.10 (see also Mendelsohn and Rosa 1985).

7.42 The following result is considered as a part of mathematical folklore (see, for instance Harary 1969, Theorem 9.1).

Theorem. *The complete graph K_v can be decomposed into 1-factors if and only if $v \geq 2$ is even.*

Proof. Obviously, v must be even and ≥ 2. We show this is also a sufficient condition. Let $v \geq 2$ be even. Set $e = v/2 - 1$. Then, by Lemma

7.41, the graph K_{2e+1} has a decomposition into subgraphs isomorphic to L_{2e}. It follows that K_{2e+1} can be decomposed into factors with all vertices of degree 1 except for one vertex of degree 0. If in each factor this vertex is joined to a new vertex denoted, e.g., ∞, then a decomposition of K_v into factors isomorphic to L_{2e+2} is obtained. This completes the proof.

Figure 7.42.1 depicts an L_{20}-decomposition of K_{20} based on Lemma 7.35 and one based on Lemma 7.41. For each of these two decompositions two factors are exhibited, the remaining ones are obtained applying rotations. It is readily seen that the second construction is much simpler (cf. R 11.67).

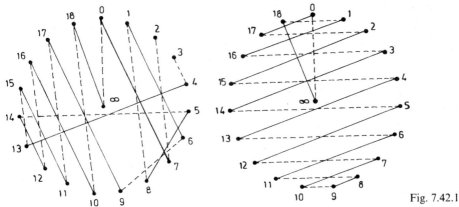

Fig. 7.42.1

The situation becomes more difficult if we require the decomposition of K_v into 1-factors to be cyclic. In Hartman and Rosa (1985) it is shown that such a decomposition exists if and only if v is even and $v \in \{8, 16, 32, 64, ...\}$ (see also Exercise 7.50).

Exercises

7.43 [Golomb 1972]. Determine grac G (7.11) for all graphs of order ≤ 5. Which of them are graceful?

7.44 Construct a graceful labelling for the graph of the 5-prism.

7.45 The *Dutch n-windmill* is defined as a graph consisting of n triangles with one vertex in common. Show that the Dutch n-windmill is graceful if and only if $n \equiv 0$ or 1 (mod 4).

7.46 The *French n-windmill* is defined as a graph consisting of n copies of K_4 with one vertex identified. For which $n \in \{0, 1, 2, 3, 4\}$ is the French windmill graceful?

7.47 For which m and n is the graph consisting of m disjoint copies of K_n graceful?

7.48 Find an Euclidean model of the Petersen graph.

7.49 [Cockayne and Hartnell 1976]. Construct a cyclic decomposition of the complete 9-partite graph $K(9 \times 5)$ into cycles of order 4.

7.50 [Hartman and Rosa 1985]. Construct a cyclic decomposition of the graph K_v with vertices $0, 1, ..., v - 1$ into 1-factors for $v = 4, 6$ and 10. Does there exist such a decomposition for $v = 8$?

7.51 [Rosa 1967a]. Find a tree with no α-labelling and the smallest possible order.

7.52 [Maheo 1980]. Construct an α-labelling for A. the 2-cube Q_2; B. the 3-cube Q_3.

7.53 For which n does there exist an α-labelling h of the n-cube that does not satisfy the condition that $\alpha(h) = (n - 1) 2^{n-2}$?

7.54 Grac $K_5 = 11$ (see 7.11).

7.55 [Kárteszi 1976]. For each $k \in \{1, 2, 3, 4, 5, 6, 8, 9, 10, 12\}$ there exists a ϱ-labelling of K_k.

7.56 [Kárteszi 1976]. For each $k \{1, 2, 3, 4, 5, 6, 8, 9, 10, 12\}$ and $v = k^2 - k + 1$ there exists a cyclic K_k-decomposition of K_v.

7.57 [Hall 1967, Table 1, Scheme 16]. There exists a cyclic K_7-decomposition of $^3K_{15}$ and a K_8-decomposition of $^4K_{15}$.

Chapter 8

Block designs and decompositions of graphs into isomorphic complete subgraphs

8.1 There is a close connection between graph decompositions and combinatorial structures known as designs.

A *design* is a system of (not necessarily distinct) subsets of a fixed finite set. The subsets are called *blocks* and their elements are called *points*. The number of elements in a block is its *order*. The *order of the design* is the number of all points of the design.

As an example consider the 4-point set $\{1, 2, 3, 4\}$ together with the system of 8 blocks $(\{1, 2, 3\}, \{1, 2, 4\}, \{1, 3\}, \{1, 4\}, \{2, 3\}, \{2, 4\}, \{3, 4\}, \{3, 4\})$. Its order is 4 and the orders of its blocks are either 2 or 3. The block $\{3, 4\}$ is said to be repeated; it is considered as two blocks.

Initiation to the study of designs came from mathematical statistics because of their applications to design of experiments in agriculture and other areas. Later they became a subject of independent interest as important combinatorial structures with tight links to algebra, number theory and finite geometries.

In a natural way to each design there corresponds a decomposition of a certain finite undirected loopless graph into complete graphs and vice versa. In this correspondence, vertices of the graph correspond to points and subgraphs of the decomposition to blocks. For instance, to design of the previous example there corresponds a decomposition of the graph $^2K_4$ into eight complete subgraphs of order 2 or 3.

A design is said to be *balanced* if each of its points belongs to the same number of blocks; more exactly, *r-balanced* if this number is r. It is easily seen that a balanced graph decomposition (6.1) and a balanced design correspond to each other. In the above example, the design is clearly not balanced; hence the corresponding decomposition of $^2K_4$ is not balanced.

A design is said to be *symmetric* (or sometimes, *projective* or *square*) if the number of its blocks is the same as the number of its points. It is easily seen that a symmetric graph decomposition (7.7) and a symmetric design correspond to each other. Obviously, the design of the above example is not symmetric $(4 \neq 8)$.

Of particular interest are (v, K, λ)-designs. Let K be an arbitrary (possibly infinite) set of positive integers, and let v and λ be positive integers. A (v, K, λ)-*design* is a design of order v whose blocks have orders in the set K and for any two distinct points u and U there exist exactly λ blocks containing both u and U. To each (v, K, λ)-design there corresponds a decomposition of the graph ${}^{\lambda}K_v$ into complete subgraphs the orders of which belong to K, and vice versa. The above-mentioned design can be considered as a $(4, K, 2)$-design with $K = \{2, 3\}$ or, say, $K = \{1, 2, 3, 4, 5\}$ (not every element of K has to be the order of some block).

If K is the set of all positive integers ($K = N$) we obtain a (v, N, λ)-*design* which is also called a *pairwise balanced design*, see, for instance, Stinson (1982). A balanced (v, N, λ)-decomposition is called a *regular pairwise balanced design*; if it is r-balanced then it is called an (r, λ)-*design* of order v. A survey of results concerning (r, λ)-designs is given in Deza et al. (1981). A (v, N, λ)-design corresponds to a decomposition of ${}^{\lambda}K_v$ into non-empty complete graphs; an (r, λ)-design corresponds to an r-balanced decomposition of ${}^{\lambda}K_v$ into non-empty complete graphs; a $(v, K, 1)$-design corresponds to a decomposition of the complete graph of order v into complete subgraphs of orders belonging to K. Sometimes they are called *finite linear spaces* (see, for instance, Mullin and Wallis 1980, Stinson 1983).

The concept of a (v, K, λ)-design will be generalized in 8.40.

8.2 An important special case of a (v, K, λ)-design is that when $K = \{k\}$, i.e., K consists of a single integer k. If positive integers v, k and λ are given, a (v, k, λ)-*design* is a system of (not necessarily distinct) subsets, called blocks, of a fixed set with v elements, called points, in which every block has k points any two distinct points belong to exactly λ blocks. (v, k, λ)-Designs are also known as (v, k, λ)-*configurations, balanced incomplete block designs*, briefly BIB-*designs*, **BIBD**, or simply *block designs*.

Our considerations in 8.1 imply:

Lemma [Bermond and Sotteau 1976, Chung and Graham 1981, Horton 1983b]. *Let v, k and λ be positive integers. Then a K_k-decomposition of ${}^{\lambda}K_v$ exists if and only if there exists a (v, k, λ)-design.*

We generalize the concept of a (v, k, λ)-design in 8.32 and 8.40.

Example. The blocks $\{1, 2, 4\}$, $\{2, 3, 5\}$, $\{3, 4, 6\}$, $\{4, 5, 1\}$, $\{5, 6, 2\}$, $\{6, 1, 3\}$, $\{1, 2, 3\}$, $\{3, 4, 5\}$, $\{5, 6, 1\}$, and $\{2, 4, 6\}$ form a $(6, 3, 2)$-design. Clearly, to this design there corresponds a K_3-decomposition of the 2-fold complete graph ${}^2K_6$ of order 6 into 10 triangles. The numbers 1, 2, ..., 6 (the points of the design) are to be considered as the vertices of ${}^2K_6$.

8.3 Let a decomposition of a graph of order v into b complete subgraphs [a design with v points and b blocks] be given. We can define its *incidence matrix* **A** of type $v \times b$ as follows: the (i, j)-entry $A(i, j) = 1$ if the i-th vertex [point] belongs to the j-th subgraph [block]; otherwise we set $A(i, j) = 0$. Of course, we assume that the vertices [points] and subgraphs [blocks] are numbered in some way by the numbers 1, 2, ..., v and 1, 2, ..., b, respectively. As usual, such a numbering will be tacitly assumed.

For example, the K_3-decomposition of $^2K_6$ [(6, 3, 2)-design] from 8.2 has incidence matrix

$$\mathbf{A} = \begin{bmatrix} 1 & 0 & 0 & 1 & 0 & 1 & 1 & 0 & 1 & 0 \\ 1 & 1 & 0 & 0 & 1 & 0 & 1 & 0 & 0 & 1 \\ 0 & 1 & 1 & 0 & 0 & 1 & 1 & 1 & 0 & 0 \\ 1 & 0 & 1 & 1 & 0 & 0 & 0 & 1 & 0 & 1 \\ 0 & 1 & 0 & 1 & 1 & 0 & 0 & 1 & 1 & 0 \\ 0 & 0 & 1 & 0 & 1 & 1 & 0 & 0 & 1 & 1 \end{bmatrix}.$$

Lemma. *Let* $\mathbf{A}$ *be an incidence matrix of a decomposition of the graph* K_v *into complete subgraphs* [*of a* (v, N, λ)-*design*]. *Let* r_i $(i = 1, 2, ..., v)$ *be the number of subgraphs in the decomposition* [*blocks*] *which contain the vertex* [*point*] *numbered i. Then the matrix* $\mathbf{B} = \mathbf{A}\mathbf{A}^T$ *has the following properties*:

$$\mathbf{B} = \begin{bmatrix} r_1 & \lambda & \lambda & \cdots & \lambda \\ \lambda & r_2 & \lambda & \cdots & \lambda \\ \multicolumn{5}{c}{\dotfill} \\ \lambda & \lambda & \lambda & \cdots & r_v \end{bmatrix} \tag{1}$$

and

$$\begin{aligned} \det \mathbf{B} = {} & (r_1 - \lambda)(r_2 - \lambda)\ldots(r_v - \lambda) + \\ & + \lambda(r_2 - \lambda)(r_3 - \lambda)\ldots(r_v - \lambda) + \\ & + \lambda(r_1 - \lambda)(r_3 - \lambda)\ldots(r_v - \lambda) + \ldots + \\ & + \lambda(r_1 - \lambda)(r_2 - \lambda)\ldots(r_{v-1} - \lambda). \end{aligned} \tag{2}$$

(Here $\mathbf{A}^T$ denotes the matrix transpose to $\mathbf{A}$, and $\det \mathbf{B}$ is the determinant of $\mathbf{B}$.)

Proof. It is sufficient to consider the graph version only. The (i, j)-th entry of the matrix $\mathbf{B}$ expresses the number of subgraphs in the decomposition which contain both the i-th vertex and the j-th vertex. However, this number is r_i, if $i = j$ and λ, if $i \neq j$. Thus (1) holds. Now (2) follows from (1) by computation of $\det \mathbf{B}$.

8.4 We now present an interesting result which, for $\lambda = 1$, was first derived in de Bruijn and Erdős (1948).

Theorem [Majumdar 1953, Woodall 1970, 1976]. *If there exists a decomposition of the graph* $^\lambda K_v$ *into b complete subgraphs of order* ≥ 2 [*a* $(v, \{2, 3,$

4 ...}, λ)-design with b blocks] then either this decomposition consists of $b = \lambda$ compete subgraphs [blocks], each containing all v vertices [points], or $b \geq v$.

Proof. (Graph version.) Let **A** be the incidence matrix of the decomposition and let $\mathbf{B} = \mathbf{A}\mathbf{A}^T$. Assume it holds $v > b$. Add $v - b$ zero columns to **A** obtaining thereby a square matrix $\mathbf{A}_0$ of order v. Clearly, $\mathbf{A}_0\mathbf{A}_0^T = \mathbf{A}\mathbf{A}^T = \mathbf{B}$ and 8.3.2 holds, with the same notation as in Lemma 8.3. However, at the same time we have that $\det \mathbf{B} = (\det \mathbf{A}_0) \cdot (\det \mathbf{A}_0^T) = 0$. Comparing these two expressions for $\det \mathbf{B}$, and taking into account that $r_i \geq \lambda$ for each $i = 1, 2, \ldots, v$, we deduce that among the numbers $r_1, r_2, \ldots, r_v$ at least two are equal to λ. However, if $r_i = \lambda$ for some i, then the λ edges joining the i-th vertex to any other vertex fall, one by one, into $r_i = \lambda$ subgraphs of the decomposition. Thus each of these subgraphs contains all v vertices and the union of these subgraphs includes each edge of ${}^{\lambda}K_v$. Consequently, there are no other subgraphs in the decomposition and $r_1 = r_2 = \ldots = r_v = \lambda = b$.

8.5 In our next theorem we give some important necessary conditions for the existence of a K_k-decomposition of ${}^{\lambda}K_v$. For the reader's convenience we state these conditions in both the graph version and the design version, providing them with a graph-theoretical proof. Proofs in a combinatorial or algebraic fashion can be found in monographs on combinatorics or finite geometry (e.g. in Hall 1967, 10.1—10.2, Dembowski 1968, 1.3.8 and 2.1, Raghavarao 1971, p. 64, Rybnikov 1972, § 4.5, Cameron 1976, p. 98, and in the survey paper of Hall 1978).

Theorem. *If there exists a K_k-decomposition of the graph K_v [a (v, k, λ)-design] then the following hold*:

A. *Each of the v vertices [points] is contained in the same number r of the total number b of graphs of the K_k-decomposition of ${}^{\lambda}K_v$ [blocks of the (v, k, λ)-design]. That is to say, the K_k-decomposition of ${}^{\lambda}K_v$ [(v, k, λ)-design] is r-balanced.*

B. $vr = bk$.

C. $\lambda(v - 1) = r(k - 1)$.

Moreover, if $k < v$ then:

D. $v \geq 3$, $b \geq 3$, $r \geq 2$, $k \geq 2$, $\lambda \geq 1$.

E. $\lambda < r$.

F. $v \leq b$ (*Fisher's inequality*).

G. $k \leq r$.

Proof. Assume that ${}^{\lambda}K_v$ has a K_k-decomposition. The part A of our theorem follows from Lemma 6.1. To prove C observe that every vertex of ${}^{\lambda}K_v$ has degree $\lambda(v - 1)$ and at the same time $r(k - 1)$.

The number of pairs $[u, G]$ where u is a vertex of G and G is a member of the K_k-decomposition of ${}^{\lambda}K_v$ is vr and at the same time bk, whence we get B.

Let $k < v$. Thus $v \geq 2$. But then k cannot be equal to 1. Therefore $k \geq 2$ and hence $v \geq 3$. Further, since $k < v$, r cannot be equal to 1. Thus $r \geq 2$.

Using B we see that $bk = vr > kr$ so $b > r \geq 2$. Therefore $b \geq 3$. The inequality $\lambda \geq 1$ holds by definition thus D is proved.

The condition E follows from C, D and the asssumption $k < v$. The inequality G is a consequence of B and F.

Remark. First proofs of Fisher's inequality were published in Fisher (1940) and Bose (1949). A generalization of this inequality was considered in 8.4. Other modified and dual forms of Fisher's inequality are given in Isbell (1959), Dembowski (1968), Ryser (1968, 1972), Woodall (1970, 1976), Hell and Rosa (1972), Vanstone (1979), Simonovits and Sós (1980), Stinson (1982). (v, k, λ)-Designs in which Fisher's inequality becomes equality, i.e., $v = b$, are *symmetric* (cf. 7.7, 8.1, 8.20). Symmetric (v, k, λ)-designs are also known as (v, k, λ)-*configurations*, λ-*planes*, λ-*spaces* or just simply (v, k, λ)-*designs* (cf. Ryser 1978).

A survey of results on symmetric designs can be found in Ryser (1963, 1968), Hall (1967), Dembowski (1968), Woodall (1970). Some basic results will be mentioned in 8.20 and 8.54—8.59.

It is easy to describe all symmetric $(v, k, 1)$-designs if $k = 1$ or $k = 2$. In the first case $v = b = 1$ and in the second case necessarily $v = b = 3$. Symmetric $(v, k, 1)$-designs for $k \geq 3$, termed *projective planes*, will be dealt with in 8.22—8.25, 8.44 and 8.45.

Symmetric $(v, k, 2)$-designs, called *biplanes*, are according to Cameron (1973) and Ryser (1978) "extraordinarily rare", because only few are kown. The following $(11, 5, 2)$-design with blocks $\{0, 2, 3, 4, 8\}$, $\{1, 3, 4, 5, 9\}$, $\{2, 4, 5, 6, 10\}$, $\{3, 5, 6, 7, 0\}$, ..., $\{10, 1, 2, 3, 7\}$ is an example of biplane, see Hall (1978). Some results on biplanes will be given in 8.51.

It is expected that the number of symmetric (v, k, λ)-designs for fixed $\lambda \geq 2$ is finite (see Ryser 1978).

8.6 Besides the incidence matrix of type $v \times b$, with each K_k-decomposition of $^\lambda K_v$ [(v, k, λ)-design] one can associate a matrix of type $k \times b$ whose columns correspond to the members of the decomposition [to blocks]. The following matrix

$$\begin{bmatrix} 1 & 1 & 2 \\ 2 & 3 & 3 \end{bmatrix}$$

shows that the inequalities 8.5.D—G are best possible in general. The parameters of the corresponding (v, k, λ)-design are: $v = 3$, $b = 3$, $r = 2$, $k = 2$, $\lambda = 1$.

8.7 If the parameters v, k and λ are given and $k \geq 2$, then the remaining two parameters of a (v, k, λ)-design are uniquely determined, as can be deduced from 8.5.B—C:

82

$$r = \lambda \frac{v-1}{k-1}, \tag{1}$$

$$b = \lambda \frac{v(v-1)}{v(k-1)}. \tag{2}$$

If $k = 1$ (whence $v = 1$) then $r = b$ and $b \geq 0$; the parameter λ can take any integral value ≥ 1.

8.8 As we have seen in 8.1—8.4 the study of K_k-decompositions of $^\lambda K_v$ can be reduced to the investigation of (v, k, λ)-designs and vice versa. This leads to many interesting results. Apart from asymptotic results (8.9, 8.12) we shall be interested in "exact" results (8.13—8.31) as well.

8.9. The following result settles a conjecture of M. Hall, Jr., stated in Hall (1967, p. 248).

Theorem [Wilson 1975]. *If there exists a K_k-decomposition of $^\lambda K_v$ [a (v, k, λ)-design] then the following hold*:

$$k - 1|\lambda(v-1), \tag{1}$$

$$k(k-1)|\lambda v(v-1). \tag{2}$$

Conversely, for all pairs $[v, \lambda]$ of positive integers satisfying the conditions (1) and (2) — with possibly a finite number of exceptions — there exists a K_k-decomposition of $^\lambda K_v$ [a (v, k, λ)-design].

Proof. The necessity of the conditions (1) and (2) follows from 6.14.3 and 6.14.2 for $G = K_k$. We omit the proof of the sufficiency and refer the reader to Wilson (1975).

8.10 Besides 8.9.1—2 there is another necessary condition (a trivial one) that $v \geq k$ if $v \neq 1$. However, for $k > \lambda$ even more can be proved.

Theorem. *If there exists a K_k-decomposition of $^\lambda K_v$ [a (v, k, λ)-design] where $k < v$ then*

$$v \geq 1 + k(k-1)/\lambda. \tag{1}$$

Proof. For $k = 1$ this is obvious. If $2 \leq k < v$ then Fisher's inequality 8.5.F and 8.7.2 imply that $v \leq b = \lambda v(v-1)/k(k-1)$. But then (1) holds.

8.11 Corollary. *If there exists a K_k-decomposition of the graph $^\lambda K_v$ [a (v, k, λ)-design] then either $v = 1$ or $v = k$ or*

$$v \geq \begin{cases} 1 + k(k-1)/\lambda & \text{for } k > \lambda, \\ k+1 & \text{for } k \leq \lambda. \end{cases}$$

Proof follows from Theorem 8.10 and the fact that if $k \geq v$ then either $v = k$ or $v = 1$. For $v \neq k$ and $v \neq 1$ we use 8.10.1 or the inequality $v \geq k + 1$ depending on which one is stronger.

8.12 Let K be an arbitrary set of positive integers. Let $\alpha(K)$ $[\beta(K)]$ denote the greatest common divisor of the numbers $k - 1$ $[k(k - 1)]$ where $k \in K$. The following result can be viewed as a generalization of Theorem 8.9.

Theorem [Wilson 1975]. *For any set K of positive integers there exists a number $n(K)$ such that for all integers $v \geq n(K)$ it holds: A decomposition of ${}^{\lambda}K_v$ into complete subgraphs whose orders belong to K exists if and only if*

$$\alpha(K)|\lambda(v - 1), \tag{1}$$

$$\beta(K)|\lambda v(v - 1). \tag{2}$$

Proof follows from the corresponding result on (v, K, λ)-designs proved by R. M. Wilson (see Wilson 1972a, 1975).

8.13 Although the asymptotic result from 8.9 is very elegant, the number of exceptional pairs $[v, \lambda]$ for any k might be quite large. Thus it makes sense to seek "exact" results concerning the existence of a K_k-decomposition of ${}^{\lambda}K_v$. One of the possible approaches is to start with investigating "small" values of k systematically. The cases $k = 1$ and $k = 2$ are trivial. In the first one a K_k-decomposition exists precisely when $v = 1$, and in the second one a K_k-decomposition always exists.

8.14 Theorem [Hanani 1961, 1972, 1975, Lenz 1981]. *Let $v \geq 2$ and $k \geq 5$ be integers. Then a K_k-decomposition of ${}^{\lambda}K_v$ $[a\,(v, k, \lambda)$-design] exists if and only if*

$$v \geq k, \tag{1}$$

$$k - 1|\lambda(v - 1), \tag{2}$$

$$k(k - 1)|\lambda v(v - 1), \tag{3}$$

$$[v, k, \lambda) \neq [15, 5, 2]. \tag{4}$$

Proof. The necessity of the condition (1) follows from 6.3.A. The necessity of (2) and (3) follows from the Theorem 8.9. The necessity of (4) follows from 8.51, see for instance Connor (1952) and Hall and Connor (1954). The sufficiency of (1)—(4) is clear for $1 \leq k \leq 2$; for $3 \leq k \leq 5$ it is proved in Hanani (1961, 1972, 1975) and Lenz (1981). The case $k = 3$ and $\lambda = 1$ is studied in detail in 7.29—7.39.

8.15 Corollary [Hanani 1961, Cho 1982]. *A K_3-decomposition of ${}^{\lambda}K_v$ $[a\,(v, 3, \lambda)$-design] exists if and only if $v \geq 3$ and one of the following cases comes true:*

A. $\lambda \equiv 1$ *or* 5 (mod 6), $v \equiv 1$ *or* 3 (mod 6).

B. $\lambda \equiv 2$ *or* 4 (mod 6), $v \equiv 0$ *or* 1 (mod 3).

C. $\lambda \equiv 3$ (mod 6), $v \equiv 1$ (mod 2).

D. $\lambda \equiv 0$ (mod 6).

Proof follows from 8.14 by considering the possibilities for λ with respect to the modulo 6.

8.16 Remarks. A. Corollary 8.15 allows decompositions which have several triangles with the same vertices (or equivalently, *multiple* or *repeated blocks*). To avoid this possibility it is necessary to add the condition $\lambda \leq v - 2$, as was proved in Dehon (1983). Additional information on this question can also be found there.

B. Instead of decompositions, coverings or packings can be considered. The maximum number of triangles in a packing of the graph K_v is equal to $\lfloor v \lfloor \lambda(v - 1)/2 \rfloor /3 \rfloor$ except for $v \equiv 2$, $\lambda \equiv 4$ (mod 6); $v \equiv 5$, $\lambda \equiv 1$ (mod 6); $v \equiv 5$, $\lambda \equiv 4$ (mod 6) in which case it is less by one, see Hanani (1975) and Stanton et al. (1983); for $\lambda \equiv 1$ see also Schönheim (1966), Spencer (1968), Hilton (1969) and Stanton and Rogers (1982). The minimum number of triangles that cover $^\lambda K_v$ is equal to $\lceil v \lceil \lambda(v - 1)/2 \rceil /3 \rceil$ except for $v \equiv \lambda \equiv 2$ (mod 6); $v \equiv 5$, $\lambda \equiv 2$ (mod 6) and $v \equiv \lambda \equiv 5$ (mod 6) in which case it is greater by one, see Hanani (1975) and Stanton and Rogers (1982); for $\lambda = 1$, see also Fort and Hedlund (1958) and Hilton (1969).

Packings and coverings of the graph $^\lambda K_v$ by complete graphs of order $k \geq 3$ are investigated in Schönheim (1966) and Hanani (1975); for $\lambda = 1$ see also Novák (1959, 1963, 1970a, 1978). Results concerning packings and coverings of complete graphs with a graph G are given in Roditty (1983) and references therein (cf. 9.4); for $G = C_4$ see Schönheim and Bialostocki (1975).

The usual formulation of results on coverings or packings of the graph $^\lambda K_v$ is in terms of triples or 3-element subsets of a v-element set. A modification of this problem which allows repeated triples can be found in Billington and Stanton (1982) and Billington (1982). Apart from the works cited above, packings of complete graphs with triangles and related problems are studied in Nash-Williams (1970b), Novák (1970b), Treash (1971), Lindner (1975), Andersen et al. (1980), Stanton and Goulden (1981), Colbourn et al. (1983), Mendelsohn and Rosa (1983), Severn (1984) and Colbourn (in press) using the terminology of completing or embedding of partial triple systems into Steiner triple systems. (A *partial triple system* of order v is a system S of order v such that every 2-element subset of V is contained in at most one member of S.)

C. A (v, k, λ)-design is said to be *cyclic* if its points are the elements $0, 1, 2, ..., v - 1$ and with every block $\{a_1, a_2, ..., a_k\}$ it has also the block $\{a_1 + 1, a_2 + 1, ..., a_k + 1\}$ with the same multiplicity; the addition is taken modulo v).

In Colbourn and Colbourn (1981a) it is proved that a cyclic K_3-decomposition of $^\lambda K_v$ [a cyclic $(v, 3, \lambda)$-design] exists if and only if $[v, \lambda] \notin \{[9, 1]$, $[9, 2]\}$ and one of the following conditions holds:

$\lambda \equiv 1$ or 5 (mod 6), $v \equiv 1$ or 3 (mod 6),. $\hspace{2cm}$ (1)

$\lambda \equiv 2$ or 10 (mod 12), $v \equiv 0, 1, 3, 4, 7$ or 9 (mod 12), $\hspace{1cm}$ (2)

$\lambda \equiv 3$ (mod 6), $v \equiv 1$ (mod 2), $\hspace{3cm}$ (3)

$\lambda \equiv 4$ or 8 (mod 12), $v \equiv 0$ or 1 (mod 3), $\hspace{2cm}$ (4)

$\lambda \equiv 6$ (mod 12), $v \equiv 0, 1$ or 3 (mod 4), $\hspace{2.3cm}$ (5)

$\lambda \equiv 0$ (mod 12), $v \geq 3$. $\hspace{4.5cm}$ (6)

The special case $\lambda = 1$ was given in 7.37. Further information on cyclic (v, k, λ)-designs can be found in Jimbo and Kuriki (1983).

D. A (v, k, λ)-design is called *n-rotational* if it has an automorphism with one fixed point and n cycles of length $(v - 1)/n$. The existence and enumeration of *n*-rotational Steiner triple systems is dealt with in Phelps and Rosa (1981), mainly for $n \leq 6$. Every 1-rotational (v, k, λ)-design can be represented in the following form: its points are the elements $0, 1, ..., v - 2$ and the element ∞; with each block $\{a_1, a_2, ..., a_k\}$ the design contains (with the same multiplicity) the block $\{a_1 + 1, a_2 + 1, ..., a_k + 1\}$. The addition is taken modulo $v - 1$ except for the rule $\infty + 1 = \infty$. This concept naturally carries over to K_k-decompositions of $^\lambda K_v$.

In Cho (1982) and Kuriki and Jimbo (1983) it is proved that a 1-rotational K_3-decomposition of $^\lambda K_v$ [a 1-rotational $(v, 3, \lambda)$-design] exists if and only if one of the following conditions is satisfied:

$\lambda \equiv 1, v \equiv 3$ or 9 (mod 24), $\hspace{3.5cm}$ (7)

$\lambda \equiv 1$ or 5 (mod 6), $\lambda \neq 1, v \equiv 1$ or 3 (mod 6), $\hspace{1cm}$ (8)

$\lambda \equiv 2$ or 4 (mod 6), $v \equiv 0$ or 1 (mod 3), $\hspace{2cm}$ (9)

$\lambda \equiv 3$ (mod 6), $v \equiv 1$ (mod 2), $\hspace{3.2cm}$ (10)

$\lambda \equiv 0$ (mod 6), $v \geq 3$. $\hspace{4.5cm}$ (11)

The special case $\lambda = 1$ was solved in Phelps and Rosa (1981). Comparing with C it can be seen that if there exists a K_3-decomposition of $^\lambda K_v$ [a $(v, 3, \lambda)$-design] then there also exists either a cyclic or 1-rotational K_3-decomposition of $^\lambda K_v$ [a $(v, 3, \lambda)$-design] with the same parameters v, 3 and λ, see Cho (1982) and Kuriki and Jimbo (1983).

8.17 We now concentrate on the special case $\lambda = 1$ which corresponds to a K_k-decomposition of K_v.

Lemma. *If there exists a K_k-decomposition of K_v [a $(v, k, 1)$-design] then there is an integer c such that*

$$0 \leq c < k, \tag{1}$$

$$k \mid c(c - 1), \tag{2}$$

$$v \equiv 1 + (k - 1)c \pmod{k(k - 1)}. \tag{3}$$

Proof. If there exists a $(v, k, 1)$-design then, by 8.9.1, for an integer c' it holds $v = 1 + (k - 1)c'$. Clearly, there exists c such that $c \equiv c' \pmod{k}$ and (1) is satisfied. Then $v \equiv 1 + (k - 1)c' \equiv 1 + (k - 1)c \pmod{k(k - 1)}$ which yields (3). By 8.9.2 we have $k(k - 1) \mid v(v - 1) = (1 + (k - 1)c')$. $\cdot (k - 1)c'$ whence $k \mid (1 + (k - 1)c')c' = kc'^2 - c'(c' - 1)$. Therefore $k \mid c'(c' - 1)$ and, since $c \equiv c' \pmod{k}$, we obtain (2).

8.18 Theorem [Hanani 1975, p. 303]. *If there exists a K_k-decomposition of the graph K_v and k is a power of a prime then*

$$v \equiv 1 \text{ or } k \pmod{k(k - 1)}. \tag{1}$$

Proof. Assume there exists a $(v, k, 1)$-design. Then, by Lemma 8.17, there is a number c satisfying 8.17.1—3. If k is a power of a prime then 8.17.2 implies that $k \mid c$ or $k \mid c - 1$. From 8.17.1 we deduce that either $c = 0$ or $c = 1$. Substituting these values to 8.17.3 we obtain (1).

8.19 Corollary. *For any integer k with $2 \leq k \leq 5$ a K_k-decomposition of K_v [a $(v, k, 1)$-design] exists if and only if 8.18.1 holds.*

Proof. The necessity of 8.18.1 follows from Theorem 8.18 because the numbers 2, 3, 4 and 5 are all prime powers. For the converse, if 8.18.1 holds then either $v = 1$ or $v \geq k$. In the first case the existence of a $(v, k, 1)$-design is obvious. In the second case one just takes into account that 8.18.1 implies $v - 1 \equiv 0 \pmod{k - 1}$, $v(v - 1) \equiv 0 \pmod{k(k - 1)}$, and the existence of a $(v, k, 1)$-design follows from 8.14 for $\lambda = 1$. (For $k = 3$ see also Corollary 7.39).

8.20 Lemma. *Let $k \geq 2$. A K_k-decomposition of the graph $^\lambda K_v$ [a (v, k, λ)-design] is symmetric if and only if*

$$\lambda(v - 1) = k(k - 1). \tag{1}$$

Proof follows directly from 7.7, 8.5 and 8.7.

Remark. In proving the existence of symmetric (v, k, λ)-designs *cyclic methods*, analogous to the method of cyclic decompositions, are often used.

Such decompositions were studied in Chapter 7 (see for instance 7.55—7.56 and R 7.56). Parallel to the notion of a ϱ-labelling of a complete graph is the so-called (v, k, λ)-*difference set* for $\lambda = 1$. The corresponding construction method is usually called the *Bose difference method*, see Bose (1939). Detailed and lucid treatment of this topic can be found in Hall (1967, Chap. 11).

On the other hand, there is a deep result concerning the non-existence of a (v, k, λ)-design (more exactly, a necessary condition for its existence):

Theorem [Chowla and Ryser 1950]. *Let v, k and λ be integers satisfying the equality* (1) *and such that there exists a K_k-decomposition of the graph $^\lambda K_v$ [a (v, k, λ)-design]. Then the following conditions hold:*

A. *If v is even then $k - \lambda$ is the square of some integer* [Shrikhande 1950].

B. *If v is odd then there exist integers X, Y and Z (at least one non-null) such that*

$$Z^2 = (k - \lambda) X^2 + (-1)^{(v - 1)/2} \lambda Y^2.$$

Proof. A. If v is even then the equality (1) implies that $k \geq 2$ and Lemma 8.20 implies that under these assumptions every K_k-decomposition of $^\lambda K_v$ [a (v, k, λ)-design] is symmetric. Hence, (8.5) $v = b$ and (8.5B) $r = k$. It follows that the corresponding incidence matrix (8.3) **A** is square. Setting $\mathbf{B} = \mathbf{AA}^T$ we see that, by Lemma 8.3.2 and Theorem 8.5.C it holds:

$$\det \mathbf{B} = (r + (v - 1))(r - \lambda)^{v-1} = (r + r(k - 1))(r - \lambda)^{v-1} =$$
$$= k^2(k - \lambda)^{v-1}.$$

Since $\det \mathbf{B} = (\det \mathbf{A})^2$, the number $(k - \lambda)^{v-1}$ must be the square of some integer. As v is even $k - \lambda$ too must be the square of some integer.

B. The proof is difficult and therefore omitted. It can be found, besides Chowla and Ryser (1950), also in Hall (1967, Theorem 10.3.1) and Dembowski (1968, 2.1.15, p. 63), see also Lenz (1983).

Employing this theorem one can disprove the existence of a (symmetric) K_k-decomposition of K_n [a (v, k, λ)-design], for instance, for $[v, k, \lambda] \in$ $\in \{[22, 7, 2], [29, 8, 2], [34, 12, 4], [43, 15, 5], [46, 10, 2], [67, 12, 2], [92, 14, 2], [106, 15, 2]\}$; for further triples see Kageyama and Hedayat (1982, Table 2) and Mathon and Rosa (1985). The case $\lambda = 1$ will be considered in 8.22—8.25. Further similar results can be found in the exercises 8.51—8.56.

8.21 We shall need the following result from number theory.

Lemma. *Let $n \geq 2$ be an integer. The following statements are equivalent.*

A. *There exist integers s and t such that $n = s^2 + t^2$.*

B. *There exist integers $X \neq 0$, Y and Z such that $nX^2 = Y^2 + Z^2$.*

C. *All prime factors* $\equiv 3$ (mod 4) *occurring in the canonical decomposition of the number n have even exponent.*

Proof. The equivalence $A \Leftrightarrow C$ is proved, for instance, in Hardy and Wright (1971, Theorem 366). On pp. 299—302 of that book four different proofs are given. A fairly elementary proof can be found also in Sierpiński (1964, Theorem XI.1, p. 351).

The equivalence $A \Leftrightarrow B$ follows from the equivalence $A \Leftrightarrow C$ by comparing the canonical decompositions of n and nX^2, taking into account prime factors $\equiv 3$ (mod 4).

8.22 Important examples of $(v, k, 1)$-designs are finite affine and projective planes, i.e., affine and projective planes of finite order n. Given integer $n \geq 2$, any $(n^2, n, 1)$-design is called an *affine plane* of order n (cf. 12.9 and R 12.66) and any $(n^2 + n + 1, n + 1, 1)$-design is called a *projective plane* of order n (cf. 8.45). Blocks of these $(v, k, 1)$-designs are called *lines*. Obviously, for each projective plane of order n it holds $v = b = n^2 + n + 1$. Hence, every finite projective plane is always a symmetric design (cf. 8.5 and 8.44). (Warning: the order n of an affine or projective plane and the order v of the corresponding $(v, k, 1)$-design are different numbers and have to be distinguished.)

For example, a projective plane of order 2 consists of the points 1, 2, ..., 7 and lines $126 = \{1, 2, 6\}$, 137, 145, 235, 247, 346, 567. Omitting the points 5, 6 and 7 and the line 567 we get an affine plane of order 2 with the points 1, 2, 3, 4 and lines 12, 13, 14, 23, 24, 34. The projective and the affine plane are depicted in Fig. 8.22.1, an affine plane of order 3 is in Fig. 12.16.1.

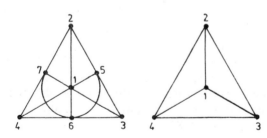

Fig. 8.22.1

The geometric background and further connections of these concepts are explained in the monographs of Dembowski (1968) and Kárteszi (1976) and in an excellent survey paper of Lorimer (1981). So far we shall content ourselves with the following facts:

8.23 Theorem. *Let $n \geq 2$ be an integer. Then the following hold*:

A. *An affine plane of order n exists if and only if there exists a projective plane of order n* [Hall 1967, Theorem 12.3.3].

B. *There exists a projective of order n if n is a power of a prime* (see for instance Hanani 1975, Theorems 2.1 and 2.2 or Street and Wallis 1977, Theorem VIII.4).

C. *A projective plane of order n does not exist if $n \in Q$, where Q is the set of all positive integers $n \equiv 1$ or 2 (mod 4) that cannot be written as $n = s^2 + t^2$, where s and t are integers. In particular there is no projective plane of order $n \equiv 6$ (mod 8)* (Bruck and Ryser 1949, Chowla and Ryser 1950, Hall 1967, Theorem 12.3.2, Dembowski 1968, 3.12.13, p. 144).

Proof. The statement A will be proved in 12.31.

B. If n is a power of a prime, then, up to isomorphism, there exists exactly one finite field of order n, the *Galois field* $GF(n)$ (see, for instance, Birkhoff and Mac Lane 1965, 15.6). On account of the statement A it is sufficient to construct an affine plane of order n, that is, an $(n^2, n, 1)$-design. The points of the affine plane will be all ordered pairs $[x, y]$, where $x, y \in GF(n)$. The lines will consist of all points $[x, y]$ which satisfy the equation $y = ax + b$ or the equation $x = c$, where a, b and c are fixed elements of the field $GF(n)$. It is clear that there are $n^2 + n$ lines and each has exactly n points. It is also easy to see that any two points $[x_1, y_1]$ and $[x_2, y_2]$ lie on exactly one line. If $x_1 = x_2$ it is the line $x = x_1$, otherwise it is the line

$$y = \frac{y_2 - y_1}{x_2 - x_1} x + \frac{y_1 x_2 - y_2 x_1}{x_2 - x_1}.$$

Thus we have a $(n^2, n, 1)$-design.

C. Assume that there exists a projective plane of order n. Set $v = n^2 + n + 1, k = n + 1$ and $\lambda = 1$. Then v is odd and by virtue of Theorem 8.20.B there exist integers X, Y and Z (at least one non-null) such that

$$Z^2 = nX^2 + (-1)^{(n+1)n/2} Y^2.$$

If $n \equiv 1$ or 2 (mod 4) then $(n+1)n/2$ is odd and $Z^2 = nX^2 - Y^2$, that is, $nX^2 = Y^2 + Z^2$. Clearly $X \neq 0$. By Lemma 8.21 there exist integers s and t such that $n = s^2 + t^2$ which contradicts the hypothesis. If $n \equiv 6$ (mod 8) then $n \equiv 2$ (mod 4); since s^2, $t^2 \equiv 0, 1$ or 4 (mod 8) we have $s^2 + t^2 = 6$ (mod 8).

8.24 Corollary. *A projective (and an affine) plane of order n exists for $n \in \{2, 3, 4, 5, 7, 8, 9, 11, 13, 16, 17, 19, 23, 25, 27, 29, ...\}$ and does not exist for $n \in \{6, 14, 21, 22, 30, ...\} = Q$.*

Proof follows from 8.23; for $n \in \{2, 3, 4, 5, 7, 8, 9, 11\}$ also from 7.56 and 12.31.

8.25 Remarks. A. The problem of the existence of a projective (and also of an affine) plane of order n for $n \in \{10, 12, 15, 18, 20, 24, 26, 28, ...\}$ is open.

B. By Lemma 8.21 the set Q from 8.23.C can be defined as the set of all integers $n \geq 2$ with $n \equiv 1$ or 2 (mod 4) the canonical decomposition of which contains at least one prime number $\equiv 3$ (mod 4) with an odd exponent.

8.26 The following theorem yields a result which corresponds to Theorem 8.14 for $k = 6$.

Theorem [Connor 1952, Hall and Connor 1954, Hanani 1975, Lenz 1981]. *Let $v \geq 2$ and $\lambda \geq 2$. Then a K_6-decomposition of the graph $^{\lambda}K_v$ [a $(v, 6, \lambda)$-design] exists if and only if*

$$v \geq 6, \tag{1}$$

$$5 | \lambda(v - 1), \tag{2}$$

$$30 | \lambda v(v - 1), \tag{3}$$

$$[v, \lambda] \neq [21, 2]. \tag{4}$$

Proof. The necessity of the conditions (1)—(4) follows from 6.3.A, 8.9.1, 8.9.2 and 8.51. The sufficiency is established in the cited literature.

8.27 The condition $\lambda \geq 2$ in Theorem 8.26 cannot be omitted. Indeed, there is no $(v, 6, 1)$-design for $v = 16$ and 21 (by Theorem 8.10) and for $v = 36$ (by 8.23—24) although the conditions (1)—(4) are satisfied.

Problem [Hanani 1975, Mills 1978a]. *Determine the set M of all integers v for which there exists a K_6-decomposition of K_v [a $(v, 6, 1)$-design].*

It is known that the set M contains the numbers 1, 6, 31, 66, 76, 91, 96, 106, 111, 121, 126, 136, 151, 156, 181 and 186. This is clear for the numbers 1 and 6, for 31 it follows from 7.56 if we set $k = 6$. A $(66, 6, 1)$-design was constructed by Denniston (see Denniston 1980b). Further $(v, 6, 1)$-designs were constructed with help of computers: for $v = 76$ see Mills (1979), for $v = 96$ and 111 see Mills (1978a), for $v = 106$ see Mills (1975a), for $v = 121$ see Mills (1975b), for $v = 126$ see Stanton and Gryte (1970), Hanani (1975) and McKay and Stanton (1978) (without computers already Bose 1963), for $v = 136$ see Mills (1978b). The case $v = 91$ was discussed in 7.4. $(v, 6, 1)$-designs for $v = 151, 156, 181$ and 186 are constructed in Wilson (1972b) and Mills (1978a).

For many numbers (for instance 46, 51, 61, 81, 141, 166, 171, 196) it is not known whether or not they belong to the set M. Mills (1978a, 1979) conjectures that it consists of all integers $v \equiv 1$ or 6 (mod 15) except for $v = 16, 21$ and 36; so far this conjecture is proved by Mills (1979) for $v > 34\,636$. The same result for v sufficiently large follows from 8.9 and 8.12.

Even less is known about K_7-decompositions of the graph $^{\lambda}K_v$ [$(v, 7, \lambda)$-designs]. A necessary and sufficient condition for their existence is known just for $\lambda = 6, 7$ and 42 (see Hanani 1975, Lenz 1981).

8.28 We now summarize the obtained necessary conditions for the existence of a K_k-decomposition of the graph K_v.

Corollary. *If there exists a K_k-decomposition of the graph K_v [a $(v, k, 1)$-design] then the following conditions hold:*

A. $k - 1|v - 1$.
B. $k(k - 1)|v(v - 1)$.
C. $v = 1$ or $v = k$ or $v \geq k^2 - k + 1$.
D. If $v = k^2 - k + 1$ then $k - 1 \notin Q$.
E. If $v = k^2$ then $k \notin Q$.

(The set $Q = \{6, 14, 21, 22, 30, ...\}$ is defined in 8.23—8.25.)

Proof. The statement A follows from 8.9.1, B from 8.9.2 and C from 8.11 for $\lambda = 1$. The statements D and E follow from the non-existence of a projective plane of order $n = k - 1$ and an affine plane of order $n = k$, see 8.23.

8.29 Known results on K_k-decompositions of K_v $(v, k, 1)$-designs for "small" values of v are summarized in the following two theorems.

Theorem. *A K_k-decomposition of the graph K_v [a $(v, k, 1)$-design] for $v \leq 140$ can exist just in the following cases:*

A. $v = 1$ or $v = k$.
B. $2 \leq k \leq 5$, $v \equiv 1$ or $k \pmod{k(k - 1)}$.
C. $k = 6$, $v \in \{31, 46, 51, 61, 66, 76, 81, 91, 96, 106, 111, 121, 126, 136\}$.
D. $k = 7$, $v \in \{49, 85, 91, 127, 133\}$.
E. $k = 8$, $v \in \{57, 64, 113, 120\}$.
F. $k = 9$, $v \in \{73, 81\}$.
G. $k = 10$, $v \in \{91, 100, 136\}$.
H. $k = 11$, $v \in \{111, 121\}$.
I. $k = 12$, $v = 133$.

Proof. Assume that A does not hold. Then $k \geq 2$. If $k \leq 5$ then, by Theorem 8.18, we have B. Let $k \geq 6$. From Corollary 8.28.C and the condition $v \leq 140$ it follows that $k \leq 12$ and $v \geq 31, 43, 57, 73, 91, 111, 133$ if $k = 6, 7, 8, 9, 10, 11, 12$, respectively.

If $k = 6$ then, by Lemma 8.17, it holds that $v \equiv 1 + 5c \pmod{30}$ where $0 \leq c < 6$ and $6|c(c - 1)$. Hence $c \in \{0, 1, 3, 4\}$ and therefore $v \equiv 1, 6, 16$ or $21 \pmod{30}$. So $v \in \{31, 36, 46, 51, 61, 66, 76, 81, 91, 96, 106, 111, 121, 126, 136\}$. Since by Corollary 8.24, there exists no affine plane of order 6, that is, a $(36, 6, 1)$-design, we get C.

Let $k \in \{7, 8, 9, 11\}$. By Theorem 8.18 it holds $v \equiv 1$ or $k \pmod{k(k - 1)}$. Since, by Corollary 8.24, there is no projective plane of order 6, that is, a $(43, 7, 1)$-design, one of the cases D, E, F or H occurs.

Let $k = 10$. By Lemma 8.17 it holds $v \equiv 1 + 9c$, where $0 \leq c < 10$ and $10|c(c - 1)$. Hence $c \in \{0, 1, 5, 6\}$ and $v \equiv 1, 10, 46$ or $55 \pmod{90}$. Thus we have G. Analogously, if $k = 12$ we have that $v \equiv 1, 12, 45$ or $100 \pmod{132}$ and I holds.

8.30 Theorem. *A K_k-decomposition of the graph K_v [a $(v, k, 1)$-design] exists in the following cases:*

A. $v = 1$ *or* k.

B. $2 \leq k \leq 5$, $v \equiv 1$ *or* k (mod $k(k-1)$).

C. $k = 6$, $v \in \{31, 66, 76, 91, 96, 106, 111, 121, 126, 136\}$.

D. $k = 7$, $v \in \{49, 91\}$.

E. $k = 8$, $v \in \{57, 64, 120\}$.

F. $k = 9$, $v \in \{73, 81\}$.

G. $k = 10$, $v = 91$.

H. $k = 11$, $v = 121$.

I. $k = 12$, $v = 133$.

Proof. A K_k-decomposition of K_v exists in the case A (obviously), in the case B (by Corollary 8.19) and also in the case C (by 8.27). The existence of affine and projective planes of order 7, 8, 9 and 11 (Corollary 8.24) implies the existence of a K_k-decomposition of K_v for the ordered pairs $[v, k] \in \{[49, 7], [64, 8], [81, 9], [121, 11]\}$ and $[v, k] \in \{[57, 8], [73, 9], [91, 10], [133, 12]\}$, respectively. A K_7-decomposition of the graph K_{91} was constructed in 7.4, a $(120, 8, 1)$-design can be found in Seiden (1963).

8.31 Remark. Comparing Theorem 8.29 with Theorem 8.30 it can be seen that the existence of a K_k-decomposition of the graph K_v [a $(v, k, 1)$-design] for $v \leq 140$ remains open in the following 11 cases:

$k = 6$; $v = 46, 51, 61, 81$;
$k = 7$; $v = 85, 127, 133$;
$k = 8$; $v = 113$;
$k = 10$; $v = 100, 136$;
$k = 11$; $v = 111$.

The results from 8.29—8.31 are extended to $v \leq 200$ in the exercises 8.48—8.50.

8.32 The concept of a (v, k, λ)-design can be generalized to a t-(v, k, λ)-design, where $v, k, \lambda \geq 1$ and $t \geq 0$. (To avoid trivial cases it is sometimes assumed that $1 < t < k < v$; we do not accept this assumption unless if it follows from the situation.) A t-(v, k, λ)-*design* is a system S of (not necessarily distinct) k-element subsets (called *blocks* or *k-tuples*) of a v-element set V (whose elements are called *points*) such that every t-element subset of V is contained in exactly λ members of S. For $t = 2$ we obtain a (v, k, λ)-design. Another important special case is obtained when $\lambda = 1$ and is called a *Steiner system of type* $S(t, k, v)$ or briefly, an $S(t, k, v)$ *system*. A system $S(k - 1, k, v)$ is called a *Steiner system of k-tuples* of order v; a system $S(2, k, v)$, also consisting of k-tuples, is nothing but a $(v, k, 1)$-design, which we have discussed in 8.17—8.31. The special case for $k = 3$, $\lambda = 1$, $t = 2$ of Steiner triple systems of order v was dealt with in 7.29—7.39.

It is easily seen (see, for instance, Hedayat and Kageyama 1980, 1.1) that if s is an integer with $0 \leq s < t \leq k$ then every t-(v, k, λ)-design is at the same time an s-(v, k, λ')-design, where

$$\lambda' = \lambda \binom{v-s}{t-s} \bigg/ \binom{k-s}{t-s}. \tag{1}$$

Therefore a necessary condition for the existence of a t-(v, k, λ)-design is that, for $s = 0, 1, 2, \ldots, t-1$,

$$\binom{k-s}{t-s} \bigg| \lambda \binom{v-s}{t-s}. \tag{2}$$

For $t = 2$ and $s = 0$ [$s = 1$] we obtain the condition 8.9.2 [8.9.1]; in this case $\lambda' = b$ [$\lambda' = r$] if $k \geq 2$; cf. 8.7.

An extensive survey of results on t-(v, k, λ)-designs is in Hedayat and Kageyama (1980) and Kageyama and Hedayat (1982).

8.33 A t-(v, k, λ)-design can be interpreted by means of a hypergraph in two ways. In the first interpretation, to a t-(v, k, λ)-design there corresponds a k-uniform hypergraph of order v (that is, each edge is incident with k vertices of the total number v of vertices) with the property that each k-tuple of distinct vertices is contained in exactly λ edges (vertices of the hypergraph correspond to points and edges to blocks of the design). In the second interpretation, to a t-(v, k, λ)-design there corresponds [Bermond et al. 1977] a decomposition of the complete λ-fold t-uniform hypergraph of order v into complete t-uniform subhypergraphs of order k (points of the design are interpreted as vertices of the hypergraph, but blocks are interpreted as subhypergraphs; for $t = 2$ we obtain the correspondence between (v, k, λ)-designs and K_k-decompositions of the graph $^{\lambda}K_v$ which we have employed in 8.2—8.31).

Minimal coverings of the complete t-uniform hypergraph of order v with complete t-uniform subhypergraphs of order k (especially in the case $t = 3$ and $v = 2k - 1$) are investigated in Todorov and Tonchev (1982).

8.34 Two designs are said to be *identical* if they have the same points and the same blocks (their order is immaterial; if there are repeated blocks then the same block must occur in both designs with the same multiplicity). Otherwise two designs are considered as different.

Two designs are said to be *isomorphic* if there exists a bijection between their sets of points which preserves blocks. Otherwise designs are non-isomorphic (essentially different).

In the spirit of 1.51 we define an *isomorphism of hypergraphs* $[V, E, I]$ and $[V', E', I']$ as a bijection $f: V \cup E \rightarrow V' \cup E'$ satisfying the conditions 1.24.1—3. Hypergraphs are said to be *isomorphic* if there is some isomorphism between them. We note that the concepts of equality and isomorphism of graph decompositions (3.9) can be naturally generalized to include hyper-graphs.

It is easily verified that two t-(v, k, λ)-designs are isomorphic if and only if the corresponding k-uniform hypergraphs or the decompositions of t-uniform hypergraphs (see 8.33) are isomorphic. The situation with the equality is different. For $\lambda \geq 2$ the same t-(v, k, λ)-design can possibly correspond to different k-uniform hypergraphs or decompositions of t-uniform hypergraphs; see Exercise 8.52 for the case $t = 2$, $v = k = \lambda$.

We shall now be interested in the number of non-isomorphic $S(t, k, v)$ systems with given t, k and v. Not much is known, especially regarding the exact values. For instance, the following values of the function $S(v)$, the number of non-isomorphic $S(2, 3, v)$ systems, that is, Steiner triple systems [K_3-decompositions of K_v], are known: $S(1) = S(3) = S(7) = S(9) = 1$, $S(13) = 2$, $S(15) = 80$ (for the first time discovered in White et al. 1919, and then many times rediscovered), $S(19) \geq 284\,407$, $S(21) \geq 2\,160\,980$, $S(25) \geq 10^{14}$, $S(27) \geq 10^{11}$. There are two asymptotical formulas known:

$$S(v) = v^{v^2(1/6 + °(1))},$$

$$S(v) \sim e^{v(v - 1)/6},$$

($e = 2.71828\ldots$) for $v \to \infty$. Details can be found in Fisher (1940), Alekseev (1974), Wilson (1974b), Doyen and Rosa (1978), Babai (1980), Phelps (1980), Mathon et al. (1983) and in references therein.

8.35 Because of the difficulties with determination of exact values of the function $S(v)$, enumeration of special types of Steiner systems was considered by many authors. For example, non-isomorphic cyclic Steiner triple systems of order v for $v \leq 45$ are computed in Colbourn (1979), Colbourn (1982a) and for $v \leq 37$ also in Bays (1923); see also 7.38.A. Further information concerning enumeration of special Steiner triple systems, that is K_3-decompositions of K_v, can be found in 7.38.A and in Colbourn and Mathon (1980), Denniston (1980a), Mathon et al. (1981, 1983), Phelps and Rosa (1981), and Mathon and Rosa (1984). Enumeration of K_3-decompositions of ${}^2K_{10}$ and C_3^0-decompositions of K_{10}^* (see 8.14 and 10.9 for the existence theorem of these decompositions), disguised as (10, 3, 2)-designs and, respectively, Mendelsohn triple systems of order 10 (named after S. Mendelsohn, see Mendelsohn 1971), can be found in Ganter et al. (1978a, b, 1979), Ivanov (1981) and Colbourn et al. (1983a). It turns out that altogether there are 960 non-isomorphic K_3-decompositions of ${}^2K_{10}$ and 566 of them contain a pair of triangles with the same vertices. A survey (including a bibliography) of the extensive field of Steiner systems can be found in Doyen and Rosa (1978), Stanton and Goulden (1981) and in the book *Topics in Steiner Systems*

(Annals of Discrete Mathematics 7, North-Holland, Amsterdam, 1980). The complete solution of the problem to construct two *Mendelsohn triple systems* with a prescribed number of common (cyclic) triples is in Hoffman and Lindner (1982); this is equivalent with finding two C_3^0-decompositions of the graph K_v^* with a prescribed number of "consistently parallel" pairs of cycles of order 3.

8.36 From the results of Wilson (1975) it follows that for $k \geq 3$ and $\lambda \geq 1$ the number of non-isomorphic K_k-decompositions of ${}^\lambda K_v$ [non-isomorphic (v, k, λ)-designs], assuming the conditions 8.9.1—2, tends to infinity with growing v. For the minimum values of the parameters, that is, for $k = 3$ and $\lambda = 1$ the number $S(v)$, defined in 8.34, is obtained.

Further results on K_k-decompositions of the graph ${}^\lambda K_k$ [(v, k, λ)-designs] can be found in Hall and Connor (1954), Hall (1967), Parker (1967), Hanani (1975), Lesniak-Foster and Roberts (1977), Hedayat and Kageyama (1980), Tran van Trung (1982), Kageyama and Hedayat (1982) and Mathon and Rosa (1985). Necessary and sufficient conditions for their existence in terms of the incidence matrices are analysed in Araoz and Tannenbaum (1979).

8.37 So far in this chapter we have dealt with the question of when the property $K_k | {}^\lambda K_v$ is satisfied. The oriented variant of this problem reads as follows: Under what conditions $K_k^\rightarrow | {}^\lambda K_v^*$, that is, when does the complete λ-fold digraph of order v admit a decomposition into acyclic tournaments of order k? The case $k = 3$ will be considered, since $K_3^\rightarrow = C_3^\rightarrow$, in 10.10—10.11. We shall see that $K_3^\rightarrow | {}^\lambda K_v^*$ if and only if $v \geq 3$ and $3 | \lambda v(v - 1)$ (Seberry and Skillicorn 1980, Colbourn and Harms 1983). In Street and Seberry (1980) it is proved (in terms of oriented designs) that $K_4^\rightarrow | {}^\lambda K_v^*$ if and only if $v \geq 4$, $3 | \lambda(v - 1)$ and $6 | \lambda v(v - 1)$. The case $k = 5$ is solved in Street and Wilson (1980), see also Colbourn and Harms (1983).

8.38 In Skillicorn (1982b) the following results (with a sketch of proof) are given in the combinatorial terminology: For each integer $v \geq 34835$ the minimum number $DN(2, 4, v)$ of acyclic (and hence transitive) tournaments of order 4 (that is, subgraphs isomorphic with $K_4^\rightarrow$) that cover the complete digraph K_v^* is equal to $\lceil (v/4) \lceil 2(v - 1)/3 \rceil \rceil$. Further, for each integer $v \geq 35781$ the maximum number $DD(2, 4, v)$ of acyclic tournaments of order 4 forming a packing of the graph K_v^* is equal to $\lfloor (v/4) \lfloor 2(v - 1)/3 \rfloor \rfloor$. Similarly defined numbers $DN(2, 3, v)$ and $DD(2, 3, v)$ will be discussed in 10.11.B.

8.39 K_k-decompositions of graphs other than ${}^\lambda K_v$ have also been investigated. Some results along this line can be found in Hell and Rosa (1971, 1972), Tomasta and Zelinka (1981) and in Wallis (1977). For $k = 3$ C_3-decompositions of H (decompositions into triangles) are obtained. Such decompositions will be discussed in 10.17 (see also 10.18). Packings of graphs with triangles have been investigated, for instance, in Sauer (1971), and coverings of graphs with complete graphs of the same order have been studied in Erdős (1970), Surányi (1970) and in Szemerédi (1970).

8.40 K_k-decompositions of complete m-partite graphs and their generalizations have often been studied (sometimes without authors being aware of this fact) in terms of group divisible designs, see, for instance, Bose and Connor (1952), Bose et al. (1953), Raghavarao (1971), Wilson (1972a, 1975), Hanani (1974), Seberry (1978), Mielants (1979), Engel (1980), Vanstone (1980) and Burr and Grossman (1982). A *group divisible design*, more precisely an $(n_1, n_2, ..., n_m; K; \lambda_1, \lambda_2)$-*design* (where $m, n_1, n_2, ..., n_m$ are positive integers, K is a set of positive integers and λ_1 and λ_2 are non-negative integers) is defined to be a system of not necessarily distinct subsets (called *blocks*), with cardinalities in K, of a given set (elements of which are called *points*) provided with a decomposition into classes (called *groups*) of cardinalities n_1, $n_2, ..., n_m$ such that an arbitrary pair of distinct points belonging to the same group [respectively, to different groups] is contained in exactly λ_1 [respectively, λ_2] blocks. (As usual, if $K = \{k\}$ we write k instead of $\{k\}$). For $m = 1$, $n_1 = v$, $1 \le \lambda = \lambda_1$ and arbitrary λ_2 we obtain a (v, K, λ)-design defined in 8.1; for $K = \{k\}$ we have a (v, k, λ)-design (8.2). Now, Lemma 8.2 can obviously be generalized as follows.

Lemma. *The complete (λ_1, λ_2)-fold m-partite graph $K_m(n_1, n_2, ..., n_m; \lambda_1, \lambda_2)$ (see 1.17) admits a decomposition into complete subgraphs whose orders belong to a given set K of positive integers if and only if there exists an $(n_1, n_2, ..., n_m; K; \lambda_1, \lambda_2)$-design.*

8.41 An important special case corresponding to decompositions of a λ-fold complete m-partite graph ${}^\lambda K_m(n_k, n_2, ..., n_m)$ is obtained for $\lambda_1 = 0$, $\lambda_2 = \lambda$. If, moreover, $n_1 = n_2 = ... = n_m = n$ we briefly speak of a (*group divisible*) $(m \times n, K, \lambda)$-*design*, and in the case when $K = \{k\}$, of an $(m \times n, k, \lambda)$-*design*. The corresponding concept in graph theory is a K_k-decomposition of the λ-fold complete m-partite graph ${}^\lambda K(m \times n)$. For $\lambda = 1$ this correspondence was pointed out in Colbourn (1979).

8.42 Results on the existence of decompositions of graphs into complete subgraphs can automatically be obtained, using Lemma 8.40 from the known results about group divisible designs (see, for instance, Bose and Connor 1952, Bose et al. 1953, Sprott 1959, Raghavarao 1971, Clatworthy 1973, Seberry 1978, Mielants 1979, Bush 1981, Jungnickel 1982). As an illustration we present some basic results of this kind brought together in the following theorem.

Theorem. *Let s be a power of a prime and m, n, k and λ be positive integers. A K_k-decomposition of the graph $K(m \times n)$ [a $(m \times n, k, \lambda)$-design] exists in each of the following cases:*

A. $m = s + 1, n = s - 1, k = s$.

B. $m = s^2 + s + 1, n = s^2 - s, k = s$.

C. $m = k \le s + 1, n = s$.

D. $m = k \le s^2 + s + 1, n = \lambda = s$.

E. $m = s^2 + s + 1, n = 2, k = s^2, \lambda = (s^2 - s)/2$.

Proof. In the cases A—C λ may be arbitrary but it is sufficient to prove these statements for $\lambda = 1$. The cases A and B now follow from the existence of a group divisible $(m \times n, k, 1)$-design which was proved, for instance, in Raghavarao (1971), Corollary 8.6.2.1 and Theorem 8.6.6. The case C follows from 12.27—12.30 (see also Tomasta and Zelinka 1981, Theorem 3). An $(m \times n, k, \lambda)$-design corresponding to case D and E has been constructed in Raghavarao (1971, Theorem 8.6.5), and Seberry (1978), respectively.

8.43 Problems standing on the boundary between G-decompositions of graphs and (v, k, λ)-designs can be generalized in many different ways: instead of one parameter λ or two parameters λ_1, λ_2 any finite number c of parameters $\lambda_1, \lambda_2, \ldots, \lambda_c$ can be allowed. These parameters then correspond to colours of edges in the graph H carrying the decomposition in such a way that any two vertices of the graph H (of order v) are joined by λ_i edges of the i-th colour ($i = 1, 2, \ldots, c$). Naturally, the graph G is then also coloured with c colours. Further, undirected graphs may be replaced by directed ones, and so on. These generalizations, with a discussion of the most important special cases, are surveyed in Wilson (1977); for their applications in organizing whist tournaments see also Wilson (1972, 1975) and Hanani (1975). Further generalizations of group divisible designs can be obtained by replacing the conditions on pairs of distinct points by conditions concerning t-element subsets ($t \ge 1$); see, for instance, Stanton and Mullin (1981). Some problems of this kind can be interpreted by means of decompositions of λ-fold complete m-partite t-uniform hypergraphs with parts of cardinality n into complete hypergraphs of order k.

Enumeration of group divisible designs is considered in Engel (1980).

Exercises

8.44 For each integer $v \ge 1$ and $k \ge 3$ every symmetric $(v, k, 1)$-design is a projective plane of order $k - 1$ and vice versa.

8.45 In every finite projective plane (8.22) the following hold: A. Any two distinct points are contained in exactly one line. B. Any two lines intersect in exactly one point. C. There exist four points no three of which lay on the same line. (The statements A—C are sometimes considered as the axioms of a projective plane, see, for instance, Rybnikov 1972 and Kárteszi 1976.)

8.44 For which k does there exist a K_k-decomposition of the graph: A. K_{91}; B. K_{129}?

8.47 Find all the possibilities for which an $(n_1, n_2, ..., n_m; K; \lambda_1, \lambda_2)$-design is at the same time a (v, K, λ)-design?

8.48 Extend Theorem 8.29 to $v \leq 200$.

8.49 Extend Theorem 8.30 to $v \leq 200$, using the information of Mathon and Rosa (1985), that there exists a (169, 7, 1)-design.

8.50 Comparing the results of 8.48 and 8.49 determine the cases in which the existence of a $(v, k, 1)$-design with $v \leq 200$ is neither proved nor disproved.

8.51 The results of Hall and Connor (1954) and Shrikhande (1960) imply the following: If $k \geq 4$ and $v = 1 + \binom{k}{2}$ then a (symmetric) K_k-decomposition of the graph $^\lambda K_v$ [a $(v, k, 2)$-design] exists if and only if there exists a K_{k-2}-decomposition of the graph $^2 K_{v-k}$ [a $(v - k, k - 2, 2)$-design]. Using this result and 8.20 find some ordered pairs $[v', k']$ for which there exists no $K_{k'}$-decomposition of $^2 K_{v'}$ [a $(v', k', 2)$-design]. (Hint: set $v' = v - k$ and $k' = k - 2$.)

8.52 A. How many K_k-decompositions of the graphs $^k K_k$ are there? B. What is the number of different (k, k, k)-designs on a fixed set of k points?

8.53 How many K_k-decompositions of $^{k-1} K_{k+1}$ ($k \geq 2$) are there?

8.54 For each $k \geq 2$ there exist at least two non-isomorphic K_k-decompositions of K_r [symmetric (v, k, λ)-designs].

8.55 [Hall 1967]. If there exists a (v, k, λ)-design with $2 \leq k \leq v - 2$ then there also exists a $(v, v - k, \lambda(v - k)(v - k - 1)/k(k - 1))$-design. Moreover, if one of them is symmetric then so is the other.

8.56 Find all ordered triples $[v, k, \lambda]$, $2 \leq k \leq 8$ for which there exists a symmetric K_k-decomposition of the graph K_v [a symmetric (v, k, λ)-design].

8.57 [Ryser 1968, Woodall 1970, Kageyama and Hedayat 1982]. There exist two non-isomorphic symmetric designs of order 7 in which arbitrary two distinct blocks have exactly two points in common.

8.58 There exist two non-isomorphic $(7, \{2, 3, ...\}, 2)$-designs.

8.59 [de Bruijn and Erdős 1948]. For each integer $v \geq 2$ there exists a symmetric design of order v in which arbitrary two distinct blocks have exactly one point in common and any two distinct points are contained in exactly one block.

Chapter 9

Decompositions into isomorphic subgraphs of small order, paths, trees, forests, complete bipartite graphs and cubes

9.1 A considerable number of papers deal with the existence of a G-decomposition of a graph H for a special choice of G and H. While H is usually chosen to be the complete graph or its modifications (λ-fold complete graph, complete digraph, complete m-partite graph), for the graph G much more possibilities have been considered. In this chapter we confine ourselves to just a few types of graphs G. Namely, we shall consider graphs of order < 10 (see 9.4—9.8, 9.28, 9.31—33, 9.35, 9.36), paths (9.6—9.18, 9.36), stars (9.19—9.26), other trees (9.5, 9.9, 9.17, 9.28, 9.34), forests (9.28, 9.32), complete bipartite graphs (9.20, 9.27) and n-cubes (9.29—9.31, 9.35).

9.2 A particular attention has been paid to balanced G-decompositions of the λ-fold complete graph $^{\lambda}K_v$, i.e., such that every vertex of $^{\lambda}K_v$ is contained in the same number of graphs of the decomposition. In this case the results are often formulated in terms of graph designs (more exactly, balanced (v, G, λ)-designs) which can be considered as a generalization of (v, k, λ)-designs defined in 8.2. This approach can be found in Hell and Rosa (1972), Huang and Rosa (1973b), Huang (1974), Bermond (1975b), Rosa and Huang (1975), Bermond and Sotteau (1977), Bermond et al. (1978c) and in survey articles Bermond and Sotteau (1976) and Harary and Wallis (1977); for hypergraphs in Bermond et al. (1977). Below we outline this approach as introduced in Hell and Rosa (1971, 1972). However, we shall slightly modify the definitions in order to avoid the use of matrices. The basic idea resides in introducing a structure within blocks.

While in (v, k, λ)-designs from 8.2 all points of the same block were in some sense equivalent now they will be bound with a binary relation J "a point u is *joined* with a point U" denoted as uJU. We shall assume that the relation J is *antireflexive* (there is no u such that uJu) and *symmetric* (uJU implies UJu).

A block together with such a relation, more exactly, an ordered pair $[V_0, J]$ where V_0 is a non-empty finite set and J an antireflexive symmetric binary relation on V_0 will be called a *structural block*; the elements of V_0 are its *points*.

Assign to each structural block $[V_0, J]$ an ordinary graph with vertex-set

V_0 in which two vertices u and U are adjacent if and only if uJU. This graph will be called the *graph of a structural block* $[V_0, J]$. A structural block the graph of which is isomorphic to G is called a *G-block*.

Let G a finite simple graph and v and λ positive integers. A (v, G, λ)-*design* is a system of (not necessarily distinct) G-blocks the points of which are elements of a fixed v-element set V and any two distinct points are joined in exactly λ blocks. If in addition each point (element of V) belongs to the same number of G-blocks, the (v, G, λ)-*design* is said to be *balanced*. Obviously a (v, K_k, λ)-design is the same as a (v, k, λ)-design; this design is always balanced as was proved in 8.5.A. More generally we have:

Lemma. *If G is a regular graph then every (v, G, λ)-design is balanced.*

P r o o f follows from Lemma 6.1.

9.3 The terminology of (v, G, λ)-designs is used especially when G is a path or cycle. Due to obvious correspondences

(v, G, λ)-design $\leftrightarrow$ G-decomposition of the graph $^{\lambda}K_v$,

balanced (v, G, λ)-design $\leftrightarrow$ balanced G-decomposition of the graph $^{\lambda}K_v$,

the theory of G-decompositions of $^{\lambda}K_v$ could be built up in two parallel interpretations similarly as was done in Chapter 8 for the correspondence

(v, k, λ)-design $\leftrightarrow$ K_k-decomposition of the graph $^{\lambda}K_v$.

In this chapter, however, we shall content ourselves with formulations in terms of graph decompositions. Therefore we omit the modification of the concept of a (v, G, λ)-design for a directed graph G (in this case we do not assume that the relation J is symmetric and so we obtain G-decompositions of the graph $^{\lambda}K_v^*$). However, the reader interested in the literature of this area should be aware of the parallel formulation in terms of *graph designs* (which is the common term for (v, G, λ)-designs). For example, results on P_k-decompositions of the graph $^{\lambda}K_v$ (9.11—9.13) have often been formulated in terms of balanced (v, P_k, λ)-designs known also as handcuffed designs (Hung and Mendelsohn 1974, 1977, Lawless 1974a, b, Huang and Rosa 1978), usually in connection with the problem of handcuffed prisoners (12.35—12.40); in Tarsi (1983) the term handcuffed design is used also for (v, P_k, λ)-designs which are not balanced. A similar kind of designs is investigated in Hwang (1976).

9.4 In a systematic study of the existence problem for G-decompositions of the complete graph of order v it was natural to begin with graphs of "small order". In Erdős and Schönheim (1975) and Bermond and Schönheim (1977) this problem has been solved for graphs G with less than 5 vertices. (In Roditty 1983 an analogous problem for packings and coverings, and in

Bermond et al. 1977 for decompositions of complete 3-uniform hypergraphs is considered.) Graphs of order 5 have been systematically studied in Bermond et al. (1980) and in most cases the final result has been found. For example, a G-decomposition of the graph K_v for the graph from Fig. 9.4.1 exists if and only if $v \neq 4$ and $v \equiv 0, 1, 4$ or $9 \pmod{12}$. In proving the necessity in similar cases the relations 6.14.1—3 are used. The sufficiency is then proved by construction.

Fig. 9.4.1

9.5 A necessary and sufficient condition for the existence of a T-decomposition of the graph K_v for all trees T of order < 10 has been found in Huang and Rosa (1978). For instance, for an arbitrary tree T of order 5 a T-decomposition of K_v exists if and only if $v \equiv 0$ or $1 \pmod 8$. Necessary and sufficient conditions for the existence of a balanced T-decomposition of the graph K_v for most trees of order < 10 are also given.

9.6 Theorem [Yavorskii 1978, Caro and Schönheim 1980]. *A P_3-decomposition of an ordinary finite graph H exists if and only if every component of H has even size.*

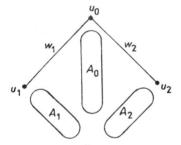

Fig. 9.6.1

Proof. The necessity is obvious. We shall establish the sufficiency by induction on the size e of the graph H. For $e = 0$ and $e = 2$ the assertion is obvious. Therefore assume that e is an even number ≥ 4 and that the assertion is true for each even number $e' \geq 0$ such that $e' < e$. Without loss of generality we may assume that H is connected. Choose in H two adjacent edges w_1 and w_2. Denote their common end-vertex by u_0 and the other end-vertices of w_1 and w_2 by u_1 and u_2, respectively (see Fig. 9.6.1). Let A_i $(i = 0, 1, 2)$ be the component of the graph $H - \{w_1, w_2\}$ containing u_i. Note that A_0, A_1 and A_2 need not be mutually distinct. If all the components A_i have even size we can apply the induction hypothesis. If A_i ($i = 1$ or 2), has odd size then the induction hypothesis applies to graphs $A_i \cup H\{\{w_i\}\}$ and $H - (E(A_i) \cup \{w_i\})$, that is, to A_i with the edge w_i added, and to the graph which arises from H by deleting the edges of the first graph. If A_0 has odd size, then so has A_1 or A_2 and we can proceed as described above.

9.7 Corollary [Huang and Rosa 1978]. *A P_3-decomposition of the graph K_v exists if and only if $v \equiv 0$ or 1 (mod 4).*

Proof. Use 9.6 with $H = K_v$.

9.8 Corollary [Chartrand et al. 1973, Sumner 1974, Yavorskii 1978; in the general form Nebeský 1979]. *Let H be a finite ordinary graph. Then the line graph $L(H)$ has an 1-factor if and only if every component of the graph $L(H)$ has even order.*

Proof. Clearly, $L(H)$ has an 1-factor if and only if the graph H has a P_3-decomposition. According to 9.6 this is equivalent with the fact that every component of H has even size, that is, every component of $L(H)$ has even order.

9.9 Let H be a finite ordinary $(2n - 1)$-regular graph ($n \geq 1$). The paper of Kotzig (1979a) deals with the question of when $P_{2n}|H$ or $D_{2n}|H$, where P_{2n} is the path of order $2n$ (1.44) and D_{2n} is the double star of order $2n$ [1.20]. In the case $n = 1$ the answer is trivial ($P_2 = D_2 = K_2$, and a K_2-decomposition exists for every graph). For $n = 2$ we have again $P_4 = D_4$ and the question is easy to answer, see Kotzig (1979a) and Bouchet and Fouquet (1983). For $n \geq 3$ we have $P_{2n} \neq D_{2n}$. The general result for D_{2n} reads as follows:

Theorem [Jaeger et al. 1983]. *A finite ordinary $(2n - 1)$-regular graph H has a D_{2n}-decomposition if and only if H has an 1-factor.*

Proof. I. Let v be the order of H. Its size is then $v(2n - 1)/2$, whence a D_{2n}-decomposition of H consists of $v/2$ subgraphs. Picking up in each of them the edge joining the two vertices of degree n we obviously get a 1-factor of H.

II. For the converse we refer the reader to Jaeger et al. (1983). The special case $n = 2$ is in 9.36.

9.10 We shall now present an important result special cases of which have been discovered in Hell and Rosa (1972) ($k = v$, see 11.4), in Bermond and Sotteau (1976) ($k = 4$, $\lambda = 1$), and Huang and Rosa (1978) ($k = 3$ and $\lambda = 1$, see 9.7).

Theorem [Tarsi 1983]. *A P_k-decomposition of the graph $^\lambda K_v$ (where $v \geq 2$ and $k \geq 2$) exists if and only if*

$$\lambda v(v - 1) \equiv 0 \pmod{2k - 2}, \tag{1}$$

$$v \geq k. \tag{2}$$

Proof. The necessity follows from 6.14.1—2. The sufficiency of these conditions is established in Tarsi (1983).

9.11 The following general result was also preceded by some of its special cases, see Hell and Rosa (1971, 1972), Hung and Mendelsohn (1974), Lawless (1974a, b), Bermond (1975a) and Huang and Rosa (1978).

Theorem [Hung and Mendelsohn 1977]. *A balanced P_k-decomposition of the $^\lambda K_v$ exists if and only if*

$$\lambda k(v - 1) \equiv 0 \pmod{2k - 2}, \tag{1}$$

$$v \geq k. \tag{2}$$

Proof. The necessity of these conditions follows from Corollary 6.14.1 and 6.14.4. Their sufficiency is proved in Hung and Mendelsohn (1977), however, in a different formulation.

9.12 Corollary [Hung and Mendelsohn 1977, Huang and Rosa 1978]. *A balanced P_k-decomposition ($k \geq 2$) of the graph K_v exists if and only if one of the following conditions holds*:
  A. *k is even and $v \equiv 1 \pmod{k - 1}$.*
  B. *k is odd and $v \equiv 1 \pmod{2k - 2}$.*

Proof. Theorem 9.11 for $\lambda = 1$ implies: A balanced P_k-decomposition of the graph K_v exists if and only if $2(k - 1) | (v - 1)k$. For k even this condition is equivalent with the condition $v \equiv 1 \pmod{k - 1}$ while for k odd with the condition $v \equiv 1 \pmod{2k - 2}$. (See also Corollary 7.27.B for $n = k - 1$.)

9.13 For some values of v, k and λ there may exist both a balanced and a non-balanced P_k-decomposition of the graph $^\lambda K_v$. For example, for $v = 5$, $k = 3$ and $\lambda = 1$ the complete graph with vertices 1, 2, 3, 4, 5 has a balanced P_3-decomposition (124, 235, 341, 452, 513) and a non-balanced P_3-decomposition (125, 235, 341, 542, 513).

9.14 The oriented version of the previous problem is also of interest. Namely, we ask under what conditions does the relation $P_k^\rightarrow | ^\lambda K_v^*$ hold. Attention has been paid especially to *balanced decompositions* (each vertex is contained in the same number of members of the decomposition). The complete solution is not even known in the case $\lambda = 1$; the final result is known just for k even.

9.15 Theorem [Bermond 1975a]. *A balanced $P_k^\rightarrow$-decomposition of the graph K_v^* for k even exists if and only if $k - 1 | v - 1$.*

Proof. From the condition 6.3.H it follows that if there exists a balanced $P_k^\rightarrow$-decomposition of the graph K_v^* then $k - 1 | k(v - 1)$. However, since $k - 1$ and k are coprime we also have $k - 1 | v - 1$. (This implication clearly holds also for k odd.) The converse can be found in Bermond (1975a).

9.16 Of a special interest is the case $k = v$ of the above theory, more exactly the study of the relations $P_v|^\lambda K_v$ and $P_v^{\rightarrow}|^\lambda K_v^*$. In other words, this is the problem of decomposing the λ-fold complete graph or digraph into Hamiltonian paths. Similar questions will be studied in Chapter 11.

9.17 Besides complete graphs and their modifications, decompositions of other graphs into paths and other types of trees were also considered. For instance, in Caro and Schönheim (1980), Alon (1981), Alon and Caro (1982) and Ajtai et al. (1982) decompositions of trees into isomorphic trees were studied. To state some results of this kind we have to define the concept of a branch.

A *branch* of a connected graph G associated with a cut-vertex $u \in V(G)$ is any subgraph of G which arises from a component K of $G - u$ by adding the vertex u and all edges joining u with a vertex of K. Moreover, each loop with the incident vertex is itself a branch.

Now we can formulate the promised result.

9.18 Theorem [Caro and Schönheim 1980]. *A finite tree T has a P_k-decomposition ($k \geq 3$) if and only if for every cut-vertex of T the number of associated branches with i (mod $k - 1$) edges is equal to the number of branches with $k - 1 - i$ (mod $k - 1$) edges (for $i = 1, 2, ..., k - 2$), and, in case k is odd, the number of branches with $(k - 1)/2$ (mod $k - 1$) edges is even.*

Proof. We omit the proof of this and of the following theorem. The interested reader is referred to the original source or to the papers of Alon (1981), Alon and Caro (1982) and Ajtai et al. (1982), where these results are generalized.

9.19 Recall that $S_k = K(1, k - 1)$ denotes the *star* of order k and size $k - 1$, that is the tree of diameter ≤ 2 with k vertices and $k - 1$ edges.

Theorem [Caro and Schönheim 1980]. *A finite tree T has an S_k-decomposition ($k \geq 3$) if and only if for every cut-vertex the number of edges of any branch is 0 (mod $k - 1$) or 1 (mod $k - 1$) and the number of branches with 1 (mod $k - 1$) edges is a multiple of $k - 1$.*

9.20 Naturally, S_k-decompositions, and more generally $K(n_1, n_2)$-decompositions, can be investigated also for graphs other than trees. In 9.21—9.27 we give some typical results. Further information on this topic can be drawn from Rosa (1967a), Huang and Rosa (1973b), Cain (1974), Huang (1974), Hogarth (1975), Kurek and Petrenyuk (1977), Caro and Schönheim (1980), Chung and Graham (1981) and Ushio (1981). The problem whether or not $K(n_1, n_2)\|^\lambda K_v$ is dealt with in Hung and Mendelsohn (1973), Huang and Rosa (1973b), Huang (1974), Bermond and Sotteau (1976), Kurek and Petrenyuk (1977) and Tarsi (1981). The origins of this problem are described in Yamamoto et al. (1975a), and a generalization to hypergraphs is considered in Yamamoto and Tazawa (1980).

9.21 Theorem [Tarsi 1979]. *The λ-fold complete graph $^\lambda K_v$ of order v has an*

S_k-decomposition if and only if the following conditions hold:

    A. $2(k-1)|\lambda v(v-1)$.

    B. $v = 1$, or else

$$v = \begin{cases} 2(k-1) & \text{for } \lambda = 1, \\ k & \text{for } \lambda \text{ even}, \\ k + (k-1)/\lambda & \text{for odd } \lambda \geq 3. \end{cases}$$

Proof. Omitted.

9.22 Theorem [Ushio et al. 1978]. *The complete m-partite graph $K(m \times n)$ has an S_k-decomposition if and only if the following conditions are satisfied*:

    A. $2(k-1)|m(m-1)n^2$.

    B. $mn = 1$ or $mn \geq 2(k-1)$.

Proof. Again omitted.

A generalization of this result to $K_m(n_1, n_2, ..., n_m)$ is given in Tazawa (1979) and Truszczyński (1983); the case $m = 2$ was solved in Yamamoto et al. (1975b) and the case $m = 3$ in Truszczyński (1983). A modification of this problem with the additional condition that all stars have their vertices in mutually different parts of $K_m(n_1, n_2, ..., n_m)$ is dealt with in Tazawa et al. (1978). Decompositions and balanced decompositions of complete bipartite (and even m-partite) graphs into complete bipartite graphs have been studied in Ushio (1981).

9.23 Corollary [Yamamoto et al. 1975b, Tarsi 1979, 1981]. *The complete graph K_v has an S_k-decomposition if and only if the following conditions are satisfied*:

    A. $2(k-1)|v(v-1)$.

    B. $v = 1$ or $v \geq 2(k-1)$.

Proof. Put $\lambda = 1$ in Theorem 9.21 or $m = v$ and $n = 1$ in Theorem 9.22.

9.24 Corollary [Yamamoto et al. 1975b]. *The complete bipartite graph $K(n, m)$ has an S_k-decomposition if and only if $k-1|n^2$ and $n \geq k-1$.*

Proof. Put $m = 2$ in Theorem 9.22.

9.25 We shall now turn our attention to balanced decompositions. First we shall give a theorem concerning decompositions of the graph $^\lambda K(m \times n)$ into stars of order k. Special cases of this result were known prior to its general form (for $n = \lambda = 1$ see Huang and Rosa 1973b, Theorem 4.1; Huang and Rosa 1978, Theorem 3.5; for $\lambda = 1$ see Ushio 1980; the case $n = 1$ will be stated in 9.26; some further cases are solved in Huang and Rosa 1973b and Huang 1974).

Theorem [Ushio 1982]. *Let $k \geq 3$ and $m \geq 2$. The following three statements are equivalent*:

    A. *There exists a balanced S_k-decomposition of the graph $^\lambda K(m \times n)$.*

    B. *There exists a cyclic S_k-decomposition of the graph $^\lambda K(m \times n)$.*

C. *The conditions* (1) *and* (2) *hold*:

$$2(k-1)|\lambda(m-1)n, \tag{1}$$

$$(m-1)n \geq k-1. \tag{2}$$

Proof. B ⇒ A. This implication follows from Lemma 7.2.

A ⇒ C. The degree of the star S_k cannot be greater than the number of vertices adjacent with an arbitrary vertex of $^\lambda K(m \times n)$, whence we get (2). Let b be the number of graphs in a S_k-decomposition of $^\lambda K(m \times n)$ and let r of them contain a fixed vertex of this graph. By 6.3.F, $bk = mnr$. Further, 6.3.B implies that $(k-1)b = \lambda \binom{m}{2} n^2$. Hence

$$b = \lambda m(m-1)n^2/(2k-2),$$

$$r = bk/(mn) = \lambda(m-1)nk/(2k-2).$$

Choose a vertex u in $^\lambda K(m \times n)$ and let s and t denote the number of stars in the decomposition which contain u as a vertex of degree $k-1$ or k, respectively. Then

$$s + t = r,$$

$$(k-1)s + t = \lambda(m-1)n$$

(the second equation is obtained by computing the degree of u in two ways). Subtracting the first equation from the second one we get

$$(k-2)s = \lambda(m-1)n - r = \lambda(m-1)n - \lambda(m-1)nk/(2k-2) =$$
$$= \lambda(m-1)n(k-2)/(2k-2),$$

and consequently

$$s = \lambda(m-1)n/(2k-2). \tag{3}$$

Thus (1) holds too.

C ⇒ B. Denote the vertices of $^\lambda K(m \times n)$ by the numbers $0, 1, 2, ..., mn-1$. Let the vertices i and j $(i \neq j)$ be adjacent if and only if $i-j$ is not divisible by m; if they are adjacent let they be joined by λ edges. Then the m parts of $^\lambda K(m \times n)$ are given by the sets

$$\{i, i+m, i+2m, ..., i+(n-1)m\}$$

for $i = 0, 1, 2, ..., m-1$.

Form the following (infinite) sequence U of the vertices of $^{\lambda}K(m \times n)$:
$$U = [1, 2, \ldots, m - 1, m + 1, m + 2, \ldots, 2m - 1, 2m + 1, 2m + 2, \ldots,$$
$$\ldots, nm - 1, 1, \ldots, m - 1, m + 1, \ldots, 2m - 1, 2m + 1, \ldots, nm - 1, 1, \ldots].$$
In order to obtain a cyclic S_k-decomposition of $^{\lambda}K(m \times n)$, form a base system (7.3) of graphs $G_1, G_2, \ldots, G_s$ isomorphic with S_k. (Note that s is a positive integer by (1), (2) and (3).) In each of them the vertex 0 has degree $k - 1$. In G_1 the vertex 0 is adjacent to the first $k - 1$ members of the sequence U, in G_2 the vertex 0 is adjacent to the next $k - 1$ members of U and so on. We take care to make the graphs $G_1, G_2, \ldots, G_s$ edge-disjoint; this is indeed possible since all the edges in $^{\lambda}K(m \times n)$ have multiplicity λ. It is easy to verify that edges of length $mn/2$ in $^{\lambda}K(m \times n)$ exist if and only if m is even and n is odd. If they exist then their number in the base graph is $\lambda/2$ altogether. The total number of edges with length $< mn/2$ in the base graphs is always λ (consider λ even and odd separately). The application of mn rotations to the base graphs gives rise to mns subgraphs of $^{\lambda}K(m \times n)$ which form its cyclic S_k-decomposition.

9.26 Corollary [Huang 1974, Ushio 1982]. *Assume that $v \geq 2$ and $k \geq 3$. Then the following three statements are equivalent:*
  A. *There exists a balanced S_k-decomposition of the graph $^{\lambda}K_v$.*
  B. *There exists a cyclic S_k-decomposition of the graph $^{\lambda}K_v$.*
  C. *The conditions (1) and (2) hold:*

$$2(k - 1) | \lambda(v - 1), \tag{1}$$

$$k \leq v. \tag{2}$$

Proof. In Theorem 9.25 put $m = v$ and $v = 1$.
9.27 Now we shall be interested in decompositions of complete graphs into complete bipartite graphs. In Corollary 7.27.C we have shown that if $v \equiv 1 \pmod{2\, n_1 n_2}$ then there exists a balanced $K(n_1, n_2)$-decomposition of K_v. The following theorem says that the converse statement is almost always true, too. However, we shall prove it only in the case $n_1 = n_2$. It is pity that just in this case there is a gap between the necessary and the sufficient condition.

Theorem [Huang and Rosa 1973b]. *Let v, n_1 and n_2 be integers for which there exists a balanced $K(n_1, n_2)$-decomposition of the graph K_v. Then the following hold:*
  A. *If $n_1 = n_2 \equiv 1 \pmod 2$ and $v \equiv 0 \pmod 2$ then $v \equiv n_1 n_2 + 1 \pmod{2 n_1 n_2}$.*
  B. *In the remaining cases $v \equiv 1 \pmod{2 n_1 n_2}$.*
Proof. Assume that $n_1 = n_2 = n$. The conditions 6.14.2—3 for $\lambda = 1$, $G = K(n, n)$, $e(G) = n^2$, $D(G) = n$ imply that if $K(n, m) \| K_v$ then

$$2n^2 | v(v - 1), \tag{1}$$

$$n | v - 1.$$

$$(2)$$

Clearly,

$$\text{g.c.d.} \{n, v\} = 1.$$

$$(3)$$

If v is even then, according to (2), n must be odd. Since

$$n^2 | (v/2)(v - 1)$$

and

$$\text{g.c.d.} \{n, v/2\} = 1,$$

it holds that $n^2 | v - 1$. That is, $v \equiv 1 \pmod{n^2}$ or, in other words, $v \equiv 1$ or $n^2 + 1 \pmod{2n^2}$. The first possibility is absurd because v is even. Thus the statement A holds. If v is odd then, according to (1), we have

$$n^2 | v(v - 1)/2.$$

Taking (3) into account we get

$$n^2 | (v - 1)/2.$$

Thus $v \equiv 1 \pmod{2n^2}$ and, in the case $n_1 = n_2$, the statement B is established. The case $n_1 \neq n_2$ is omitted and the reader is referred to Huang and Rosa (1973b, Theorem 3.5).

9.28 So far we have dealt with decompositions of graphs into paths and stars. Decompositions into other trees or forests have been studied mainly for complete graphs. Several results of this kind are included in Huang and Rosa (1978); for the forest $L_4 = 2K_2$ see Exercise 9.32. Decompositions of complete graphs into trees have been considered also in Chapter 7 of this book.

In Huang and Rosa (1978) the following is proved: If $v - 1 | n_1$ and $v - 1 | n_2$ then for each caterpillar (7.22) T of order v it holds that $T | K(n_1, n_2)$.

Recently remarkable results were obtained concerning the relation $L_{2t} | H$, where $L_{2t} = tK_2$ (that is, the 1-regular graph of order $2t$ and size t) and H is a finite simple graph. First we give slightly more general results.

If a finite simple graph H of size $e(H)$ with maximum degree $\Delta(H)$ has a decomposition into subgraphs of the same order t with maximum degrees $\leq p$ then

$$t | e(H),$$

$$(1)$$

$$\Delta(H) \le pe(H)/t. \tag{2}$$

Indeed, the condition (1) is obvious and (2) follows from the fact that no vertex of H can have degree greater than p-times the number of subgraphs in the decomposition, i.e., $pe(H)/t$.

It is interesting that, in some sense, the converse is also true.

Theorem [Lonc and Truszczyński, Preprint Ia]. *For every ordered pair $[p, t]$ of positive integers the number of types of finite simple graphs H that satisfy (1) and (2) and have no decomposition into subgraphs of size t with maximum degree $\le p$ is finite.*

Proof. See the cited paper.

Setting $p = 1$ we obtain the conditions for the relation $L_{2t}|H$. From our theorem it follows that the number of types of exceptional (finite simple) graphs H satisfying (1) and (2) but having no L_{2t}-decomposition is finite for each positive integer t. Meanwhile this number is exactly known just for $t \le 3$. For $t = 1$ there is no exception. For $t = 2$ the only exception is the graph $H = K_2 \vee K_3$ [Caro, Preprint]. For $t = 3$ there are exactly 26 exceptional types of graphs H and they are determined in Bialostocki and Roditty (1982). In Alon (1983) it is proved that for $e(H) > (8/3) t^2 - 2t$ the relation $L_{2t}|H$ holds if and only if (1) and (2) are satisfied (for $p = 1$).

Decompositions of graphs into subgraphs of size 2 or 3 are studied in Lonc and Truszczyński (Preprint Ib).

9.29 Now we shall be interested in the existence of a decomposition of the complete graph K_v into isomorphic n-cubes Q_n (for $v \ge 2$). From Corollaries 6.14.2 and 6.14.4 we obtain the following necessary conditions for the existence of a Q_n-decomposition of K_v:

$$n2^n|v(v - 1). \tag{1}$$

$$n|v - 1. \tag{2}$$

Using them we shall prove the following:

9.30 Theorem [Kotzig 1981]. *Let v and n be positive integers. Then:*

A. *If $v \equiv 1 \pmod{n2^n}$ then $Q_n|K_v$.*

B. *If $Q_n|K_v$ and v is odd (or n even) then $v \equiv 1 \pmod{n2^n}$.*

C. *If $Q_n|K_v$, where v is even and n is odd then $v \equiv 1 \pmod{n}$ and $v \equiv 0 \pmod{2^n}$.*

Proof. A. This statement has been proved in Corollary 7.29.D.

B. Let v be odd and $Q_n|K_v$. From (2) we deduce that g.c.d. $\{n, v\} = 1$. Since g.c.d. $\{2^n, v\} = 1$ we also have g.c.d. $\{n2^n, v\} = 1$. Therefore the condition (1) implies that $n2^n|v - 1$ that is $v \equiv 1 \pmod{n2^n}$. Starting from the assumption that n is even we see that, according to (2), v is odd and we can repeat the previous arguments.

C. If $Q_n|K_v$ then (2) implies that $v \equiv 1 \pmod{n}$. From (1) it follows that $2^n|v(v-1)$. Now, if v is even then $2^n|v$ which means that $v \equiv 0 \pmod{2^n}$.

9.31 Remarks. A. The following result of Maheo (1980) completes the previous theorem: $Q_3|K_v$ if and only if $v \equiv 1$ or $16 \pmod{24}$. The case $v = 16$ is in Exercise 9.35.

B. If v is odd (or if n is even) then the statements 9.30.A—B yield a necessary and sufficient condition for the complete graph K_v to have a Q_n-decomposition. If v is even and n is odd we only have a necessary condition 9.30.C; this condition is sufficient for $n = 1$ (which is obvious) and for $n = 3$ (by Remark A). Whether or not this condition is sufficient for $n \geq 5$ is not yet known.

C. If n is odd then g.c.d.$\{2^n, n\} = 1$ and, by the Chinese Remainder Theorem [Sierpiński 1964, p. 31] there exists $v_0 \in \{0, 1, ..., n2^n - 1\}$ with the property that the two congruences in 9.30.C are equivalent with one congruence $v \equiv v_0 \pmod{n2^n}$. This is particularly interesting if n is an odd prime. In this case Fermat's Little Theorem [Sierpiński 1964, p. 202] implies that $2^{n-1} \equiv 1 \pmod{n}$ in which case $v_0 = (n+1)2^{n-1}$. Thus for an odd prime n the two congruences in 9.30.C can be replaced by the congruence $v \equiv (n+1)2^{n-1} \pmod{n2^n}$ (cf. Kotzig 1979b).

D. Since $Q_2 = C_4$, Theorem 9.30.A—B for $n = 2$ yields the following statement:

$$C_4|K_v \qquad \text{if and only if } v \equiv 1 \pmod 8.$$

This has been proved in Kotzig (1965b) (cf. 10.5—6).

Exercises

9.32 The complete graph K_v has a $2K_2$-decomposition if and only if $v \equiv 0$ or $1 \pmod 4$.

9.33 Let e be a power of a prime and λ be not divisible by this prime. Let G be a graph of size e such that $G|^\lambda K_v$. Then the following hold:
A. If e is odd then $v \equiv 0$ or $1 \pmod e$.
B. If e is even then $v \equiv 0$ or $1 \pmod{2e}$.

9.34 [Huang and Rosa 1978]. Let k be a positive integer such that $k-1$ is a power of a prime and let T be a tree of order k satisfying the property $T|K_v$. Then the following hold:
A. If k is even then $v \equiv 0$ or $1 \pmod{k-1}$.
B. If k is odd then $v \equiv 0$ or $1 \pmod{2k-2}$.

9.35 [Kotzig 1979b, Maheo 1980]. The complete graph of order 16 can be decomposed into 3-cubes, i.e., $Q_3|K_{16}$.

9.36 [Kotzig 1979a]. Every simple cubic graph which has a 1-factor has also a P_4-decomposition.

Chapter 10

Decompositions into isomorphic cycles

10.1 Many papers deal with the following question: under what conditions does there exist a C_k-decomposition of the graph K_v or, more generally, of the λ-fold complete graph $^\lambda K_v$? (Briefly, when $C_k|^\lambda K_v$?) It is easy to see that the following hold:

Lemma [Bermond and Thomassen 1981]. *If $C_k|^\lambda K_v$ where $v \geq 2$ and $k \geq 2$ then*

$$k \leq v, \tag{1}$$

$$2k|\lambda v(v-1), \tag{2}$$

$$2|\lambda(v-1). \tag{3}$$

Proof follows from 6.14.1—3 for $G = C_k$.

10.2 We need not consider the case $k = 2$ since $C_2|^\lambda K_2$ if and only if λ is even. Therefore we can restrict ourselves to the case $k \geq 3$. Our point of departure will be the following working conjecture.

Conjecture [Bermond et al. 1978c]. *Let $v \geq 2$ and $k \geq 3$ be integers. Then $C_k|^\lambda K_v$ if and only if the conditions 10.1.1—3 hold.*

10.3 The most remarkable result in direction to settling this conjecture is the following theorem.

Theorem [Rosa and Huang 1975, Bermond and Faber 1976, Bermond and Sotteau 1977, Bermond et al. 1978c]. *Suppose that $v \geq 2$ and $k \in \{3, 4, 5, 6, 7, 8, 10, 12, 14, 16\}$. Then $C_k|^\lambda K_v$ if and only if the conditions 10.1.1—3 hold.*

Proof. The necessity follows obviously from Lemma 10.1. For $k = 3$ the sufficiency follows from 8.14 since $C_3 = K_3$. For $4 \leq k \leq 8$ the theorem is proved in Bermond and Faber (1976), for $4 \leq k \leq 6$ also in Reid and Beineke (1978) using a different method. The cases $k = 8, 10, 12, 14, 16$ are proved in Bermond et al. (1978c).

10.4 Further results on C_k-decompositions of the graph $^\lambda K_v$ can be found in Bermond and Sotteau (1977), Bermond et al. (1978c) and Hering (1980). The special case $k = v$ will be considered in 11.1—11.6. Applications of this topic in serology are discussed in Rees (1967) and Hwang and Lin (1976).

10.5 Conjecture 10.2 is not proved even in the case when $\lambda = 1$. The sufficiency of the conditions 10.1.1—3 for $\lambda = 1$ has been proved in the following cases:

A. $k = 3$ (the case of Steiner triple systems, see 7.29—7.39, 8.14—8.19, 8.32).

B. $k \in \{5, 7, 9\}$ [Bermond and Sotteau 1977].

C. $k \in \{4, 6, 8, 10, 12, 14, 16\}$ [Bermond et al. 1978c].

D. $k \in \{4, 8, 16, 32, ...\}$ [Alspach and Varma 1980]. This result follows from results of the papers Kotzig (1965b) and Rosa (1966b).

E. $k/2$ is a power of a prime [Alspach and Varma 1980].

F. $k = v$ (see 11.5).

10.6 We shall now give some special results on C_k-decompositions of the graph K_v. Further results can be found for instance in Keedwell (1982).

Theorem. *Assume that $3 \le k \le v$. Then $C_k | K_v$ in the following cases:*

A. $k \equiv 1 \pmod 2$ *and* $2k | v - k$.

B. $2k | v - 1$.

C. $k \equiv 6 \pmod 8$ *and* $v = 3k/2$.

D. $k \equiv 2 \pmod 8$ and $v = 5k/2$.

E. $k = 20, v = 25$.

Proof. The statement A has been proved in Rosa (1966a) in a different formulation. The statement B follows by combining the results of Kotzig (1965b) and Rosa (1966a, b) (for $k \equiv 0 \pmod 4$ see also Corollary 7.27.A; for $k = 4$ also 9.31.D). The statements C and D are proved in Alspach and Varma (1980) and E in Rosa (1967b).

10.7 The oriented version of the problem investigated in 10.1—6 is also interesting. It reads as follows: under what conditions does there exist a decomposition of the λ-fold complete graph of order v into cycles of order k (briefly $C_k^0 |^\lambda K_v^*$)? Again there are obvious necessary conditions:

Lemma [Bermond and Sotteau 1977, Bermond et al. 1978c]. *Assume that $C_k^0 |^\lambda K_v^*$, where $v \ge 2$ and $k \ge 2$. Then*

$$k \le v, \tag{1}$$

$$k | \lambda v(v - 1). \tag{2}$$

Proof follows from 6.16.1—2 for $G = C_k^0$.

10.8 The problem of sufficiency of the conditions 10.7.1—2 seems to be even more difficult than the analogous problem for the conditions 10.1.1—3. The relationship between these two problems is expressed by the following obvious implications

$$C_k |^\lambda K_v \Rightarrow C_k^0 |^\lambda K_v^* \Rightarrow C_k |^{2\lambda} K_v .$$

We shall deal with the relation $C_k^0 |^\lambda K_v^*$ in 10.9 and 10.12—16.

10.9 Theorem [Mendelsohn 1971, Bermond 1974, Bruck, Preprint]. *For any positive integer v, $C_3^\circ | K_v^*$ if and only if $3\,|\,v(v-1)$ and $v \neq 6$.*
also Bermond and Faber 1976 for C_3°-decompositions of K_{16}^* and K_{18}^* found $k = 3$ and $\lambda = 1$; the necessity of the second one from R 10.22. The sufficiency was proved independently in Mendelsohn (1971), Bermond (1974), Bruck (Preprint); a short proof can be found also in Bermond (1975a, b). (See also Bermond and Faber 1976 for C_3°-decompositions of K_{16}^* and K_{18}^* found with a computer.)

C_3°-decompositions of the graph K_v^* are discussed also in 8.35.

10.10 We shall now give a very similar result for the quasicycle $C_3^\rightarrow$ (i.e., acyclic tournament $K_3^\rightarrow$) of order 3 which can be obtained from the oriented cycle C_3° by reversing the orientation of one edge.

Theorem [Hung and Mendelsohn 1973]. *For any positive integer v, $C_3^\rightarrow | K_v^*$ if and only if $3\,|\,v(v-1)$.*

P r o o f. The necessity of the condition $C_3^\rightarrow | K_v^*$ follows from 6.16.2. The proof of the sufficiency is in Hung and Mendelsohn (1973), for a simple proof see also Bermond (1975a, b).

10.11 R e m a r k s. A. Theorem 10.9 and 10.10 solve the problem of decomposing the complete digraph into isomorphic semicycles of order 3. In fact there exist only two types of such semicycles: the oriented cycle C_3° and the quasicycle $C_3^\rightarrow$.

It is easy to see that there exist exactly four non-isomorphic semicycles of order 4: C_4°, $C_3^\rightarrow$ and two semicycles which arise from $C_4^\rightarrow$ by reversing the orientation of some edge. Necessary and sufficient conditions for the existence of a G-decomposition where G is an arbitrary (fixed) semicycle of order 4 are derived in Harary et al. (1978).

There are again four semicycles of order 5: C_5°, $C_5^\rightarrow$ and two semicycles which can be obtained from $C_5^\rightarrow$ by reversing the orientation of some edge. The following result holds (Alspach et al. 1979): Let G be an arbitrary semicycle of order 5. Then the complete digraph K_v^* has a G-decomposition if and only if $v \equiv 0$ or 1 (mod 5).

B. From Theorem 10.10 it follows that the minimum number $DN(2, 3, v)$ of quasicycles of order 3 that cover the complete digraph K_v^* is equal to $v(v-1)/3$ if $v \equiv 0$ or 1 (mod 3). In the remaining case $v \equiv 2$ (mod 3), $v \neq 2$, this number is equal to $\lceil v(v-1)/3 \rceil + 1$ (Skillicorn 1981) while for $v = 2$ the number $DN(2, 3, v)$ is obviously not defined. From this result it can be derived that the minimum number of triangles (not necessarily distinct) such that every edge of K_v belongs to at least two of them is equal to the number $D(2, 3, v)$ (Skillicorn 1981). (Such systems of triangles are called bicoverings.)

Analogously, the maximum number $DD(2, 3, v)$ of quasicycles of order 3 that can be packed into the complete digraph K_v^* is equal $v(v-1)/3$ for $v \equiv 0$

or 1 (mod 3). For $v \equiv 2$ (mod 3) it holds that $DD(2, 3, v) = \lfloor v(v-1)/3 \rfloor$ (see Skillicorn 1982a).

C. Denoting by $D(v)$ the number of non-isomorphic (3.9, 8.34) $C_3^{\rightarrow}$-decompositions of the graph K_v^* we have the following: $D(1) = D(3) = 1$, $D(4) = 3$, $D(6) = 32$, $D(7) = 2\,368$ as was proved in Colbourn and Colbourn (1981b) in terms of directed triple systems.

D. Theorem 10.10 can be generalized as follows: For any integer $v \geq 3$, $C_3^{\rightarrow}|K_v^*$ if and only if $3|\lambda v(v-1)$. The proof can be found in Seberry and Skillicorn (1980) and Colbourn and Harms (1983) in terms of directed blocks designs (more exactly, directed balanced incomplete block designs). Theorem 10.9 allows a similar generalization, even for some values of k different from 3:

10.12 Theorem [Bermond and Sotteau 1977, Bermond et al. 1978c]. *Let v, k and λ be positive integers with $k \leq v$, $k|\lambda v(v-1)$ and $k \in \{2, 3, 4, 5, 6, 7, 8, 10, 12, 14, 16\}$. Then $C_k^{\circ}|^{\lambda}K_v^*$ except for the following cases:*

A. $v = 6, k = 3, \lambda = 1$.
B. $v = k = 6, \lambda = 1$.
C. $v = k = 4, \lambda \equiv 1$ (mod 2).

Proof. The case of even k is solved in Bermond et al. (1978), for $\lambda = 1$ also in Bermond (1975a) and Bermond and Faber (1976) and for $k = 4$ and $\lambda = 1$ in Schönheim (1975). The result for odd k follows from Bermond (1975a) and Bermond and Sotteau (1977); for $\lambda = 1$ also from Mendelsohn (1971), Bermond (1974, 1975a, b) and Bruck (Preprint) ($k = 3$), from Bermond (1975a, b), Bermond and Faber (1976) and Merriel (Preprint) ($k = 5$) and from Merriel (Preprint) ($k = 7$). The necessity of the exceptions follows from R.10.22 and 11.9.B.

10.13 Further direction of research could lead to new values of k (with possible new exceptions) for which Theorem 10.12 would be true. Certain arguments suggest that in two important cases there are no new exceptions; that is to say, that the following two conjectures are true.

10.14 Conjecture [Bermond 1975a, b]. *If $2 \leq k \leq v$ then $C_k^{\circ}|K_v^*$ if and only if $k|v(v-1)$ except for $v = 6, k = 3$; $v = k = 6$ and $v = k = 4$.*

10.15 Conjecture [Bermond et al. 1978c]. *Let $k \leq v$ and k be even. Then $C_k^{\circ}|^{\lambda}K_v^*$ if and only if $k|\lambda v(v-1)$ with the following exceptions:*

A. $v = k = 6, \lambda = 1$.
B. $v = k = 4, \lambda \equiv 1$ (mod 2).

10.16 Proofs of special cases of these conjectures and further related results can be found in the sources mentioned in 10.12 as well as in Hartnell and Milgram (1975), Hartnell (1975), Bermond and Sotteau (1976, 1977), Sotteau (1980), Tillson (1980), Alspach et al. (1981) and Keedwell (1982) and in 11.9.B. It is known that Conjecture 10.14 is true at least in the following cases:

A. v is a prime number [Hartnell and Milgram 1975].

B. v is a power of a prime and $k|v - 1$ [Hartnell and Milgram 1975], or k is odd [Sheehan 1976].

C. k is a power of a prime [Bermond 1975a, Sotteau 1976].

D. k is odd and $v \equiv 0$ or $1 \pmod{k}$ [Bermond and Faber 1976, Sotteau 1976].

E. $v = 2k$ [Bermond 1975a].

F. $v = k + 1$ [Bermond and Faber 1976, Sotteau 1976, Alspach et al. 1981].

G. $v = k$ (see 11.10).

H. $v > 2k$, k is even an $k|v - 1$ [Köhler 1976].

From Theorem 10.12 it follows that Conjecture 10.14 is true for $k \leq 8$, $k = 10, 12, 14$ and 16. From 10.16.C it follows that the smallest open case is $k = 15$ (in Conjecture 10.15 even $k = 18$).

Instead of decompositions of the graph K_v^* into cycles C_k^o one can consider, more generally, packings and coverings of the graph K_v^* by cycles. Results of this kind can be found in Bermond (1975a).

10.17 One can investigate decompositions into cycles also for graphs different from complete graphs or digraphs. For instance, in Vasiliev (1967), C_{2n}-decompositions of the n-cube for $n = 2^q$ are investigated. In Graham (1970) and Nash-Williams (1970b) decompositions of an arbitrary graph into triangles, and in Ajtai et al. (1982) packings ("close" to coverings) of a random graph with triangles, or possibly other isomorphic graphs, are studied. We present one sample:

Problem [Nash-Williams 1970b]. *Let G be a loopless graph of order $v \geq 15$, size $e \equiv 0 \pmod{3}$ and multiplicity $\leq \lambda$ in which every vertex has even degree $\geq 3\lambda v/4$. Does then G have a K_3-decomposition?* (See also Exercise 10.24.)

We shall now restrict ourselves to complete bipartite graphs and digraphs. The reader interested in decompositions of complete m-partite graphs into isomorphic cycles should consult the survey articles of Cockayne and Hartnell (1976, 1977). Some results can also be found in Exercises 7.49, 10.23 and R 10.23.

10.18 Theorem [Sotteau 1981]. *The complete bipartite graph $K(n_1, n_2)$ has a C_k-decomposition if and only if $k \geq 4$, $k \equiv n_1 \equiv n_2 \equiv 0 \pmod{2}$, $k \leq 2n_1$, $k \leq 2n_2$ and $k|n_1 n_2$.*

Proof. Obviously each of these conditions is necessary. The sufficiency of these conditions is proved in Sotteau (1981).

10.19 Theorem [Sotteau 1981]. *The graph $K^*(n_1, n_2)$ has a C_k^o-decomposition if and only if $k \equiv 0 \pmod{2}$, $k \leq 2n_1$, $k \leq 2n_2$ and $k|2n_1 n_2$.*

Proof. If $C_k^o|K^*(n_1, n_2)$ then clearly k is even. By 6.3.C it holds that $k|2n_1 n_2$. Further, every cycle of length k has exactly $k/2$ vertices in each of the parts of the vertex-set of $K^*(n_1, n_2)$. This is only possible if $k \leq 2n_1$ and $k \leq 2n_2$.

The proof of the converse implication is in Sotteau (1981); for $k|2n_2$ also in Köhler (1976).

10.20 In infinite undirected graphs one can also consider infinite analogues of cycles — two-way infinite paths (that is, infinite connected regular graphs of degree 2) C_∞ (cf. 1.11, 1.44). Characterizations of locally finite graphs decomposable into a prescribed (or minimum or maximum) number of subgraphs isomorphic to C_∞ are derived in Nash-Williams (1963) and Zelinka (1974).

Another modification of our problem is obtained by studying G-decompositions of the graph K_v where G is a 2-regular graph of order v. In fact, these decompositions coincide with decompositions into isomorphic 2-factors. The celebrated Oberwolfach problem asks, essentially, when do such decomposition exist. It was formulated by G. Ringel at a meeting in Oberwolfach in 1967 as follows: An odd number v of guests has to be placed around s round tables that have $k_1, k_2, \ldots, k_s$ places ($k_1 + k_2 + \ldots + k_s = v$) for l different dishes in such a way that each of guests will have any other guest as a neighbour exactly once. Basic results about this problem can be found in Köhler (1973), Hell et al. (1975) and Huang et al. (1979). The special case $k_1 = k_2 = \ldots = k_s = 3$ will be treated in 12.1—12.8 in a slightly different interpretation as the problem of strolling schoolgirls, the case $s = 1$, $k_1 = v = 2l + 1$ will be mentioned in 11.5—11.6 as the problem of knights.

Exercises

10.21 $C_4^0|^\lambda K_4^*$ if and only if λ is even.

10.22 Under what conditions $C_k^0|K_6^*$?

10.23 [Cockayne and Hartnell 1976]. If $C_k|K(m \times n)$ then:

A. $k \leq mn$.

B. $2|(m-1)n$.

C. $2k|m(m-1)n^2$.

10.24 [Graham 1970]. The fraction 3/4 in Problem 10.17 cannot be replaced by a smaller number.

10.25 Find three non-isomorphic $C_3^\rightarrow$-decompositions of the graph K_4^* (see 10.11.C) with the vertices 1, 2, 3, 4.

10.26 What is the number of non-isomorphic C_3^0-decompositions of the graph K_4^{*}?

Chapter 11

Decompositions into Hamiltonian cycles (Hamiltonian paths) and operations on graphs

11.1 In this chapter we shall be dealing with decompositions of graphs into Hamiltonian cycles or paths. These problems are related mostly to finite graphs. Thus, throughout the chapter, all the graphs considered will be finite unless stated otherwise (11.48).

Our first aim will be to investigate the existence of a decomposition of the λ-fold complete graph of order v into Hamiltonian cycles or paths. In doing this, the following lemma (part A of which has been generally known — cf. Harary 1969, p. 91) will prove useful.

    Lemma [Hell and Rosa 1972].

A. *If v is odd then $P_v|K_v$.*

B. *If v is even then $P_v|^2K_v$.*

Proof. In both cases we shall make use of cyclic decompositions (7.1—7.3). The base system will consist of a single Hamiltonian path. Let us denote the vertices of K_v by $0, 1, 2, ..., v - 1$. The base system path for the case A will be of the form $[0, v - 1, v - 2, 2, ..., v/2 - 1, v/2]$. It is obvious that every possible edge length occurs twice there except for the length $v/2$ which occurs just once. The $v/2$ Hamiltonian paths forming a decomposition of K_v are obtained by i-fold rotations for $i = 0, 1, 2, ..., v/2 - 1$. In the case B, the base system path will be $[0, v - 1, 1, v - 2, 2, ..., (v + 1)/2, (v - 1)/2]$. Here, every edge length occurs twice. The decomposition of ^2K_v into v Hamiltonian paths is now obtained by i-fold rotations, $i = 0, 1, 2, ..., v - 1$. For $v = 6$ or 7, the base system path is depicted in Fig. 11.1.1.

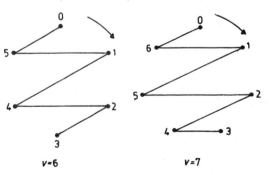

v=6   v=7   Fig. 11.1.1

11.2 The following lemma will be helpful in simplifying the investigation of decompositions of graphs into Hamiltonian paths [cycles].

Lemma. A. *Each decomposition of an arbitrary graph into Hamiltonian paths [Hamiltonian cycles] is balanced.*

B. *For each decomposition of $^\lambda K_v$ [or $^\lambda K_v^*$] into Hamiltonian paths, each vertex of $^\lambda K_v$ [$^\lambda K_v^*$] is the endpoint [initial point as well as terminal point] of exactly λ paths of the decomposition.*

Proof. The part A follows from the fact that each vertex of the graph belongs to all paths [cycles] of the decomposition. As regards the part B, it is sufficient to notice that in a balanced decomposition of $^\lambda K_v$ [$^\lambda K_v^*$] into Hamiltonian paths, every vertex is the endpoint [initial, as well as terminal point] of the same number of paths. As the number of paths in such a decomposition is $\lambda \binom{v}{2} \Big/ (v-1) = (\lambda v)/2$ $[\lambda v(v-1)/(v-1) = \lambda v]$, the total number of endpoints [initial points; terminal points] of the paths is λv, which implies B.

11.3 Now, it is easy to reach the aim marked out in 11.1. Its essential part (namely, the equivalence of 11.3.A and C) has been proved by Hell and Rosa (1972) in terms of balanced P-designs; a generalization of this result concerning paths was given in 9.10—9.11. The importance of the topic may be illustrated e.g. by the fact that decompositions of complete λ-fold graphs into Hamiltonian cycles have been used to construct so-called neighbour-designs, with applications in serology (Rees 1967).

Theorem. *For $n \geq 2$ the following assertions are equivalent:*

A. $P_n | ^\lambda K_n$.

B. $C_{n+1} | ^\lambda K_{n+1}$.

C. $2 | \lambda n$.

Proof. A $\Rightarrow$ B. Let $P_n | ^\lambda K_n$. According to Lemma 11.2.B for $v = n$, every vertex of $^\lambda K_n$ is the endpoint of exactly λ paths of the decomposition. Thus, adding a new vertex to $^\lambda K_v$ and joining it to both end-vertices of each Hamiltonian path of the decomposition of $^\lambda K_v$ we obtain a decomposition of $^\lambda K_{n+1}$ into Hamiltonian cycles.

B $\Rightarrow$ C. If B is true then from 6.3.3.C we get

$$n + 1 | \lambda \binom{n+1}{2},$$

implying $2 | \lambda n$.

C $\Rightarrow$ A. If n is even, then (according to Lemma 11.1.A) $P_n | K_n$. Since obviously $K_n | ^\lambda K_n$, we have $P_n | ^\lambda K_n$. If n is odd, λ must be even. From 11.1.B we get $P_n | ^2 K_n$, and thus $P_n | ^\lambda K_n$ for λ even.

11.4 Corollary. *For $v = 2$ the following holds*:

A. *There is a decomposition of $^\lambda K_v$ into Hamiltonian paths if and only if λv is an even number* (Hell and Rosa 1972).

B. *There is a decomposition of $^\lambda K_v$ into Hamiltonian cycles if and only if the number $\lambda(v - 1)$ is even.*

Proof. See Theorem 11.3.

11.5 The following assertion has become a part of "mathematical folklore".

Corollary [Lucas 1883, p. 161, Narasimhamurti 1940, Abramson 1967, Harary 1969, pp. 88—92, Berge 1970, pp. 233—234, Tillson 1980]. *There exists a decomposition of the complete graph of order $v \geq 2$ into Hamiltonian path [cycles] if and only if v is even [odd].*

This follows directly from 11.4. Some modifications of the result can be found in exercises 11.56—11.57.

11.6 The proof of Theorem 11.3 yields also a method of constructing a C_v-decomposition of the graph K_v for odd v. It is illustrated in Fig. 11.6.1 for $v = 7$; the symbol ∞ stands for the added vertex. The $(v - 1)/2$ Hamiltonian cycles are obtained by successive rotations, as indicated in Fig. 11.6.1. Recall that a P_v-decomposition of K_v (v even) was described in the proof of Lemma 11.1.A.

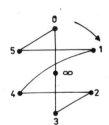

Fig. 11.6.1

According to Berge (1958, or 1970, p. 232), the problem of decomposing K_v into Hamiltonian cycles had already been investigated by T. P. Kirkman in the 19th century in solving the *problem of knights*: How many ways are there for v knights to sit around a table so that none of them sits twice next to the same neighbour? Equivalently, what is the maximum number of Hamiltonian cycles in a packing of K_v? The answer is $\lfloor (v - 1)/2 \rfloor$. For v odd it follows from 11.5, and for v even from Exercise 11.57 (see also Berge 1970, p. 233). Another kind of the "problem of knights" can be found in 11.20.

11.7 Results analogous to those of 11.3—11.5 for directed graphs are not so easy, as regards both their formulation and their proofs. Among others, the following deep result of Tillson (1980) which settled a problem of Bermond and Faber [cf. Bermond and Faber 1976, Bermond and Sotteau 1976] will be used in our considerations.

Lemma. *If $v \geq 8$ is even then the complete digraph K_v^* is decomposable into Hamiltonian cycles.*

We omit the rather complicated proof of this lemma and refer the reader to the original paper of Tillson (1980).

11.8 Theorem. *The following assertions are equivalent:*

A. $P_n^{\rightarrow} |^\lambda K_n^*$.

B. $C_{n+1}^0 |^\lambda K_{n+1}$.

C. *Neither of the cases C1 and C2 holds:*

    C1. $n = 3$ *and* λ *is odd.*

    C2. $n = 5$ *and* $\lambda = 1$.

Proof. A $\Rightarrow$ C. We are to show that the graphs $^\lambda K_3^*$ (λ odd) and K_5^* do not admit a decomposition into Hamiltonian paths.

Assume the contrary. Consider a decomposition of $^\lambda K_3^*$ with vertices 1, 2, 3 into Hamiltonian paths. Denote by a, b, c the number of paths in the decomposition of the form 123, 231, 312, respectively. Every arc emanating from 1 and terminating at 2 has to belong to the path of the form 123 or 312, thus $a + c = \lambda$. Similar considerations lead to equations $a + b = \lambda$ and $b + c = \lambda$. Adding these equations yields $2a + 2b + 2c = 3\lambda$, which contradicts the fact that λ is odd.

Now, suppose the graph K_5^* with vertices 1, 2, 3, 4, 5 to have a decomposition into Hamiltonian paths. According to 11.2.B ($\lambda = 1$), every vertex has to be the initial point of exactly one Hamiltonian path of the decomposition, and the same is true when considering terminal points. Without loss of generality we may suppose the decomposition to be either of the form $1u_1u_2u_32$, $2u_4u_5u_61$, $3u_7u_8u_94$, $4u_{10}u_{11}u_{12}5$, $5u_{13}u_{14}u_{15}3$, or $1u_1u_2u_32$, $2u_4u_5u_63$, $3u_7u_8u_94$, $4u_{10}u_{11}u_{12}5$, $5u_{13}u_{14}u_{15}1$, where $u_i \in \{1, 2, 3, 4, 5\}$, $1 \leq i \leq 15$. But it can be shown by a systematical examination of all cases that this is impossible (cf. Mendelsohn 1968, a computer-aided examination has been carried out by Bankes; see also Bermond and Faber 1976).

C $\Rightarrow$ B. If C is true then one of the following cases has to occur:

C3. n is even.

C4. $n = 1$.

C5. $n = 3$ and λ is even.

C6. $n = 5$ and λ is even.

C7. $n = 5$ and $\lambda \geq 3$ is odd.

C8. $n \geq 7$, n is odd.

In the case C4, B is clearly true. The cases C3, C5 and C6 can be handled easily, using Theorem 11.3. Namely, if each edge of $^\lambda K_{n+1}$ is replaced by two oppositely directed arcs, the desired decomposition of $^\lambda K_{n+1}^*$ is obtained from that of $^\lambda K_{n+1}$ in a straightforward manner. In the case C8 it is sufficient to apply Lemma 11.7 for $v = n + 1$. Thus, the only one possibility that remains to be considered is C7.

Let 0, 1, 2, 3, 4, 5 be the vertices of the graph $^\lambda K_6^*$. If $\lambda = 3$, it can be verified that the cycles 0134250, 0152340, 0154230, 0214350, 0243510, 0253140, 0321450, 0324510, 0354120, 0415320, 0425310, 0452130, 0512340, 0524130, 0543120 form a decomposition of $^3K_6^*$ into Hamiltonian cycles. If $\lambda \geq 5$ and λ is odd, it suffices to pick in $^\lambda K_6^*$ a subgraph isomorphic to $^3K_6^*$, decompose it as indicated above, and handle the rest in accordance with C6. Thus, B is true in either case.

B $\Rightarrow$ A. Suppose there is a decomposition of $^\lambda K_{n+1}^*$ into Hamiltonian cycles. If we delete a vertex of $^\lambda K_{n+1}^*$ (and thus the same vertex from all Hamiltonian cycles) we obtain a decomposition of $^\lambda K_n^*$ into Hamiltonian paths.

11.9 Let us restate the preceding result in a more compact form.

Corollary. A. *The graph $^\lambda K_v^*$ is decomposable into Hamiltonian paths if and only if neither $v = 3$ and $\lambda \equiv 1 \pmod 2$, nor $v = 5$ and $\lambda = 1$.*

B. *The graph $^\lambda K_v^*$ is decomposable into Hamiltonian cycles if and only if neither $v = 4$ and $\lambda \equiv 1 \pmod 2$, nor $v = 6$ and $\lambda = 1$.*

11.10 J. C. Bermond and V. Faber [or E. G. Strauss] (see Bermond and Faber 1976) asked when the graph K_v^* is decomposable into Hamiltonian cycles [paths]. Now, the answer is obtained immediately from 11.9.

Corollary [Tillson 1980]. *There exists a decomposition of K_v^* into Hamiltonian cycles [paths] if and only if $v \neq 4$, 6 [$v \neq 3$, 5].*

Some modifications of this result can be found in exercises 11.58—11.60.

11.11 Remarks. A. The equivalence of decompositions of K_v^* into Hamiltonian paths and K_{v+1}^* into Hamiltonian cycles has been proved by Bermond and Faber (1976). Some partial results concerning Corollary 11.10 can be found in Mendelsohn (1968), Wang (1973), Keedwell (1974), Bermond (1975) and Bermond and Faber (1976). These problems are closely related to the so-called horizontally complete Latin squares, as well as to sequentiable groups. The reader interested in this topic is referred to the book by Dénes and Keedwell (1974, Chap. 2 and 9).

B. Colbourn (1982b) proposed to study the number $f(v)$ of non-isomorphic (3.9) decompositions of the graph K_v into Hamiltonian cycles and proved that $f(3) = f(5) = 1$, $f(7) = 2$, $f(9) = 122$, $f(11) \geq 3\,140$. Some results in this direction had already been obtained by Levi (1940).

11.12 It is clear that if a directed graph G is decomposable into k Hamiltonian cycles then G is finite, loopless (if its order is $\neq 1$), strongly connected and k-diregular. However, these conditions are by far not sufficient (see the examples given in 11.13, R 11.62; for $k = 2$ confer also Skupień (1981), where for each $v \geq 7$ a 2-diregular non-Hamiltonian digraph of order v is con-

122

structed in which every vertex is the initial (and terminal) vertex of a Hamiltonian path; for $k = 3$ see also Aubert and Schneider (1982a). For this reason, many authors have been dealing with sufficient conditions for the existence of a decomposition of a digraph into Hamiltonian cycles. But still, as we shall see, the current status of the matter lies, for the most part, in the sphere of problems and conjectures.

Conjecture [Jackson 1980a]. *If a complete bipartite graph can be directed so as to produce a k-regular digraph G, then G is decomposable into k Hamiltonian cycles.*

11.13 J. A. Bondy proposed to investigate the existence of the decomposition above in the case when the order of the digraph is limited. From this point of view the following problem seems to be interesting.

Problem [Bondy 1978, Thomassen 1979, Bermond and Thomassen 1981]. *Find the set S of all those k for which every k-diregular digraph of order $2k + 1$ can be decomposed into k Hamiltonian cycles.*

(Note that such digraphs are a fortiori finite and strongly connected.)
Obviously $1 \in S$. Examples given in Figs 11.13.1—3 show that 2, 3, $5 \notin S$.

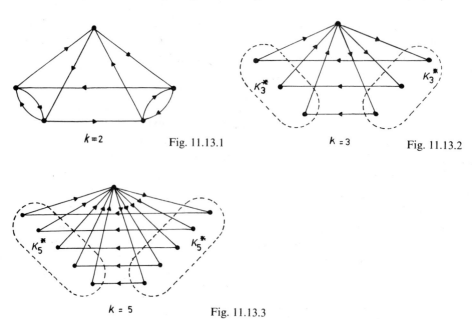

$k = 2$ Fig. 11.13.1 $k = 3$ Fig. 11.13.2

$k = 5$ Fig. 11.13.3

(For $k = 3$ or 5 this is also a consequence of 11.10.) There are several authors who believe that S is infinite, and some even think each sufficiently large number belongs to S. Let us note that, according to Thomassen (1981), up to two exceptions every k-diregular digraph of order $2k + 1$ is Hamiltonian (the exceptions can be found in R 11.62).

11.14 A problem analogous to that of 11.13 for tournaments had even been posed earlier. In its formulation, the condition concerning the order of the digraph is omitted because it is fulfilled automatically.

Problem [Moon 1968, Kotzig 1969, Koh 1977, Bermond and Thomassen 1981]. *Find the set S' of those k for which every k-diregular tournament can be decomposed into k Hamiltonian cycles.*

This problem is claimed to originate from the work of P. Kelly (cf. Moon 1968). Some related results are due to Kotzig (1969) and Thomassen (1980, 1982). Most authors believe that $S' = N$ — it is the so-called *Kelly conjecture*. It is known to be true for $k \leq 4$; see Thomassen (1982).

It is easy to verify that the following two assertions are equivalent [Thomassen 1982]:

A. Every k-diregular tournament (of order, of course, $2k + 1$) is decomposable into k Hamiltonian cycles.

B. Every tournament of order $2k$ in which every vertex has outdegree k or $k - 1$ is decomposable into k Hamiltonian paths.

11.15 To simplify the language, from now on a decomposition of a graph G into k Hamiltonian cycles will be called *Hamiltonian decomposition* (of cardinality k) of G. Similarly, a decomposition of G into k Hamiltonian cycles and one 1-factor will be called *quasi-Hamiltonian decomposition* (of cardinality k) of G. (Some authors, e.g. Alspach 1980, omit here the prefix "quasi".) Clearly, every graph which admits a Hamiltonian [quasi-Hamiltonian] decomposition of cardinality k [$k + 1$] is $2k$-regular [$(2k + 1)$-regular]. The Corollaries 11.4.B and 11.9.B provide us an answer to the question of the existence of Hamiltonian decompositions of complete λ-fold graphs and digraphs. The solution to analogous problems for quasi-Hamiltonian decompositions is postponed to exercises 11.67—11.68.

Let us consider now Hamiltonian and quasi-Hamiltonian decompositions of complete m-partite graphs. According to Berge (1958, or 1970, p. 232), E. Lucas is the author of the following *dancing rounds problem*: In how many different ways can six boys and six girls create a dancing round by catching at their hands in such a way that none of the boys catches the same girl twice at hand? The correct answer is 3 because there is a Hamiltonian decomposition of the graph K(6, 6). We now show how this can be generalized (11.16—11.17).

11.16 Theorem [Hetyei 1975, Laskar and Auerbach 1976]. *The complete m-partite graph $K(m \times n)$, $m \geq 2$ has a Hamiltonian [quasi-Hamiltonian] decomposition if and only if the number $(m - 1)n$ is even [odd].*

Proof. The necessity is a consequence of the fact that $K(m \times n)$ is a regular graph of degree $(m - 1)n$. The proof of the sufficiency can be found in the papers quoted above or in Bermond (1978); for $m = 2$ see also Dirac (1972), or Ninčák (1974).

11.17 Corollary. *The complete m-partite graph* $K(n_1, n_2, ..., n_m)$, $m \geq 2$ *has a Hamiltonian [quasi-Hamiltonian] decomposition if and only if* $n_1 = n_2 = = ... = n_m$ *and the number* $(m - 1)n_1$ *is even [odd]. In particular, the complete bipartite graph* $K(n_1, n_2)$ *has a Hamiltonian [quasi-Hamiltonian] decomposition if and only if* $n_1 = n_2 \equiv 0 \ [\equiv 1]$ (mod 2).

This follows easily from 11.16.

11.18 Theorem [Bermond and Faber 1976, Sotteau 1981]. *The complete bipartite digraph* $K^*(n_1, n_2)$ *has a Hamiltonian decomposition if and only if* $n_1 = n_2$.

Proof. If a digraph has a Hamiltonian decomposition then it must be diregular, which means for $K^*(n_1, n_2)$ that $n_1 = n_2$. On the other hand, the existence of such decomposition follows from Theorem 10.19 putting $k = 2n_1$ and $n_1 = n_2$.

11.19 Let H_1, H_2 be subgraphs of a given graph G. H_1 and H_2 are said to be *orthogonal* if they share at most one edge. Further, H_1 and H_2 are said to be *locally orthogonal* (shortly, *l-orthogonal*) if for each vertex of G there is at most one edge e incident with that vertex such that e belongs to both H_1 and H_2.

Two decompositions R_1 and R_2 of the same graph G are called *orthogonal* [*l-orthogonal*] if for any $H_1 \in R_1$, $H_2 \in R_2$ the subgraphs H_1 and H_2 are orthogonal [l-orthogonal].

Let us note that so far there seemed to be no need to distinguish the concepts of orthogonality and l-orthogonality, and both of them have been referred to as orthogonality.

A factor F of a graph G will be called an *almost n-factor* if it has an isolated vertex and all the remaining vertices are of degree n. A decomposition of G into almost n-factors will be an *almost n-factorization* of G. As a consequence of Lemma 7.41 we get the fact that a complete graph of odd order always admits an almost 1-factorization (as well as a Hamiltonian decomposition — see Corollary 11.5). If the latter two decompositions are orthogonal, they give rise to the so-called *Kotzig factorization* (see Colbourn and Mendelsohn 1982).

Theorem [Horton 1983a]. *For every odd v the graph* K_v *has a Kotzig factorization.*

Proof — see Horton (1983a).

11.20 *The problem of knights.* Dudeney (1917) (cf. also Meally 1971, Huang and Rosa 1973a) posed the following problem: Is it possible (for each $v \geq 3$) to find $(v - 1)(v - 2)/2$ round-table arrangements of v knights in such a way that each of the knights would sit between arbitrary two other knights exactly once? Dudeney himself presented a solution for $v = 3, 4, ..., 12$ and claimed to have found one for $v = 13, 14, ..., 25$ as well as for $v = 33$.

To give a more detailed account on the knights problem let us introduce some new notions. A set S of mutually l-orthogonal Hamiltonian cycles of a graph G will be called *maximal* if there is no Hamiltonian cycle C in G, $C \notin S$ such that C is l-orthogonal to every cycle in S.

For $v \geq 3$ let $m(v)$ $[m'(v)]$ denote the largest [smallest] possible cardinality of a maximal set of l-orthogonal Hamiltonian cycles in the complete graph K_v. It is obvious that

$$1 \leq m'(v) \leq m(v) \leq \binom{v-1}{2}.$$

The Dudeney problem of knights can now be restated in the following lucid way:

Conjecture [Dudeney 1917, Meally 1971, Huang and Rosa 1973a]. *For every $v \geq 3$ it holds*:

$$m(v) = \binom{v-1}{2}.$$

As indicated before, the conjecture was verified for $v \leq 12$ by Dudeney in 1917. Huang and Rosa (1973a) have settled it for all v such that $v - 1$ is a prime; a computer confirmed its validity for $v = 3, 5, 9, 11, 13$ and 15. According to Nakamura et al. (1980), the conjecture is true also for all v such that $v - 1$ is a prime power.

11.21 As far as the function m' is concerned, only the values $m'(3) = 1$, $m'(4) = 3$, $m'(5) = 2$ and $m'(6) = 6$ are known (see Huang and Rosa 1973a). This is too little for stating a conjecture on $m'(v)$ in general.

11.22 A set M of mutually l-orthogonal Hamiltonian decompositions of a graph G is said to be *maximal* if there is no Hamiltonian decomposition D of G, $D \notin M$ such that D is l-orthogonal to every member of M.

Let v be an odd number, $v \geq 3$. Denote by $M(v)$ $[M'(v)]$ the largest [smallest] size of a maximal set of l-orthogonal Hamiltonian decompositions of the complete graph K_v. It is easily seen that for any odd $v \geq 3$ it holds:

$$1 \leq M'(v) \leq M(v) \leq (v-2)! \tag{1}$$

The first inequality follows from 11.5, the second one is trivial, and the third one is a consequence of the fact that none of the $(v - 1)!/2$ Hamiltonian cycles of K_v can appear in two l-orthogonal decompositions.

So far, the only non-trivial result on M and M' says that, for $v \equiv 3$ (mod 4),

$$M'(v) \leq \frac{v-1}{2} \leq M(v) \tag{2}$$

(see Huang and Rosa 1973a and also Exercise 11.63).

126

11.23 In the study of graph decompositions into isomorphic subgraphs, various graph operations play an important role. We shall confine ourselves to some basic results. Let us start with definitions.

For any two simple graphs G_1, G_2 we define the following four new simple graphs:

$G_1 \square G_2$ — *the Cartesian product of G_1 and G_2*,

$G_1 \times G_2$ — *the direct product of G_1 and G_2*,

$G_1 \boxtimes G_2$ — *the strong product of G_1 and G_2*,

$G_1 [G_2]$ — *the composition of G_1 and G_2*.

The vertex set of all of these four graphs is the Cartesian product $V(G_1) \times V(G_2)$. The edge sets are defined as follows:

Let x, $X \in V(G_1)$ and y, $Y \in V(G_2)$. Then the vertices $[x, y]$ and $[X, Y]$ are adjacent

A. in the graph $G_1 \square G_2$ if and only if $x = X$ and $yY \in E(G_2)$, or $y = Y$ and $xX \in E(G_1)$;

B. in $G_1 \times G_2$ if and only if $xX \in E(G_1)$ and $yY \in E(G_2)$;

C. in the graph $G_1 \boxtimes G_2$ if and only if they are adjacent in $G_1 \square G_2$ or in $G_1 \times G_2$;

D. in $G_1 [G_2]$ if and only if either $x = X$ and $yY \in E(G_2)$, or $xX \in E(G_1)$.

For $G_1 = P_2$ and $G_2 = P_3$ the resulting four products are depicted in Fig. 11.23.1.

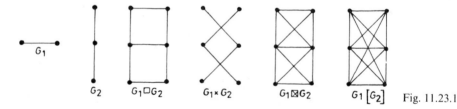

G_1 G_2 $G_1 \square G_2$ $G_1 \times G_2$ $G_1 \boxtimes G_2$ $G_1 [G_2]$ Fig. 11.23.1

Let us remark that neither the terminology nor the notation concerning graph operations has been unified. Our notion of the *Cartesian product* is occasionally used also for the *direct product*, and the latter is also referred to as *categorical product, tensor product,* and *conjunction of graphs*. Instead of the term strong product some authors use *strong direct product* or *normal product*. The composition is sometimes called *lexicographic product* or *wreath product*. In place of the symbol $\square$, the symbols $\times$, $+$, $\dotplus$ or $\oplus$ have also been used. A similar role for our symbol $\times$ [$\boxtimes$] has been played by symbols $\wedge$, $\otimes$ and $.$ [., $\times$ and $\bar{\times}$].

Graph operations (in particular, the Cartesian product) provide an important model of linking computers. In order to synchronize the work of the whole system it is necessary to search for Hamiltonian paths and cycles in the network. Thus, some results on Hamiltonian paths and cycles in Cartesian products of graphs can be applied in computer network design (see Zaretskii 1966).

11.24 There is a number of results on the relations between decompositions of graphs $G_1, G_2, \ldots$ and decompositions of those graphs which arise from $G_1, G_2, \ldots$ by means of graph operations. As an example let us state a result of Hell and Rosa (1972).

Theorem. *Let F, G, H be simple graphs. Then*:

A. *If both G and H have an F-decomposition as defined in 6.1, then the similar holds for the graph $G \square H$.*

B. *If both G and H are decomposable into 1-factors then the similar is true for $G \square H$, $G \times H$ and $G \boxtimes H$.*

C. *If both G and H are decomposable into 2-factors then the similar holds for $G \square H$, $G \times H$ and $G \boxtimes H$.*

Proof. For the Cartesian product $\square$ the assertions A—C follow from the representation

$$E(G \square H) = \bigcup_x (x \square H) \cup \bigcup_y (G \square y),$$

where $x\,[y]$ runs over the vertices of G [or H]. It is clear that each of the graphs $x \square H$ [$G \square y$] is isomorphic to H [or G].

To prove B and C for the direct product $\times$ it is sufficient to consider the following fact. If both G and H are decomposable into r-factors ($r = 1, 2$) then $G \times H$ is decomposable into factors of the form $F_1 \times F_2$, where F_1 [or F_2] is an r-factor of G [or H]. However, it is easily verified that each such factor $F_1 \times F_2$ can be decomposed into two r-factors ($r = 1, 2$).

Since $G \boxtimes H = (G \times H) \cup (G \square H)$, the assertions B and C are true also for the strong product. (For the validity of A for the direct product see Exercise 11.64.)

11.25 For further results of this type we refer the reader to papers of Ringel (1955), Vizing (1963), Kotzig (1965a, 1978, 1979c), Nash-Williams (1965), Zaretskii (1966), Hell and Rosa (1972), Bermond (1975b), Hell et al. (1975), Bermond et al. (1979a), Koh et al. (1979c), Liouville (1981), Wallis (1981), Himelwright et al. (1982), Mohar (1982), Mohar and Pisanski (1982), Mohar et al. (1982), Pisanski et al. (1982), Teichert (1982, 1983b), Witte et al. (1983) and Hartman and Rosa (1985). Now we turn the attention to Hamiltonian decompositions of graph operations, and start with two conjectures [A, p] and [B, p] depending on the parameter $p \in N$.

Conjecture [A, p] [Kotzig 1973a]. *If q is arbitrary and if each of the simple graphs $G_1, G_2, \ldots, G_p$ is decomposable into q Hamiltonian cycles, then the graph $G_1 \square G_2 \square \ldots \square G_p$ has a decomposition into pq Hamiltonian cycles.*

Conjecture [B, p]. *The Cartesian product of arbitrary p cycles is decomposable into p Hamiltonian cycles.*

The reader might be interested in a modification of [B, p], where the cycle is replaced by a two-way infinite path (1.44, 10.20). Such a modification was shown to be true for every p by Nash-Williams (1960).

11.26 Using the method of Kotzig (1973a) we first show that the above conjectures are equivalent.

Theorem. *Let p be a natural number. Then the conjectures [A, p] and [B, p] are equivalent.*

Proof. [B, p] clearly follows from [A, p] by taking cycles in place of graphs G_1, G_2, ..., G_p and putting $q = 1$. Now suppose [B, p] to be true. Let the graph G_i ($i = 1$, 2, ..., p) be a union of mutually disjoint Hamiltonian cycles H_{i1}, H_{i2}, ..., H_{iq}. Then we have:

$$G_1 \,\square\, G_2 \,\square\, ... \,\square\, G_p = (H_{11} \cup H_{12} \cup ... \cup H_{1q}) \,\square\, (H_{21} \cup H_{22} \cup ... \cup H_{2q}) \,\square$$

$$\square\, ... \,\square\, (H_{p1} \cup H_{p2} \cup ... \cup H_{pq}) = (H_{11} \,\square\, H_{21} \,\square\, ... \,\square\, H_{p1}) \cup$$

$$\cup (H_{12} \,\square\, H_{22} \,\square\, ... \,\square\, H_{p2}) \cup ... \cup (H_{1q} \,\square\, H_{2q} \,\square\, ... \,\square\, H_{pq}).$$

The rest is a consequence of [B, p].

11.27 Theorem [Myers 1972a, Kotzig 1973a, Foregger 1978]. *The conjecture* [B, p] *is valid for* $p \leq 3$.

Proof. This is trivial for $p = 1$, for $p = 3$ we refer to Foregger (1978). For the case $p = 2$ we use Kotzig's method (1973a) [Myers 1927a, b considered the case $p = 2$ only in the special case of two copies of the same graph]:

Figures 11.27.1—3 illustrate decompositions into two Hamiltonian cycles in three typical cases: $C_9 \,\square\, C_9$, $C_{10} \,\square\, C_9$, and $C_{10} \,\square\, C_{10}$; the diagrams are to be considered as drawings on a torus. One of the Hamiltonian cycles is always indicated in heavy lines. It is clear that these diagrams provide a method for constructing a Hamiltonian decomposition in the general case for $C_m \,\square\, C_n$.

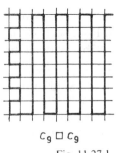

$C_9 \,\square\, C_9$

Fig. 11.27.1

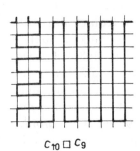

$C_{10} \,\square\, C_9$

Fig. 11.27.2

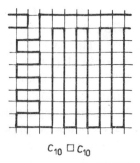

$C_{10} \,\square\, C_{10}$

Fig. 11.27.3

11.28 As shown in the next theorem, the conjecture [A, p] (or, according to 11.26 [B, p]) reduces to prime numbers p.

Theorem [Kotzig 1973a]. *The validity of* [A, p] *and* [A, r] *implies* [A, pr].

Proof. Let $G_1, G_2, ..., G_{pr}$ be graphs decomposable into q Hamiltonian cycles. Let us consider the product

$$H = G_1 \square G_2 \square ... \square G_{pr} = (G_1 \square G_2 \square ... \square G_p) \square$$

$$\square (G_{p+1} \square G_{p+2} \square ... \square G_{2p}) \square ... \square (G_{p(r-1)+1} \square$$

$$\square G_{p(r-1)+2} \square ... \square G_{pr}) = F_1 \square F_2 \square ... F_r,$$

where, according to [A, p], each of the graphs $F_1, F_2, ..., F_r$ is decomposable into pq Hamiltonian cycles. Applying [A, r] to the graph $H = F_1 \square F_2 \square \square ... \square F_r$ we see that it can be decomposed into $(pq)r = (pr)q$ Hamiltonian cycles. Thus we have [A, pr].

11.29 Corollary [Foregger 1978]. *The conjectures* [A, p] *and* [B, p] *are true for each p of the form $2^a 3^b$ for non-negative integers a and b.*

Proof — see 11.27, 11.26 and 11.28.

11.30 The conjecture [A, p] can be generalized in the following way:

Conjecture [C, p] [Chung and Graham 1981; for $p = 2$ see also Bermond 1978, Aubert and Schneider 1982b]. *Let $q_1, q_2, ..., q_p$ be arbitrary positive integers. Let $G_1, G_2, ..., G_p$ be simple graphs such that G_i is decomposable into q_i Hamiltonian cycles. Then the graph $G_1 \square ... \square G_p$ admits a decomposition into $q_1 + q_2 + ... + q_p$ Hamiltonian cycles.*

In view of Theorem 11.26, the relationship between the above three conjectures is given by

$$[C, p] \Rightarrow [A, p] \Leftrightarrow [B, p]. \tag{1}$$

Obviously [C, p] holds for $p = 1$. It is not known whether it is true for $p = 2$, whereas [A, 2] is clearly true (it is sufficient to put $q = 1$, $b = 2$ in Corollary 11.29). The result of 1.29 supports [C, p] in the case $p = 2^a 3^b$, $q_1 = q_2 = ...q_p$. Another important case is treated in the following theorem.

Theorem [Aubert and Schneider 1982b]. *Let q_1 and q_2 be natural numbers, $q_2 \leq q_1 \leq 2q_2$. Let G_1, G_2 be simple graphs having a decomposition into q_1 and q_2 Hamiltonian cycles, respectively. Then the graph $G_1 \square G_2$ is decomposable into $q_1 + q_2$ Hamiltonian cycles.*

Proof: Omitted.

11.31 As the author of this book was informed by J. C. Bermond, the preceding theorem yields an easy proof to the following result that had been predicted by Ringel (1955); for some of the partial results see also Kotzig (1973a) and Bermond (1978).

130

Corollary. *For any p the $2p$-cube Q_{2p} is decomposable into p Hamiltonian cycles.*

Proof. The assertion holds trivially for $p = 1$. Suppose it holds for all $p \leq n$. The graph Q_{2n+2} is isomorphic to both $Q_{n+2} \square Q_n$ and $Q_{n+1} \square Q_{n+1}$. Applying Theorem 11.30 either to the first or the second product (according to the parity of n) it is clear that the assertion holds also for $p = n + 1$.

11.32 An easy consequence of the definition of (repeated) Cartesian product $\square$ is the following result.

Lemma. *For every $p \geq 2$, the statements* [C, 2] *and* [C, p] *imply* [C, p + 1].

Thus, combining this with 11.30.1, to establish [A, p], [B, p] and [C, p] for arbitrary p it is sufficient to prove [C, 2].

Our next theorem shows that [C, 2] is true at least for complete graphs, as predicted by Bermond (1978).

11.33 Theorem [Aubert and Schneider 1981, for $m = n$ see also Myers 1972b]. *The graph $K_m \square K_n$ is decomposable into $(m + n - 2)/2$ Hamiltonian cycles if and only if $m + n$ is even. The same graph can be decomposed into $(m + n - 3)/2$ Hamiltonian cycles and one 1-factor if and only if $m + n$ is odd.*

Proof: Too long and therefore omitted.

11.34 Remark. The impossibility of decomposing the directed graph $(K_2 \square C_m)^*$ into three Hamiltonian cycles was proved by Aubert and Schneider (1982b).

11.35 Theorem. *Let $m \geq 3$ and $n \geq 3$. The graph $C_m \times C_n$ can be decomposed into two Hamiltonian cycles if and only if at least one of the numbers m, n is odd.*

Proof. If $m \equiv n \equiv 0 \pmod 2$ then $C_m \times C_n$ cannot have any Hamiltonian decomposition since it is disconnected. Now suppose that n is odd. Let us represent the vertices of $C_m \times C_n$ by ordered pairs $[x, y]$, $x = 0, 1, \ldots, m - 1$; $y = 0, 1, \ldots, n - 1$. Two vertices $[x, y]$ and $[X, Y]$ are adjacent if and only if $|x - X| \in \{1, m - 1\}$ and $|y - Y| \in \{1, n - 1\}$. One of the Hamiltonian cycles of $C_m \times C_n$ consists of all the edges joining either the vertices $[x, y]$ and $[x + 1, y + 1]$ for $x = 0, 2, 4, \ldots, m - 1$, $y = 0, 1, \ldots, n - 1$, or the vertices $[x, y]$ and $[x - 1, y + 1]$ for $x = 1, 3, \ldots, m - 2$, $y = 0, 1, \ldots, n - 1$ (the operations on the coordinates being taken modulo m or n, respectively). It is a matter of routine to check that all the remaining edges form another Hamiltonian cycle in $C_m \times C_n$. For $m = 8$ and $n = 5$ these cycles are depicted on Figs 11.35.1—2; the opposite sides on each diagram are to be identified.

11.36 Theorem 1.18 of Bermond (1978) states erroneously that there always exists a Hamiltonian decomposition of $C_m \times C_n$; the corresponding proof, however, works only for $n = 3$. For similar reasons, the Corollary 1.19 of that theorem is also in error. We present here a corrected version.

131

11.37 Corollary. *Suppose that the simple graphs G_1 and G_2 both have a Hamiltonian decomposition. If at least one of them has odd order then $G_1 \times G_2$ has a Hamiltonian decomposition.*

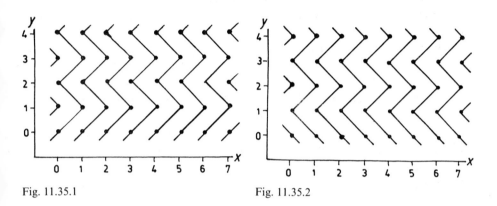

Fig. 11.35.1   Fig. 11.35.2

Proof. Assume that G_1 [or G_2] has a Hamiltonian decomposition $(H_i | i \in I)$ [or $(H_j' | j \in J)$]. Then by means of the distributive law we obtain

$$G_1 \times G_2 = (\cup (H_i | i \in I)) \times (\cup (H_j' | j \in J)) = \cup (H_i \times H_j' | i \in I, j \in J).$$

By 11.35, each of the graphs $H_i \times H_j'$ has a Hamiltonian decomposition. Thus, the same is true for $G_1 \times G_2$.

11.38 The reader interested in Hamiltonian decompositions of compositions of graphs is referred to Bermond (1978), Laskar (1978), Baranyai and Szász (1981) and Teichert (1983a).

11.39 If an undirected graph G admits a decomposition into n Hamiltonian cycles then obviously G must be a connected finite $2n$-regular graph. It is natural to ask about the validity of the converse statement, i.e. whether each finite connected $2n$-regular undirected graph has a decomposition into n Hamiltonian cycles. Since this holds trivially for $n = 1$, let us deal with the next smallest case, $n = 2$. Unfortunately, any finite connected 4-regular graph with a cutvertex provides a counterexample (see e.g. Fig. 11.39.1).

11.40 Let us now pose a similar question as above for finite 2-connected 4-regular undirected graphs. Although for some classes of graphs the answer is affirmative (see Exercise 11.65), in general it is negative. There are even planar finite 3-connected graphs without any Hamiltonian cycle (cf. Zaks 1976). Consider first the graph depicted in Fig. 11.40.1. It is a matter of routine to check that this planar 3-regular graph of order 38 is 3-connected (and hence 3-edge-connected) and contains no Hamiltonian cycle. (Further four examples of planar non-Hamiltonian graphs having the same parameters as that in Fig. 11.40.1 were given by: Bosák 1967, Grünbaum 1967, Lederberg 1967 and Klee 1971.) Let us now convert this graph to a 4-regular graph by means of the transformation depicted in Fig. 11.40.2 (edges of the new graphs are indicated by dotted lines). It can be shown that the graph thus constructed is a non-Hamiltonian planar 3-connected 4-regular graph; its order is 161. This method was described by Capobianco and Molluzo (1978, pp. 172—173), and was used by Sachs (1967) for the construction of an analogous example of order 207. A similar relationship between Hamiltonian cycles in 3-regular graphs and related Hamiltonian decompositions of certain 4-regular graphs was studied by Kotzig (1957).

11.41 Nash-Williams (1971b) conjectured that each finite 4-connected 4-regular undirected graph is Hamiltonian. A counterexample (Fig. 11.41.1)

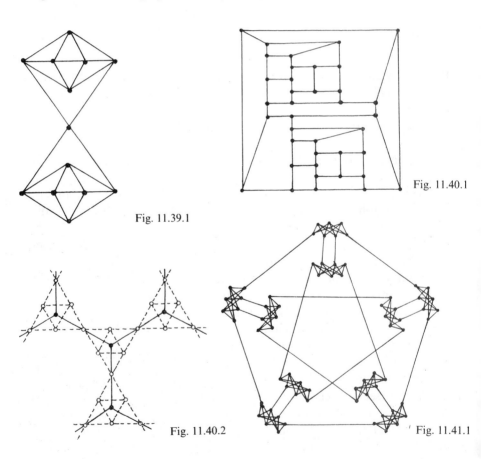

Fig. 11.39.1

Fig. 11.40.1

Fig. 11.40.2

Fig. 11.41.1

was found by Meredith (1973). It is easy to check that his graph is 4-regular, 4-connected, of order 70, and without any Hamiltonian cycle; the last claim follows from the fact that the Petersen graph is non-Hamiltonian.

11.42 Advantages of both of the preceding examples cannot be put to-gether to obtain a non-Hamiltonian 4-connected planar graph. As it was shown by Tutte (1956), each finite undirected planar 4-connected graph is Hamiltonian. However, among them there are many that are 4-regular and not decomposable into two Hamiltonian cycles. Such a simple example of order 18 (Fig. 11.42.1) was given by Grünbaum and Malkevitch (1976); further examples can be found either in their paper or in Bollobás (1978, pp. 145—146) and Martin (in press). It should be noted that, originally, the question of existence of such graphs was raised by Grünbaum and Zaks (1974).

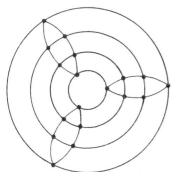

Fig. 11.42.1

A simple necessary condition for the existence of a Hamiltonian decom-position of a planar 4-regular graph was proved by Bondy and Häggkvist (1981).

Zaks and Erdős (cf. Zaks 1982) have asked the following questions:

Problem. A. *Does there exist a number l such that each planar finite undirected 4-connected 4-regular graph G contains l Hamiltonian cycles whose union is G?*

B. *Is there a function l(n) such that each finite undirected planar 4-con-nected graph G of minimum degree at least n contains l(n) Hamiltonian cycles covering G?*

11.43 In 11.39—11.42 we dealt with the question whether an undirected r-regular graph satisfying certain connectivity conditions is necessarily Ha-miltonian, or even Hamilton-decomposable. Since most results in this direc-tion are negative, it is natural to accept a further assumption — namely that of the existence of a Hamiltonian cycle — in hope of some positive results.

For $r \geq 2$ let $f(r)$ denote the smallest number of Hamiltonian cycles in an undirected r-regular Hamiltonian graph. Clearly $f(r) \geq 1$ and $f(2) = 1$. As mentioned by Tutte (1946), it was proved by Smith that $f(3) \geq 3$. Since there are 3-regular graphs containing exactly three Hamiltonian cycles (e.g. K_4) we have $f(3) = 3$. In addition, Bosák (1967), constructed for each $n \geq 4$ an undirected 3-regular planar 3-connected triangle-free graph with exactly n Hamiltonian cycles (see Exercise 11.66). However, nearly nothing is known on $f(r)$ for $r \geq 4$.

Problem [Sheehan 1976, in press]. *Determine $f(r)$ for all $r \geq 4$. In particular, is there an $r \geq 4$ for which $f(r) = 1$?*

So far it is even not known whether $f(4) = 1$, i.e. whether there is an undirected 4-regular graph containing exactly one Hamiltonian cycle. But Entringer and Swart (1980) found an example of a graph having all but one [two] vertices of degree 3, the exceptional vertex [vertices] being of degree 4, which contains exactly two [one] Hamiltonian cycles.

11.44 Now let us pass to the study of graphs having a Hamiltonian decomposition.

Suppose that an undirected graph G is decomposable into two Hamiltonian cycles (thus, G is 4-regular). Sloane (1969) proved that such G has at least three Hamiltonian cycles. This result was improved by Ninčák (1975) to 6 Hamiltonian cycles, and the author conjectured 12 to be the best lower bound for simple graphs. A further substantial progress in this direction was made by Thomason (1978). Some of his results are presented below (11.45 —11.51).

11.45 For any pair of edges w, w' of a graph G let us denote by $P_G(w, w')$ [or $Q_G(w, w')$] the set of all Hamiltonian decompositions of G in which the edges w and w' belong to the same cycle [to different cycles] of the decomposition. Further let P_G denote the set of all Hamiltonian decompositions of G. (For the sake of convenience we shall omit the subscript G in the notations defined above whenever possible.) Clearly

$$|P| = |P(w, w')| + |Q(w, w')| = |P(w, w)|$$

for arbitrary edges w, w' of G.

11.46 Lemma [Thomason 1978]. *Let G be a finite undirected 4-regular graph of order at least 3. Then:*

    A. *The number $|P(w, w')|$ is even for each w, $w' \in E(G)$.*

    B. *$|Q(w, w')|$ is an even number for any w, $w \in E(G)$.*

    C. *The number $|P|$ of Hamiltonian decompositions of G is even.*

Proof. I. First we prove that if $|P(w, w')|$ is even for arbitrary adjacent edges w, w' then C is true. (In particular, A implies C.) Assume that w_1, w_2, w_3, and w_4 are edges incident with a common vertex of G. Then obviously

$$|P| = |P(w_1, w_2)| + |P(w_1, w_3)| + |P(w_1, w_4)|,$$

and our assertion follows.

II. We prove that B is a consequence of A. Indeed, it suffices to consider the fact that A implies C and the relation

$$|Q(w, w')| = |P| - |P(w, w')| .$$

Thus, to prove the lemma it is sufficient to prove A.

III. The assertion A will be proved by induction on n, the order of G. The case $|P| = 0$ is trivial. If $|P| \geq 1$ then G is a connected loopless graph. For $n = 3$ is isomorphic to the complete 2-fold graph $^2K_3$, thus $|P(w, w')| \in$ $\in \{0, 2, 4\}$ for any w, w' (as it can be easily checked).

Now suppose that A holds for all graphs of order n where $3 \leq n \leq s$. We prove the validity of A also for all graphs G of order $s + 1$. Let w, w' be edges of such a G.

(a) Consider first the case when the edges $w = w_1$ and $w' = w_2$ are adjacent. Let w_1, w_2, w_3, and w_4 be incident with a common vertex u; let $w_i = uu_i$ ($i = 1, 2, 3, 4$) (Fig. 11.46.1). Now, delete from G the vertex u together with the four edges w_i, and add two new edges $w_5 = u_1u_2$, $w_6 = u_3u_4$. In this way we obtain a new 4-regular graph G' of order s. It is easy to see that

$$|P(w, w')| = |P_G(w_1, w_2)| = |Q_{G'}(w_5, w_6)| \equiv 0 \;(\text{mod } 2) .$$

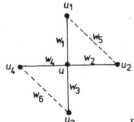

Fig. 11.46.1

(The first equality is obvious, the second follows from a 1-1 correspondence between the Hamiltonian decompositions of G and G' in question, and the congruence is a consequence of the induction hypothesis and part II of our proof.)

(b) Assume now that $w = w'$. Then, according to IIIa and I we have

$$|P(w,w')| = |P(w, w)| = |P| \equiv 0 \;(\text{mod } 2) .$$

(c) At last, let w, w' be two non-adjacent edges of G. Since G is connected, there is a sequence of edges $w = W_0, W_1, \ldots, W_{t-1}, W_t = w'$ such that any two consecutive edges are adjacent in G. For any three edges x, y, z of G there is a simple identity

$$Q(x, y) = P(x, z)\,\Delta P(z, y)$$

(Δ stands for the symmetric difference of sets). The properties of our sequence of W_i's and the parity of $|P|$ enable us to write down the following series of congruences:

$$|P(w, W_{i+1})| = |P| - |Q(w, W_{i+1})| \equiv |Q(w, W_{i+1})| =$$
$$= |P(w, W_i)\,\Delta P(W_i, W_{i+1})| \equiv |P(w, W_i)| + |P(W_i, W_{i+1})| \equiv$$
$$\equiv |P(w, W_i)| \pmod 2 .$$

Therefore

$$|P(w, w')| = |P(w, W_t)| \equiv |P(w, W_{t-1})| \equiv \ldots \equiv |P(w, W_1)| =$$
$$= |P(W_0, W_1)| \equiv 0 \pmod 2 .$$

IV. According to III, the assertion A is true. As both B and C are consequences of A, Lemma 11.46 follows.

11.47 Theorem [Thomason 1978]. *Let G be an undirected 4-regular graph of order at least 3 with at least one Hamiltonian decomposition. Then*

(a) *G contains at least 8 Hamiltonian cycles,*

(b) *G has an even number (≥ 4) of Hamiltonian decompositions,*

(c) *every edge of G lies in at least 4 Hamiltonian cycles.*

Proof. Let $R_1 = (F_1, F_2)$ be a Hamiltonian decomposition of G. By Lemma 11.46.C, there is another Hamiltonian decomposition $R_2 = (F_3, F_4)$ of G. Obviously, there is an edge w which belongs to both F_1 and F_3, and an edge w' in F_1 which is not in F_3. But, according to Lemma 11.46.A, the number $|P(w, w')|$ is even, $R_1 \in P(w, w')$ and $R_2 \notin P(w, w')$. Thus there exists a third Hamiltonian decomposition $R_3 \in P(w, w')$. Using Lemma 11.46.C once again, the number $|P|$ is even and therefore there must be a fourth Hamiltonian decomposition $R_4 \in P$. Hence $|P| \geq 4$; G has at least 8 Hamiltonian cycles and each edge is contained in at least four of them (since there are at least four Hamiltonian decompositions).

11.48 It might be interesting to investigate the following statement analogous to Theorem 11.47 for infinite graphs:

Problem. *Suppose that an infinite undirected 4-regular graph has a decomposition into two-way infinite Hamiltonian paths. What is then the minimum number of such decompositions?*

Decompositions of specific graphs into two-way infinite Hamiltonian paths have been investigated by Nash-Williams (1960); see 11.25.

11.49 Although some examples might indicate that the number of Hamiltonian decompositions of a 4-regular graph could grow rapidly with respect to the order of the graph, it is not true in general: Thomason (1978) has constructed for each $n \geq 10$ a 4-regular graph order n with precisely 32 Hamiltonian decompositions.

Problem. *What is the smallest number s for which there are infinitely many 4-regular graphs with exactly s Hamiltonian decompositions?*

We know only that s is even and $4 \leq s \leq 32$ (11.47, 11.49).

11.50 Corollary [Thomason 1978]. *Let G be an undirected 2n-regular graph of order ≥ 3 ($n \geq 2$) which has a Hamiltonian decomposition. Then:*

A. *Each edge of G is contained in at least $3n - 2$ Hamiltonian cycles of G.*

B. *There are at least $n(3n - 2)$ Hamiltonian cycles in G.*

C. *G has an even number of Hamiltonian decompositions, and the number is at least $(3n - 2)(3n - 5) \ldots 7.4$.*

Proof. For $n = 2$ the assertions follow from Theorem 11.47. We shall proceed by induction and assume that A—C hold for all $n < s$, where $s \geq 3$; let G be a $2s$-regular graph with a Hamiltonian decomposition $(F_1, F_2, ..., F_s)$.

A. Let $w \in F_1$ be an edge of G. According to our induction hypothesis, w is contained in at least $3s - 5$ Hamiltonian cycles in the graph $F_1 \cup F_2 \cup ... \cup F_{s-1}$. By Theorem 11.47, the graph $F_1 \cup F_s$ has at least 4 Hamiltonian cycles, each containing the edge w; however, only one of those cycles (namely F_1) is a factor of the graph $F_1 \cup F_2 \cup ... \cup F_{s-1}$. Thus, G contains at least $(3s - 5) + (4 - 1) = 3s - 2$ Hamiltonian cycles passing through the edge w.

B. By the induction hypothesis, the graph $F_1 \cup F_2 \cup ... \cup F_{s-1}$ has at least $(s - 1)(3s - 5)$ Hamiltonian cycles. Each of the graphs $F_1 \cup F_s$, $F_2 \cup F_s$, ..., $F_{s-1} \cup F_s$ contains at least 8 Hamiltonian cycles; among them one belongs to $F_1 \cup F_2 \cup ... \cup F_{s-1}$ and another one, F_s, is common. Therefore G contains at least

$$(s - 1)(3s - 5) + (8 - 2)(s - 1) + 1 = s(3s - 2)$$

different Hamiltonian cycles.

C. Let w be an edge of G. We know (A) that w belongs to at least $3s - 2$ Hamiltonian cycles of G. By deleting one of these Hamiltonian cycles from G we obtain a graph which, according to our induction hypothesis, contains at least $(3s - 5)(3s - 8) \ldots 7.4$ Hamiltonian decompositions (and, of course, their number is an even one). Thus, G has again an even number of such decompositions, bounded from below by $(3s - 2)(3s - 5) \ldots 7.4$.

11.51 Corollary. *Suppose that an undirected graph G of order ≥ 3 contains n pairwise edge-disjoint Hamiltonian cycles. Then there are at least $n(3n - 2)$ Hamiltonian cycles in G.*

Proof. It is sufficient to consider the subgraph of G induced by the n Hamiltonian cycles, and then to apply Corollary 11.50.B.

11.52 There is a great number of further papers dealing with Hamiltonian decompositions of graphs and related questions. For example, Camion (1978) studied certain 6-regular, toroidal, Hamilton-decomposable graphs in the context of automorphism groups and the Diophantine equation $x^2 + dy^2 = n^2$.

A graph G is *vertex-transitive* if its automorphism group acts transitively on $V(G)$, i.e. if for any pair of vertices u, v of G there is an automorphism of G sending u to v. It is known (Turner 1967, Alspach 1980, 1981) that a simple, connected, and vertex-transitive graph of order n has either a Hamiltonian or a quasi-Hamiltonian decomposition if n is prime, or if $n \equiv 6$ (mod 8) and $n/2$ is prime.

11.53 Neumann-Lara (1975) proved the following assertions:

A. For any $k \geq 2$, $g \geq 3$, $n \geq g$ there is a graph decomposable into k Hamiltonian cycles whose girth is equal to g and whose order is divisible by n.

B. For any $k \geq 2$, $g \geq 4$, $n > g$ ($g \equiv n - 1 \equiv 0$ (mod 2)) there exists a bipartite graph decomposable into k Hamiltonian cycles whose girth is equal to g and whose order is a multiple of n.

11.54 The maximum number of pairwise disjoint Hamiltonian cycles (or paths) in undirected graphs was investigated by Nash-Williams (1969, 1970a, 1971a), Grünbaum and Zaks (1974), Grünbaum and Malkevitch (1976), Zaks (1976), Bondy (1978), Jackson (1978, 1979), Biggs (1979), Owens (1980), Chung and Graham (1981). A similar problem for digraphs was studied by Nash-Williams (1969), Ninčák (1973), Alspach et al. (1976), Jackson (1978, 1979, 1980b) and Thomassen (1982). For example, Zaks (1976) has constructed a 5-connected 5-regular planar graph without two edge-disjoint Hamiltonian cycles, answering thereby a question posed in Grünbaum and Zaks (1974). Further sophisticated constructions of graphs having similar properties can also be found in the paper of Akiyama et al. (1980).

11.55 Hamiltonian decompositions of hypergraphs have been studied by Bermond (1978) and Bermond et al. (1978b).

Exercises

11.56 For every odd v the graph K_v can be decomposed into $(v - 1)/2$ Hamiltonian paths and one almost 1-factor (for the definition see 11.19).

11.57 If v is even then the graph K_v is decomposable into $v/2 - 1$ Hamiltonian cycles and one 1-factor (Harary 1969, p. 89).

11.58 For every odd v the complete digraph K_v^* has a decomposition into $v - 1$ Hamiltonian paths and $(v - 1)/2$ disjoint 2-cycles (Bigalke and Köhler 1979).

11.59 For every even v, K_v^* can be decomposed into $v - 2$ Hamiltonian cycles and $v/2$ disjoint 2-cycles.

11.60 An antipath is a directed semi-path in which every two adjacent edges are oppositely directed. The complete digraph K_v^* is always decomposable into Hamiltonian antipaths (Bigalke and Köhler 1979).

11.61 Find a decomposition of K_7^* into Hamiltonian paths (Mendelsohn 1968).

11.62 Is there a non-Hamiltonian k-diregular digraph of order $2k + 1$ for $k = 1, 2, 3$? (Thomassen 1981)

11.63 Let M and M' be as in 11.22. Show that $M(3) = M'(3) = 1$ and $M(5) = M'(5) = 2$.

11.64 Let F, G, H be simple graphs. If both G and H have an F-decomposition, what can be said for the product $G \times H$? (Cf. 11.23—11.24.)

11.65 For every $v \geq 5$ there exists a simple 4-regular graph of order v such that each of its decompositions into two 2-factors contains at least one Hamiltonian cycle (Doutre and Kotzig 1980).

11.66 For $n = 0$ and for each $n \geq 4$ there is an undirected 3-regular 3-connected planar triangle-free graph which contains exactly n Hamiltonian cycles (Bosák 1967).

11.67 The graph ${}^\lambda K_v$ has a quasi-Hamiltonian decomposition if and only if v is even an λ odd.

11.68 For which values of v and λ does the digraph ${}^\lambda K_v^*$ admit a quasi-Hamiltonian decomposition?

Chapter 12

Kirkman's schoolgirl problem, resolvable designs and graph decompositions

12.1 Kirkman (1847, 1850a, b) is the author of the famous *schoolgirl problem*: $6n + 3$ schoolgirls, grouped in $2n + 1$ triples, go for walks during each of $3n + 1$ days. In order to give every two schoolgirls an equal chance to become friends, a scheduling for $3n + 1$ days is to be found in which each pair of schoolgirls would belong to the same triple on exactly one of the $3n + 1$ days. Is such an arrangement always possible?

The first and most widely known version of the problem was a puzzle concerned with 15 schoolgirls. In Kirkman's original setting, "fifteen young ladies in a school walk out three abreast for seven days in succession: It is required to arrange them daily, such that no two shall walk twice abreast".

It is fairly easy to find a solution to the schoolgirl problem for small values of n (Rosa 1963a, Gardner 1980, Mathon et al. 1981); for the sake of convenience the schoolgirls will be denoted 1, 2, ..., $v = 6n + 3$:

$n = 0$:

1st day: (1 2 3).

$n = 1$:

1st day: (1 2 3,   4 5 6,   7 8 9).
2nd day: (1 4 7,   2 5 8,   3 6 9).
3rd day: (1 5 9,   2 6 7,   3 4 8).
4th day: (1 6 8,   2 4 9,   3 5 7).

$n = 2$:

1st day: (1 2   3,   4 5 6,   7 8 9,   10 11 12,   13 14 15).
2nd day: (1 4 12,   2 7 10,   3 8 14,   5 9 13,   6 11 15).
3rd day: (1 9 15,   2 6 13,   3 4 10,   5 8 12,   7 11 14).
4th day: (1 6   7,   2 5 11,   3 12 15,   4 8 13,   9 10 14).
5th day: (1 5 14,   2 9 12,   3 11 13,   4 7 15,   6 8 10).
6th day: (1 8 11,   2 4 14,   3 6 9,   5 10 15,   7 12 13).
7th day: (1 10 13,   2 8 15,   3 5 7,   4 9 11,   6 12 14).

$n = 3$:

1st day: (1 2 4, 8 9 11, 15 16 18, 3 10 17, 5 12 19, 6 13 20, 7 14 21).
2nd day: (2 3 5, 10 11 13, 18 19 21, 1 9 17, 4 12 20, 6 14 15, 7 8 16).
3rd day: (3 4 6, 12 13 8, 21 15 17, 1 10 19, 2 11 20, 5 14 16, 7 9 18).
4th day: (4 5 7, 14 8 10, 17 18 20, 1 11 21, 2 12 15, 3 13 16, 6 9 19).
5th day: (5 6 1, 9 10 12, 20 21 16, 2 13 17, 3 14 18, 4 8 19, 7 11 15).
6th day: (6 7 2, 11 12 14, 16 17 19, 1 13 18, 3 8 20, 4 9 21, 5 10 15).
7th day: (7 1 3, 13 14 9, 19 20 15, 2 8 21, 4 10 16, 5 11 17, 6 12 18).
8th day: (1 8 15, 2 10 18, 3 12 21, 4 14 17, 5 9 20, 6 11 16, 7 13 19).
9th day: (2 9 16, 3 11 19, 4 13 15, 5 8 18, 6 10 21, 7 12 17, 1 14 20).
10th day: (4 11 18, 5 13 21, 6 8 17, 7 10 20, 1 12 16, 2 14 19, 3 9 15).

12.2 Solving the schoolgirl problem is equivalent to finding, for each $v \equiv 3$ (mod 6), a Steiner triple system of order v in which it is possible to resolve all triples into classes in such a way that each class contains each element exactly once; more precisely, each class corresponds to a decomposition of the v elements into 3-element subsets. This leads naturally to some new definitions, relevant for $(v, k\ \lambda)$-designs in general.

By a *replication* of a (v, k, λ)-design we mean any system of blocks (of the design) containing each element exactly once. A *resolution* of a (v, k, λ)-design is any collection of its replications such that the number of occurrences of any block in the design is equal to the sum of the numbers of occurrences of the block in all replications. For example, the (4, 2, 2)-design (12, 12, 13, 13, 14, 14, 23, 23, 24, 24, 34, 34) has a resolution ((12, 34), (12, 34), (13, 24), (13, 24), (14, 23), (14, 23)) comprising 6 replications. (Here, as well as in many other places, we simplify the notation of a block by putting $u_1 u_2 \dots u_k$ instead of $\{u_1, u_2, \dots, u_k\}$.)

A (v, k, λ)-design is *resolvable* (Hanani et al. 1972, Hanani 1974) if it admits a *resolution*. In a resolvable (v, k, λ)-design, every element is contained in exactly r blocks, where r is the number of classes of each resolution. Thus, for resolvable designs we get the result of 8.5A saying that every such design is balanced. A similar result holds for resolvable $(n_1, n_2, \dots, n_m; k; \lambda_1, \lambda_2)$-designs in general (cf. 8.40, the concept of resolvability can there be defined analogously). Some results in this direction can be found in Vanstone (1980); other generalizations are mentioned in 12.41.

A resolvable $(v, k, 1)$-design will also be called a *Kirkman system of k-tuples* of order v; in particular, for $k = 3$ we get a *Kirkman triple system* of order v, or a *Kirkman design* of order v.

Now, the *schoolgirl problem* can be restated as follows. For which numbers v does there exist a Kirkman design of order v?

12.3 The concept of resolvability can be transferred in a natural way to corresponding K_k-decompositions of the graph $^\lambda K_v$, or even to the more general G-decompositions of a graph H.

Let F, G, H be graphs. We shall say that F is a graph of type G [Dénes and Gergely 1977], if each component of F is isomorphic to G. If F is of type G and, at the same time, F is a factor of H, we shall say that F is a G-factor of H. A decomposition of a graph H into G-factors will be called G-factorization of H. In particular, we shall speak about K_k-factors and K_k-factorizations of a graph H if all components of the factors are complete graphs of order k.

A G-decomposition R of a graph H will be called *resolvable* if it admits a *resolution*, i.e., a G-factorization R' of H such that the set of all components of the factors in R' is the same as the set of all components of the graphs in R.

For example, a resolvable K_3-decomposition of K_v is a decomposition of K_v into triangles which can be grouped so as to form 2-factors decomposing K_v. The problem of its existence constitutes the "graphical" version of the schoolgirl problem. In general, a resolvable K_k-decomposition of K_v exists if and only if there is a resolvable (v, k, λ)-design, and there is a natural correspondence between both resolved structures.

As we have seen, the schoolgirl problem (existence of a resolvable K_3-decomposition of K_v) can be generalized to the problem of the existence of resolvable K_k-decompositions of $^\lambda K_v$. Another possibility of generalizing the original problem can be obtained by realizing that $K_3 = C_3$; the question concerning the existence of resolvable C_k-decompositions of K_v was handled by Hell et al. (1975). However, the Oberwolfach problem is even more general (see 10.20).

12.4 Now, a natural question arises: For which parameters v, k, λ does there exist a K_k-factorization of the graph $^\lambda K_v$? Unfortunately, this question is far from being answered completely. In what follows (12.4, 12.11, 12.13) we state some of the necessary conditions which we summarize in Theorem 12.5 and Corollary 12.14. Sufficient conditions are presented in 12.6 and 12.43—12.48.

The first two necessary conditions are indeed very simple.

Lemma [Bose 1942]. *Let there exist a K_k-factorization of the graph $^\lambda K_v$ (or, equivalently, a resolvable (v, k, λ)-design). Then*

$$k - 1 | \lambda(v - 1), \tag{1}$$

$$k | v. \tag{2}$$

Proof. (1) is a consequence of 8.9.1, (2) follows from the resolvability.

12.5 Theorem. *If there is a K_k-factorization of the graph $^\lambda K_v$, then*

$$k \geq 2, \tag{1}$$

$$v \equiv k \pmod{k(k-1)/s}, \tag{2}$$

where $s = $ g.c.d. $\{k - 1, \lambda\}$.

144

Proof. From 12.4.1 we get (1) and the fact that $k - 1|s(v - 1)$. Thus, $k - 1|s(v - 1) - s(k - 1)$, or

$$k - 1|s(v - k). \tag{3}$$

From 12.4.2 we obtain $k|v - k$, and hence

$$k|s(v - k). \tag{4}$$

Combining (3) and (4) we see that $k(k - 1)|s(v - k)$, and also $k(k - 1)/s|v - k$, which proves (2).

Remark. It is now clear that both (1) and (2) are consequences of 12.4.1—2. For the converse see Exercise 12.67.

12.6 The question of 12.4, restricted to $k = 3$ and $\lambda = 1$, is precisely Kirkman's schoolgir problem. It took over 120 years until it was completely solved:

Theorem [Ray-Chaudhuri and Wilson 1971]. *A K_3-factorization of the graph K_r (or, equivalently, a Kirkman design of order v) exists if and only if* $v \equiv 3 \pmod 6$.

Proof. The necessity follows from 12.5.2, setting $k = 3$ and $\lambda = s = 1$. For the sufficiency we refer to Ray-Chaudhuri and Wilson (1971).

12.7 Once we are given a resolvable (v, k, λ)-design, we can create many others by relabelling the elements. However, in such a way we cannot get "substantially different" designs, because they all are isomorphic in the sense of the following definition:

Two resolvable (v, k, λ)-designs [equivalently, resolvable K_k-decompositions of $^\lambda K_v$] are:

A. *equal* if they have equal resolutions; else they are *different*;

B. *isomorphic* if there is a bijection between the elements of both design preserving blocks and replications.

The reader may have noticed that the concept just introduced extends those defined in 3.9 or 8.34.

As an example, consider a Steiner triple system of order 15. Starting from it we can get two different Kirkman designs of order 15:

```
((1  2  3,  4  5  6,  7  8  9,  10 11 12,  13 14 15),
 (1  4  7,  2  5 10,  3  6 13,  8 11 14,   9 12 15),
 (1  8 10,  2  6  9,  3  7 14,  4 11 15,   5 12 13),
 (1  6 11,  2  7 12,  3 10 15,  4  8 13,   5  9 14),
 (1 12 14,  2 11 13,  3  5  8,  4  9 10,   6  7 15),
 (1  9 13,  2  8 15,  3  4 12,  5  7 11,   6 10 14),
 (1  5 15,  2  4 14,  3  9 11,  6  8 12,   7 10 13))
```

and

```
((1  2  3,   4  5  6,   7  8  9,   10 11 12,   13 14 15),
 (1  4  7,   2  5 10,   3  6 13,    8 11 14,    9 12 15),
 (1  9 13,   2  7 12,   3  5  8,    4 11 15,    6 10 14),
 (1  8 10,   2 11 13,   3  4 12,    5  9 14,    6  7 15),
 (1 12 14,   2  6  9,   3 10 15,    5  7 11,    4  8 13),
 (1  6 11,   2  8 15,   3  7 14,    4  9 10,    5 12 13),
 (1  5 15,   2  4 14,   3  9 11,    6  8 12,    7 10 13)).
```

These designs are obviously different, and Read (1963) proved that they are even non-isomorphic.

12.8 As regards the number $K(v)$ of pairwise non-isomorphic Kirkman designs of order v, it is known that $K(3) = K(9) = 1$, $K(15) = 7$ (see Mulder 1917, Cole 1922, Rosa 1963b), $K(21) \geq 78$ (Mathon et al. 1981, Mathon and Rosa 1984), $K(27) \geq 661$ (Janko and Tran van Trung 1981).

It can be expected that the function $K(v)$ grows very fast. Thus, similarly as in the case of Steiner triple systems, it has been tried to enumerate at least some restricted classes of Kirkman designs. For example, Mathon et al. (1981) and Janko and Tran van Trung (1981) counted the number of pairwise non-isomorphic Kirkman designs of order 21 and 27, respectively, which can be created starting from a corresponding Steiner triple system of order 21, resp. 27, which has an automorphism consisting of tree disjoint cycles of length 7 (an automorphism of order 13, respectively). Another result in this direction can be found in Mathon and Rosa (1984).

For further modifications of the original Kirkman's problem the reader is referred to Ahrens (1901), Smith (1977), Rosa (1980a), Vanstone (1980); the relation of Kirkman designs to quasi-groups was studied by Rosa (1963a).

12.9 A resolution of a (v, k, λ)-design will be called *affine* if any two blocks belonging to different resolution classes intersect in the same number of elements. A (v, k, λ)-design admitting an affine resolution will be called *affine resolvable*. Finally, a (v, k, λ)-design is said to be *affine* if it is affine resolvable and, moreover, each of its resolution is affine.

The affine plane of order 3 ($v = 9$, $k = 3$, $\lambda = 1$; see 12.16) is an example of an affine design; this holds true also for other finite affine planes, as we shall see in 12.17.

At a first glance it may sound surprising that the concepts of affine design and affine resolvable design coincide. In addition, they depend only upon the existence of a resolution and the parameters v, b, r, k, λ:

12.10 Theorem [Bose 1942, Raghavarao 1971]. *Let v, b, r, k, λ be natural numbers such that $1 < k < v$, $b = \lambda v(v - 1)/(k(k - 1))$, $r = \lambda(v - 1)/(k - 1)$. Then, for a resolvable (v, k, λ)-design X, the following statements are equivalent.*

A. *The design X is affine resolvable.*
B. *X is an affine design.*
C. $v = k + k(k - 1)/\lambda$.
D. $b = v + r - 1$.

Proof. Clearly $C \Leftrightarrow D$.

Suppose now that we are given a resolved K_k-decomposition of the graph K_v (corresponding to a resolved (v, k, λ)-design) consisting of r factors F_i $(i = 1, 2, ..., r)$, each of them being a union of $n = v/k$ copies of k-cliques S_{i1}, $S_{i2}, ..., S_{in}$. Set $q_{ij} = |V(S_{11}) \cap V(S_{ij})|$, $i = 2, 3, ..., r, j = 1, 2, ..., n$. Obviously we have

$$\sum_{i=2}^{r} \sum_{j=1}^{n} q_{ij} = k(r - 1) = \frac{k(\lambda v - \lambda - k + 1)}{k - 1}. \tag{1}$$

Let us form the arithmetic mean of the numbers q_{ij}:

$$q = \frac{\sum_{i=2}^{r} \sum_{j=1}^{n} q_{ij}}{n(r - 1)} = \frac{k}{n} = \frac{k^2}{v}. \tag{2}$$

Comparing two ways of determining the number of edges not in S_{11} but joining two vertices in S_{11} we get

$$\sum_{i=2}^{r} \sum_{j=1}^{r} \binom{q_{ij}}{2} = \binom{k}{2}(\lambda - 1).$$

Using this equality we obtain

$$\sum_{i=2}^{r} \sum_{j=1}^{n} q_{ij}^2 = \sum_{i=2}^{r} \sum_{j=1}^{n} q_{ij}(q_{ij} - 1) + \sum_{i=2}^{r} \sum_{j=1}^{n} q_{ij} =$$
$$= k(k - 1)(\lambda - 1) + k(\lambda v - \lambda - k + 1)/(k - 1). \tag{3}$$

Clearly,

$$\sum_{i=2}^{r} \sum_{j=1}^{n} q^2 = \frac{k^4}{v^2}(r - 1)n = \frac{k^3(\lambda v - \lambda - k + 1)}{v(k - 1)}. \tag{4}$$

Finally, from (1)—(4) we get:

$$\sum_{i=2}^{r} \sum_{j=1}^{n} (q_{ij} - q)^2 = \sum_{i=2}^{r} \sum_{j=1}^{n} q_{ij}^2 - 2q \sum_{i=2}^{r} \sum_{j=1}^{n} q_{ij} + \sum_{i=2}^{r} \sum_{j=1}^{n} q^2 =$$

$$= k(k-1)(\lambda-1) + \frac{k(\lambda v - \lambda - k + 1)}{k-1} -$$

$$- \frac{2k^3(\lambda v - \lambda - k + 1)}{v(k-1)} + \frac{k^3(\lambda v - \lambda - k + 1)}{v(k-1)} =$$

$$= \frac{k(v-k)(\lambda v - \lambda k - k^2 + k)}{v(k-1)} = 0$$

if and only if $v = k + k(k-1)/\lambda$.

Since in our considerations the set S_{11} can be replaced by any other clique S_{ij}, the proof is complete.

12.11 Corollary. *Let v, k, λ be natural numbers, $1 < k < v$. If there exists an affine (v, k, λ)-design then*:

A. $v = k + k(k-1)/\lambda$.

B. $\lambda | (k-1)$.

C. $v | k^2$.

D. $(\lambda + k - 1)|\lambda k$.

E. *Any two blocks belonging to different classes of the same resolution of the design have exactly $\lambda k/(\lambda + k - 1) = k^2/v$ elements in common.*

Proof. A is a consequence of Theorem 12.10. If $s = \text{g.c.d.}\{k-1, \lambda\}$ then clearly $s \le \lambda$ and

$$k < v = k + k(k-1)/\lambda \le k + k(k-1)/s.$$

By virtue of Theorem 12.5 we have $\lambda = s$; thus, B is true. The rest follows from 12.10.2.

12.12 Remark. It may happen that v, k, λ satisfy 12.11.A—D, and yet an affine (v, k, λ)-design does not exist. This occurs in the following two cases:

A. There is no (v, k, λ)-design at all, e.g. for $v = 36$, $k = 6$, $\lambda = 1$ (cf. 8.23 —8.25).

B. There are (v, k, λ)-designs but none of them admits a resolution (e.g. $v = 6$, $k = 3$, $\lambda = 2$; see 8.3, 12.14.1—2, 12.65.A—B). However, this cannot happen for $v = k^2$ and $\lambda = 1$ (cf. 12.17).

12.13 Theorem [Kageyama 1978]. *If there exists a K_k-factorization of the graph $^\lambda K_{2k}$ (equivalently, a resolvable $(2k, k, \lambda)$-design) then $k \ge 2$ and $\lambda \equiv r$ (mod 2), where $r = \lambda(2k-1)/(k-1)$.*

Proof. Clearly $k \ge 2$. Let G_1, G_2, ..., G_r, H_1, H_2, ..., H_r be complete graphs of order k such that $(G_1 \cup H_1, G_2 \cup H_2, ..., G_r \cup H_r)$ is a K_k-factorization of the graph $^\lambda K_{2k}$. Without loss of generality we may suppose that graphs G_1, G_2, ..., G_r contain a fixed vertex $u \in V(^\lambda K_{2k})$. Consider the graph $G = (G_1 \cup G_2 \cup ... \cup G_r) - u$. Let G' be the factor of $^\lambda K_{2k}$ induced by all edges of $^{\lambda}K_{2k}$ that are not in G. For two vertices u', u'' of G let $e(u', u'', G)$, $e(u', u'', G')$ denote the number of edges in G and G', respectively, joining u' to u''. Obviously,

$$\lambda = e(u', u'', G) + e(u', u'', G').$$

Each edge w joining u' to u'' in G' can be assigned a unique graph H_i containing w, and thus also a graph G_i containing neither u' nor u'', $i \in \{1, 2, ..., r\}$. However, the number of such G_i's is equal to

$$e(u', u'', G') = r - \lambda - \lambda + e(u', u'', G).$$

where we subtracted both the number of graphs containing u' and the number of those containing u'', and by the inclusion — exclusion principle we added the number of graphs containing both u' and u''. Hence,

$$\lambda = 2e(u', u'', G) + r - 2\lambda,$$

and thus $\lambda \equiv r \pmod 2$.

12.14 Corollary. *If there exists a resolvable (v, k, λ)-design then*

$$v = k + k(k - 1)/\lambda \;\Rightarrow\; v | k^2, \tag{1}$$

$$v = 2k \;\Rightarrow\; 2(k - 1) | \lambda k, \tag{2}$$

$$v = k^2, \lambda = 1 \;\Rightarrow\; k \notin Q, \tag{3}$$

where Q is as defined in 8.23—8.25.

Proof. (1) is a consequence of 12.10 and 12.11.C, (2) follows from 12.13, and (3) follows from 8.22—8.23.

Applications of this result can be found in 12.65.

12.15 Interesting examples of affine designs are finite affine planes; we prove this in 12.17. Here we first give a preliminary result resembling the 5th axiom of Euclid. For the sake of convenience, blocks will now be called *lines*. If a line (= block) H contains the point (= element) u, we say that the line H *passes through* u, and u *lies on* H. Two lines are *parallel* if they are equal or else if they have no common point.

The reader who is interested in a deeper insight in combinatorial aspects of the concept of parallelism is referred to Cameron (1976) or Di Paola and Nemeth (1979).

Lemma [Hall 1967]. *Let u be a point and H a line of an $(n^2, n, 1)$-design, $n \geq 2$. Then there is exactly one line parallel to H and passing through u.*

Proof. Our design has parameters $v = n^2$, $k = n$, $\lambda = 1$, $b = n(n + 1)$,

$r = n + 1$ (see 8.7). Therefore, each line contains exactly n points, exactly $n + 1$ lines pass through each fixed point, and each two different points lie on exactly one line. Consider a line $H = \{u_1, u_2, ..., u_n\}$ and a point u. If $u \in H$ then there is nothing to prove. If not, then consider the lines passing through u and u_i ($i = 1, 2, ..., n$). We get n different lines passing through u. Since there are exactly $n + 1$ different lines containing u, there exists exactly one line passing through u and parallel to H.

12.16 In Fig. 12.16.1, a model of an affine plane of order 3 is depicted; its lines are represented by segments joining the points. It can be easily seen that this affine plane has an affine resolution (grouping) consisting of four replications, each containing 3 lines:

(123, 456, 789),
(147, 258, 369),
(159, 267, 348),
(168, 249, 357).

What we are going to show now is that affine resolvability is imanent to each finite affine plane.

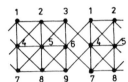

Fig. 12.16.1

12.17 Theorem [Hall 1967]. *Each affine plane of order $n \geq 2$ is an affine $(n^2, n, 1)$-design in which every two lines are either parallel or they intersect in precisely one point.*

Proof. The relation of "being parallel lines", as defined in 12.15, is obviously reflexive and symmetric. Since any affine plane of order n is an $(n^2, n, 1)$-design we can use Lemma 12.15 to see that the relation is also transitive. Thus, the set of all lines of the affine plane splits into classes of mutually parallel lines. As a further consequence of Lemma 12.15 we have that such a decomposition of lines yields a resolution of the $(n^2, n, 1)$-design. The rest follows from Theorem 12.10 and Corollary 12.11 for $v = n^2$, $k = n$, $\lambda = 1$.

12.18 The graphical version of the preceding theorem is also worth mentioning.

Theorem. *If there exists a decomposition of the complete graph K_{n^2}, $n \geq 2$ into complete graphs of order n, then this decomposition is affine. In other words, it is possible to group its classes into K_n-factors of K_{n^2} which together form a K_n-factorization of K_{n^2}; moreover, in each K_n-factorization of K_{n^2}, two subgraphs belonging to different K_n-factors have exactly one common vertex.*

12.19 In our subsequent consideration we shall need some new concepts. By an *n-matrix* we shall understand a square matrix of order n whose elements come from a fixed *n*-element set M; we put $M = \{1, 2, ..., n\}$ unless stated otherwise. A *Latin square* of order n is an *n*-matrix in which no row and no column contain an element more than once (this is possible only if every row and column contain each of the elements 1, 2, ..., n).

For the sake of convenience we shall denote by $\mathbf{A}(i, j)$ the element appearing in the *i*-th row and *j*-th column of a matrix $\mathbf{A}$ (the (i, j)-th *entry* of A).

12.20 Lemma. *For every natural number n there exists a Latin square of order n.*

Proof. It is sufficient to consider the following matric $\mathbf{A}$.

$$\mathbf{A} = \begin{bmatrix} 1 & 2 & 3 & ... & n \\ 2 & 3 & 4 & ... & 1 \\ 3 & 4 & 5 & ... & 2 \\ .... \\ n & 1 & 2 & & n-1 \end{bmatrix}.$$

12.21 Theorem [König 1916]. *The complete bipartite graph $K(n, n)$ can always be decomposed into n 1-factors.*

Proof. Let $\mathbf{A}$ be a Latin square of order n. Let $\{1, 2, ..., n; u_1, u_2, ..., u_n\}$ be the bipartition of $K(n, n)$. The factor F_i ($i = 1, 2, ..., n$) will consist of all edges of the form $u_j A(i, j)$, $1 \leq j \leq n$. It is easy to see that $(F_1, F_2, ..., F_n)$ is a 1-factorization of $K(n, n)$.

12.22 Remarks. A. König (1916) proved even a stronger result: Every finite bipartite *n*-regular graph can be decomposed into n 1-factors.

B. Further information on Latin squares can be found in the monograph by Dénes and Keedwell (1974).

12.23 Two *n*-matrices $\mathbf{A}$, $\mathbf{B}$ will be said to be *orthogonal* if for any $i, j, I, J \in \{1, 2, ..., n\}$, the equality of the ordered pairs $[\mathbf{A}(i, j), (\mathbf{B}(i, j)] = [\mathbf{A}(I, J), \mathbf{B}(I, J)]$ implies $i = I$ and $j = J$.

As there are n^2 ordered pairs of elements of the set $\{1, 2, ..., n\}$, the above condition is equivalent to the following one: Each ordered pair of elements of $\{1, 2, ..., n\}$ can be uniquely expressed in the form $[\mathbf{A}(i, j), \mathbf{B}(i, j)]$.

In particular, we can speak about *orthogonal Latin squares*. For example,

$$\begin{bmatrix} 1 & 2 & 3 \\ 2 & 3 & 1 \\ 3 & 1 & 2 \end{bmatrix} \text{ and } \begin{bmatrix} 1 & 2 & 3 \\ 3 & 1 & 2 \\ 2 & 3 & 1 \end{bmatrix}$$

are orthogonal Latin squares of order 3.

12.24 A collection of n-matrices will be called *orthogonal* if any two of its distinct members are orthogonal. Again, *orthogonal sets of Latin squares* are of special importance. For example the squares

$$\begin{bmatrix} 1 & 2 & 3 & 4 & 5 \\ 2 & 3 & 4 & 5 & 1 \\ 3 & 4 & 5 & 1 & 2 \\ 4 & 5 & 1 & 2 & 3 \\ 5 & 1 & 2 & 3 & 4 \end{bmatrix}, \begin{bmatrix} 1 & 2 & 3 & 4 & 5 \\ 3 & 4 & 5 & 1 & 2 \\ 5 & 1 & 2 & 3 & 4 \\ 2 & 3 & 4 & 5 & 1 \\ 4 & 5 & 1 & 2 & 3 \end{bmatrix}, \begin{bmatrix} 1 & 2 & 3 & 4 & 5 \\ 4 & 5 & 1 & 2 & 3 \\ 2 & 3 & 4 & 5 & 1 \\ 5 & 1 & 2 & 3 & 4 \\ 3 & 4 & 5 & 1 & 2 \end{bmatrix} \quad \text{and}$$

$$\begin{bmatrix} 1 & 2 & 3 & 4 & 5 \\ 5 & 1 & 2 & 3 & 4 \\ 4 & 5 & 1 & 2 & 3 \\ 3 & 4 & 5 & 1 & 2 \\ 2 & 3 & 4 & 5 & 1 \end{bmatrix}$$

constitute an orthogonal family of Latin squares of order 5.

12.25 A multiplication table of a groupoid (or a quasigroup) of order n can be considered as an n-matrix (or a Latin square of order n). The concept of orthogonality is sometimes transferred from n-matrices to the corresponding groupoids or quasigroups. However, this interpretation is beyond the scope of this book; the interested reader is referred to Dénes and Keedwell (1974).

12.26 Let us view an n-element set A as an *alphabet* consisting of n *letters*. The ordered t-tuples of elements of A are then called *codewords* of length t over the alphabet A. An arbitrary set of such codewords is called a *code* over the alphabet A (with words of length t). The *distance between two codewords* is given by the number of occurrences of different letters on corresponding places in the words. By the *minimum distance of a code* we understand the smallest distance of codewords in the code. For example, the 12 lines of 12.16 considered as ordered triples can be viewed as a code consisting of 12 codewords of length 3 over the alphabet $\{1, 2, ..., 9\}$. Any two words there have distance 2 or 3. Thus, the minimum distance of this code is equal to 2.

Codes of large minimum distance are of particular interest since they are more suitable for detecting and correcting transmission errors.

12.27 The following theorem closely relates several assertions dependent on two parameters n and k. The equivalence of A, B, C and F was proved by Dénes and Gergely (1977), and the equivalence of D and E follows from 8.40. The assertion E was investigated by Tomasta and Zelinka (1981).

Theorem. *Let $n > 1$, $k > 1$. The following statements are equivalent*:

A. *There exists an orthogonal family of $k - 2$ Latin squares of order n.*

B. *There exists an orthogonal family of k n-matrices.*

C. *There is a code over an n-element alphabet comprising n^2 codewords of length k, with minimum distance $k - 1$.*

D. *There exists a group-divisible $(k \times n, k, 1)$-design (8.41).*

E. *There is a K_k-decomposition of the graph $K(k \times n)$.*

F. *There is a packing of k K_n-factors into the graph K_{n^2}.*

Proof. A $\Rightarrow$ F. Let $(\mathbf{A}_1, \mathbf{A}_2, ..., \mathbf{A}_{k-2})$ be a family of orthogonal Latin squares of order n. Let us define the n-matrices $\mathbf{A}_{k-1}$ and $\mathbf{A}_k$ as follows:

$$\mathbf{A}_{k-1} = \begin{bmatrix} 1 & 1 & ... & 1 \\ 2 & 2 & ... & 2 \\ \cdots\cdots\cdots\cdots \\ n & n & ... & n \end{bmatrix}, \qquad \mathbf{A}_k = \begin{bmatrix} 1 & 2 & ... & n \\ 1 & 2 & ... & n \\ \cdots\cdots\cdots\cdots \\ 1 & 2 & ... & n \end{bmatrix}.$$

Suppose that the vertex set of K_{n^2} consists of all ordered pairs $[i, j]$; $i, j = 1, 2, ... n$. The factors $F_1, F_2, ..., F_k$ of K_{n^2} will be defined in the following way: the vertices $[i, j]$ and $[I, J]$ $(i, j, I, J = 1, 2, ..., n)$ are adjacent in F_t $(t = 1, 2, ..., k)$ if and only if $\mathbf{A}_k(i, j) = \mathbf{A}_t(I, J)$. It can be checked that each F_t is a factor of type K_n and that $F_1, F_2, ..., F_k$ can be packed into K_{n^2}.

F $\Rightarrow$ B. Assume that $F_1, F_2, ..., F_k$ are edge-disjoint factors of type K_n in the complete graph K_{n^2}. Again let $V(K_{n^2}) = \{[1, 1], [1, 2], ..., [1, n], [2, 1], [2, 2], ..., [n, n]\}$. Assign to each $F_t(1 \leq t \leq k)$ an n-matrix $\mathbf{A}_t$ in such a way that $\mathbf{A}_t(i, j) = \mathbf{A}_t(I, J)$ if and only if vertices $[i, j]$ and $[I, J]$ are adjacent in F_t (the assignment is possible due to the fact that F_t comprises n components, each of them being a complete graph of order n). It is straight forward to verify that $(\mathbf{A}_1, \mathbf{A}_2, ..., \mathbf{A}_k)$ is an orthogonal family of n-matrices.

B $\Rightarrow$ D. Suppose that $(\mathbf{A}_1, \mathbf{A}_2, ..., \mathbf{A}_k)$ is an orthogonal set of n-matrices. We shall construct a group-divisible $(k \times n, k, 1)$-design whose points are all ordered pairs $[i, j]$, $i = 1, 2, ..., n; j = 1, 2, ..., k$, whereby the j-th group of the design consists of pairs having the second coordinate equal to j. The n^2 blocks of the design are defined as follows:

$$\begin{aligned} B_1 &= \{[\mathbf{A}_1(1, 1), 1], \quad [\mathbf{A}_2(1, 1), 2], \quad ..., \quad [\mathbf{A}_k(1, 1), k]\}. \\ B_2 &= \{[\mathbf{A}_1(1, 2), 1], \quad [\mathbf{A}_2(1, 2), 2], \quad ..., \quad [\mathbf{A}_k(1, 2), k]\}. \\ &\cdots\cdots\cdots\cdots\cdots\cdots\cdots\cdots\cdots\cdots\cdots\cdots\cdots\cdots\cdots \\ B_{n^2} &= \{[\mathbf{A}_1(n, n), 1], \quad [\mathbf{A}_2(n, n), 2], \quad ..., \quad [\mathbf{A}_k(n, n), k]\}. \end{aligned} \qquad (1)$$

It is obvious that $(B_1, B_2, ..., B_{n^2})$ is a $(k \times n, k, 1)$-group-divisible design.

D $\Rightarrow$ E. This implication is a consequence of Lemma 8.40 for $m = k$, $n_1 = n_2 = ... = n_m = n$, $K = \{k\}$, $\lambda_1 = 0$, $\lambda_2 = 1$.

E $\Rightarrow$ C. Let $[i, 1], [i, 2], ..., [i, n]$ be vertices constituting the i-th part of the graph $K(k \times n)$, $1 \leq i \leq k$. Then, the vertex set of each graph of a K_k-decomposition of $K(k \times n)$ can be written in the form

$$\{[1, s_1], [2, s_2], ..., [k, s_k]\},$$

where $s_i \in \{1, 2, ..., n\}$ for $1 \leq i \leq k$. Thus, to each of the n^2 subgraphs of the decomposition there corresponds a codeword $s_1 s_2 ... s_k$ over the alphabet $\{1, 2, ..., n\}$. Obviously, the minimum distance of such a code is $k - 1$.

$C \Rightarrow A$. Assume that there exists a code over an n-element alphabet, comprising n^2 words of length k, and with minimum distance $k - 1$. Then, any two different codewords differ at least on $k - 1$ places ($=$ coordinates). Thus, after a suitable rearrangement of the words and replacing the letters by symbols 1, 2, ..., n, the codewords can be written in the form

$$
\begin{aligned}
&[1, \quad 1, \quad B_1(1, 1), \quad B_2(1, 1), \quad ..., \quad B_{k-2}(1, 1)] \, . \\
&[1, \quad 2, \quad B_1(1, 2), \quad B_2(1, 2), \quad ..., \quad B_{k-2}(1, 2)] \, . \\
&\cdots\cdots\cdots\cdots\cdots\cdots\cdots\cdots\cdots\cdots\cdots\cdots\cdots\cdots\cdots\cdots\cdots\cdots\cdots\cdots\cdots\cdots \\
&[n, \quad n, \quad B_1(n, n), \quad B_2(n, n), \quad ..., \quad B_{k-2}(n, n)] \, .
\end{aligned}
\tag{2}
$$

But in this way we have defined the elements of $k - 2$ n-matrices B_1, B_2, ..., B_{k-2}, and it is a matter of routine to verify that $(B_1, B_2, ..., B_{k-2})$ is an orthogonal family of Latin squares of order n.

12.28 R e m a r k. Setting up a table of type $k \times n^2$ whose colums are codewords satisfying the condition C we get the so-called *orthogonal array* OA(n, k) — see e.g. Bose and Shrikhande (1970a), Hall (1967, Chap. 13), Laskar and Hare (1971), or Wilson (1972). Using this terminology, the equivalence of A and C has been well known (cf. Bennet et al. 1980).

12.29 Now we pass to the natural question of characterizing the pairs $[n, k]$ for which the (equivalent) assertions 12.27. A—F hold. For each $n \geq 2$ denote by $M(n)$ the largest $k \geq 2$ such that 12.27 is valid for the pair $[n, k]$.

Lemma. *The symbol $M(n)$ is well defined for each $n \geq 2$. Moreover, $3 \leq M(n) \leq n + 1$ for each $n \geq 2$.*

P r o o f. It follows from Lemma 12.20 that the assertion 12.27.A is always valid for $k = 3$. On the other hand, 12.27.F cannot hold for $k > n + 1$, since a K_k-factor has $n^2(n - 1)/2$ edges whereas K_{n^2} has size $n^2(n^2 - 1)/2$.

12.30 Clearly, the assertions of 12.27 are valid if and only if $k = M(n)$. Thus it is sufficient to investigate the number $M(n)$, or equivalently, the number $N(n) = M(n) - 2$ giving the maxium number of mutually orthogonal Latin squares of order n. For example, it is known that $N(n) = n - 1$ if n is a prime power. More generally, if $n = p^\alpha q^\beta ... r^\gamma$ is a prime factorization then $N(n) \geq \min \{p^\alpha, q^\beta, ..., r^\gamma\} - 1$, as it was shown by Mac Neish (1922). It is also known that $N(6) = 1$ (Tarry 1900, Stinson 1984), and that $N(n) \geq 2$ for $n \neq 6$ (Bose and Shrikhande 1959, 1960a, b, Sade 1980, Crampin and Hilton 1975, Lie 1982). Further estimations on $N(n)$, such as $N(12) \geq 5$, $N(15) \geq 3$, $N(14) \leq 10$, etc., can be found, e.g., in Dénes and Keedwell (1974), Mills (1978a) and Mullin et al. (1980a). The following very important (but apparently very difficult) problem is concerned with determining $N(n)$ for small values of n.

154

Problem. *Determine* $N(n)$ *for* $n =$ 10, 12, 14, 15, 18, *and* 20.

A great effort has been exerted to find the value $N(10)$, but so far without success. The only fact we know is that $2 \le N(10) \le 9$.

12.31 Let us now discuss the case when $N(n) = n - 1$, i.e. $M(n) = n + 1$. In this case, namely, the list 12.27.A—F can be enlarged by five more assertions listed below (for some of the equivalences see e.g. Hall 1967, or Dénes and Gergely 1977).

Theorem. *For each* $n \ge 2$ *the following statements are equivalent*:

A. *There exists an orthogonal set of* $n - 1$ *Latin squares of order* n.

B. *There exists an orthogonal family of* $n + 1$ *n-matrices.*

C. *There is a code over an n-element alphabet, consisting of* n^2 *words of length* $n + 1$, *with minimum distance* n.

D. *There exists a group-divisible* $((n + 1) \times n, n + 1, 1)$-*design.*

E. *There is a* K_{n+1}-*decomposition of the graph* $K((n + 1) \times n)$.

F. *There is a decomposition of the graph* K_{n^2} *into* $n + 1$ K_n-*factors.*

G. *There exists a decomposition of* K_{n^2} *into* $n(n + 1)$ *complete graphs of order* n.

H. *There is a decomposition of the graph* $K_{n^2 + n + 1}$ *into* $n^2 + n + 1$ *complete graphs of order* n.

I. *There exists a resolvable* $(n^2, n, 1)$-*design.*

J. *There exists an affine plane of order* n.

K. *There exists a projective plane of order* n.

Proof. The equivalence of assertions A—F is a special case of 12.27 for $k = n + 1$. The facts that $F \Leftrightarrow I$, $G \Leftrightarrow J$ and $H \Leftrightarrow K$ are obvious (only a different terminology is used). Thus, it is sufficient to prove that $I \Leftrightarrow J \Leftrightarrow K$. Since an affine plane is an $(n^2, n, 1)$-design we have $I \Rightarrow J$; the reverse implication was proved in Theorem 12.17. To prove $I \Rightarrow K$ recall that a resolution of an $(n^2, n, 1)$-design ($=$ affine plane of order n) has $n + 1$ classes, each comprising n mutually parallel lines containing n points. Let us now add $n + 1$ new points, forming thereby a new line. Further, let us enlarge each of the original lines by one of the new points in such a way that any two parallel lines are enlarged by the same point. Clearly we get an $(n^2 + n + 1, n + 1, 1)$-design, i.e., a projective plane of order n. Thus, $I \Rightarrow K$. The reverse implication can be proved simply by reversing the preceding argumentation.

12.32 The question for which n the statement 12.31. K is true was handled in 8.23—8.25.

The validity of 12.31.H for $n \in \{2, 3, 4, 5, 7, 8, 9, 11\}$ is also a consequence of 7.56 for $k = n + 1$.

12.33 The 15 schoolgirls problem (12.1) can be put into "higher dimension", as done by Sylvester (1861, 1892): 15 schoolgirls should go for a walk everyday during a period of 13 weeks. A schedule is to be found which, within each of the 13 weeks, satisfies the conditions of original Kirkman's schoolgirl problem, but in addition it is required that each triple of girls occurs as a "walking triple" on some of the 7 × 13 days. It has been believed for a long time that there is no solution to this problem. Finally, Denniston (1974b) using a computer, found an appropriate arrangement. From the point of view of graph theory, it is a decomposition of the graph $^{13}K_{15}$ into triangles, which is *resolvable in two stages*: At first, all the triangles are to be divided in a suitable way into 13 sets; then, each of these sets is to be shown to be a resolvable decomposition of K_{15}.

Of course, the problem can be generalized to v schoolgirls. Some results in this direction can be found in Denniston (1974a), Cameron (1976, pp. 129—130) and Di Paola and Nemeth (1979); the case $v = 9$ was investigated in Kirkman (1850b), Bays (1917), Kramer (1974) and Mathon and Rosa (1983).

It is also possible to restate the problem in terms of 1-factorizations of 3-hypergraphs of order v (cf. Baranyai 1975, Di Paola and Nemeth 1979, Mathon and Rosa 1983).

Denniston's solution of the Sylvester problem ($v = 15$) proposes the following first week schedule (schoolgirls are denoted by symbols 0, 1, 2, 3, 4, 5, 6, 7, 8, 9, 10, 11, 12, u, U):

```
((0   1   9,   2 4 12,   5 10 11,   7  8   u,    3   6 U),
 (0   2   7,   3 4  8,   5  6 12,   9 11   u,    1  10 U),
 (0   3  11,   1 7 12,   6  8 10,   2  5   u,    4   9 U),
 (0   4   6,   1 8 11,   2  9 10,   3 12   u,    5   7 U),
 (0   5   8,   1 2  3,   6  7  9,   4 10   u,   11  12 U),
 (0  10  12,   3 5  9,   4  7 11,   1  6   u,    2   8 U),
 (1   4   5,   2 6 11,   3  7 10,   8  9  12,    0   u U)).
```

The schedules for all the remaining 12 weeks are obtained from the above by "shifting" cyclically all the numbers modulo 13, and leaving the occurrences of u and U unchanged.

12.34 The resolvability introduced in 12.33 should not be confused with the so-called *double resolvability* of Kirkman designs which was studied e.g. in Denniston (1974a), Smith (1977), Fuji-Hara and Vanstone (1980, 1982), Vanstone (1980). The latter consists of considering of a pair of orthogonal resolutions. Two resolutions are called *orthogonal* if each class of the first resolution has at most one block (= subgraph) in common with each class of the second resolution. Clearly, this concept can be introduced also in more general cases, e.g. for K_k-decompositions of the graph $^\lambda K_r$ (Rosa 1978, 1980b, Horton 1981).

12.35 Another modification of Kirkman's schoolgirl problem is the so-called *handcuffed prisoners problem*: A group of v enormously dangerous prisoners should go everyday for a walk in $v/3$ triples, each pair of neighbours

156

in a triple being handcuffed. A schedule of walks is to be found for a period
of $3(v-1)/4$ days such that each two prisoners are handcuffed to each other
on exactly one day. For which v is it possible?

The original version of the problem is due to Dudeney (1917, 1967), and
was formulated for 9 *handcuffed prisoners*. A possible solution is the following one:

　1st day: (413, 276, 598).
　2nd day: (746, 519, 832).
　3rd day: (179, 843, 265).
　4th day: (124, 739, 586).
　5th day: (457, 163, 829).
　6th day: (781, 496, 253).

It is clear from the statement of the problem of handcuffed prisoners that
both $v/3$ and $3(v-1)/4$ are integers; thus we have a necessary condition
$v \equiv 9 \pmod{12}$. Hell and Rosa (1972) obtained a solution for an infinite
family of numbers v; finally, Wilson proved that the condition $v \equiv 9$
(mod 12) is also sufficient. A generalization of Wilson's result will be discussed in 12.39.

12.36 The problem of handcuffed prisoners can easily be restated in
graph-theory language: It is the problem of the existence of a P_3-factorization
of the complete graph K_v, i.e. a decomposition into factors consisting of
disjoint paths of order 3 (= length 2). For example, the schedule of 12.35
yields a P_3-factorization of K_9 (a decomposition into six P_3-factors). Since the
isomorphism of decomposition (and thus also of factorizations) has already
been defined in 3.9, it is meaningful to ask about the number of non-isomorphic solutions to the handcuffed prisoners problem for any fixed v. For $v = 9$,
this question was studied by Rosa and Huang (1971) and Ollerenshaw and
Bondi (1978). With respect to results on the number of Kirkman designs
(recall 12.8, especially the values $K(3) = K(9) = 1$, $K(15) = 7$) it is rather
surprising that the problem of 9 handcuffed prisoners has an enormous
number of non-isomorphic solutions, namely, 332.

12.37 A natural generalization of the problem of handcuffed prisoners in
graph-theoretical setting is the problem of the existence of a P_k-factorization
of the graph $^\lambda K_v$. A necessary condition can be derived quite easily:

Theorem [Hell and Rosa 1972]. *Let v, k, λ be positive integers, $v \geq 2, k \geq 2$.
If there is a P_k-factorization of the graph $^\lambda K_v$, then*

$$v \equiv k^2 \left(\mathrm{mod}\ \frac{k(2k-2)}{\mathrm{g.c.d.}\{\lambda k, 2k-2\}} \right). \tag{1}$$

Proof. Suppose there exists the required factorization. Obviously we
have

$$v \equiv 0 \,(\mathrm{mod}\ k). \tag{2}$$

Comparing the size of $^\lambda K_v$ and P_k yields

$$v(k-1)/k|\lambda\binom{v}{2}, \tag{3}$$

which implies

$$\lambda vk \equiv \lambda k \,(\mathrm{mod}\ 2k-2). \tag{4}$$

The latter is easily seen to be equivalent to

$$v \equiv 1 \left(\mathrm{mod}\ \frac{2k-2}{\mathrm{g.c.d.}\,\{\lambda k,\ 2k-2\}}\right). \tag{5}$$

It is a matter of routine to check that the congruences (2) and (5) have mutually prime moduli and a common solution $v = k^2$. Thus, (2) and (5) is equivalent to (1), q.e.d.

12.38 As regards the sufficiency of 12.37.1 we have the following:

Conjecture [Hell and Rosa 1972, Horton 1983b]. *For $v \ge 2, k \ge 2$, 12.37.1 is sufficient for the existence of a P_k-factorization of the graph $^\lambda K_v$.*

12.39 Theorem [Hell and Rosa 1972, Horton 1983]. *The preceding conjecture is true for*:

A. $k = 2$.
B. $k = 3$.
C. $k = 4,\ \lambda = 1$.
D. $v > C(k),\ \lambda = 1$, *where $C(k)$ depends on k.*

Proof. In the case A, 12.37.1 is equivalent to $v \equiv 0$ (mod 2) and we ask for a 1-factorization. Thus, A follows from Theorem 7.42 (see 12.43).

B. This was proved by Horton (1983); for $\lambda = 1$ also by Wilson, see (12.35).

C. See Horton (1983).

D. For $\lambda = 1$, 12.37.1 reduces to $v \equiv k^2$ (mod l.c.m. $\{2k - 2,\ k\}$). The existence of $C(k)$ was proved by Horton (1983). Note, that according to A —C, for $k \le 4$ it is possible to put $C(k) = 1$.

12.40 In 9.2 we introduced (v, G, λ)-designs as a generalization of (v, k, λ)-designs defined in 8.2. A (v, G, λ)-design is said to be *resolvable* if its G-blocks can be grouped into classes ($=$ *replications*) in such a way that each class represents a decomposition of the underlying set of v points. Thus, in 12.35 to 12.39 we dealt with resolvable (v, P_k, λ)-designs ($= P_k$-factorizations of $^\lambda K_v$). Similarly as pointed out in 12.3, there is a natural correspondence between resolved (v, G, λ)-designs and G-factorizations of $^\lambda K_v$. These factorizations were studied for several particular graphs G, e.g. Hell and Rosa

(1971, 1972) for $G = P_k$, Huang (1976) for G bipartite, Hell et al. (1975) for $G = C_k$).

12.41 The following simple result can often be useful.

Lemma. *If G is a simple graph, then each resolvable G-decomposition of $^\lambda K_v$ is balanced.*

P r o o f. If we have a grouping into r classes, each vertex of the graph $^\lambda K_v$ is contained in exactly r subgraphs of the G-decomposition.

12.42 We can also ask for resolvable decompositions of directed graphs, as done e.g. by Bennet et al. (1980) for $G = C_k^\circ$ or by Colbourn and Harms (1983) for $G = K_3^\rightarrow$. As a further example let us state a result on decomposition of K_v^*, the complete digraph of order v (recall that $K_3^\rightarrow$ is the transitive tournament of order 3, C_3° is the directed cycle of order 3).

Theorem [Bermond et al. 1979b]. *The following statements are equivalent:*

A. *There exists a $K_3^\rightarrow$-factorization of the graph K_v^*.*

B. *There is a C_3°-factorization of K_v^*.*

C. *$v \equiv 0 \pmod 3$, $v \neq 6$.*

P r o o f. Implications A $\Rightarrow$ C and B $\Rightarrow$ C are a consequence of resolvability and 12.14. The proof of the reverse implications is lengthy and can be found in Bermond et al. (1979b); note that for $v \equiv 3 \pmod 6$ it is possible to get a shorther proof using Theorem 12.6.

12.43 Let us turn back to K_k-factorizations of the graph $^\lambda K_v$ (= resolvable (v, k, λ)-designs). For $2 \leq k \leq 4$, the problem of existence of such factorizations is completely solved. The simplest case is, of course, $k = 2$:

Theorem. *A K_2-factorization of the graph $^\lambda K_v$, $v \geq 2$, exists if and only if v is an even number.*

P r o o f: This is a direct consequence of Theorem 7.42.

12.44 The problem of K_3-factorization of $^\lambda K_v$ was solved for $\lambda = 1$ by Ray-Chaudhuri and Wilson (1971) (see 12.6), and for $\lambda = 2$ by Hanani (1974). Here we state the general result:

Theorem. *A K_3-factorization of $^\lambda K_v$, $v \geq 2$ exists if and only if one of the following cases occurs:*

A. *$v \equiv 3 \pmod 6$.*

B. *$v \equiv 0 \pmod 6$, $v \neq 6$, $\lambda \equiv 0 \pmod 2$.*

C. *$v = 6$, $\lambda \equiv 0 \pmod 4$.*

P r o o f. The necessity of one of the A—C is a consequence of Lemma 12.4 and Corollary 12.14.2. To prove the sufficiency, we shall make use of 6.21, putting there $H = {}^\lambda K_v$, $G =$ graph of type K_3, and $F = K_v$ in case A, $F = {}^2 K_v$ in case B, and finally $F = {}^4 K_v$ for C. In any case we have $F|H$. According to 6.21 it remains to verify that $G|F$, i.e.

$$G|K_v \qquad \text{for } v \equiv 3 \pmod 6, \qquad (1)$$

$$G|{}^2 K_v \qquad \text{for } v \equiv 0 \pmod 3, \ v \neq 6, \qquad (2)$$

$G|^4K_6.$ (3)

Note that (1) follows from Theorem 12.6. To prove (2) let us direct the edges of the graph ^2K_v in order to obtain the complete digraph K_v^*. By 12.42.A there is a $K_3^{\rightarrow}$-factorization of K_v^*. "Forgetting" about the orientation, we clearly get a K_3-factorization of ^2K_v. Finally, we prove (3) by exhibiting a K_3-factorization of the graph $^4K_6$ on the vertex set $\{1, 2, ..., 6\}$ (cf. Vanstone and Schellenberger 1977): ({123, 456}, {124, 356}, {125, 346}, {126, 345}, {134, 256}, {135, 246}, {136, 245}, {145, 236}, {146, 235}, {156, 234}).

12.45 Similarly as in the preceding case, the key role in studying K_4-decompositions of $^\lambda K_v$ play the following two special cases: $\lambda = 1$, (Hell and Rosa 1972) and $\lambda = 3$ (Baker 1975, 1983, Baker and Wilson 1977), for some partial results see Pukanow and Wilczyńska (1981), Wilczyńska (1982), or Rokowska (in press).

Theorem. *A K_4-factorization of the graph $^\lambda K_v$ exists if either $v \equiv 0$ (mod 4) and $\lambda \equiv 0$ (mod 3), or else $v \equiv 4$ (mod 12).*

Proof. The necessity is an easy consequence of Theorem 12.5.2. To prove the sufficiency we again use 6.21 putting $H = {}^\lambda K_v$, G = graph of type K_4, $F = K_v$ or $F = {}^3K_v$. What we need to verify is that $G|F$, i.e.:

$G|K_v$ for $v \equiv 4$ (mod 12), (1)

$G|^3K_v$ for $v \equiv 0$ (mod 4). (2)

(1) was proved by Hell and Rosa (1972), (2) by Baker (1983).

12.46 Only a few facts on the existence of K_k-factorization of $^\lambda K_v$ are known for $k \geq 5$. For example, we can deduce from Theorem 12.5.3 that a necessary condition for the existence of a K_5-factorization of the graph K_v is $v \equiv 5$ (mod 20). According to Hanani (1979), such factorizations can easily be constructed for $v = 5^m$. Bose (1963) and Kageyama (1972) gave a construction for $v = 65$; for $v = 85$ see Colbourn (1980) and Mathon and Rosa (1985). To complete the list for small values of v it would be desirable to clarify the situation at least for two more values: $v = 45$ and 105.

12.47 A lot of published constructions of resolvable (v, k, λ)-designs seem to support the following (maybe too optimistic) conjecture:

Conjecture. *Let $v \geq 2$, $k \geq 2$, $s =$ g.c.d. $\{k - 1, \lambda\}$. If*

$v \equiv k \pmod{k(k - 1)/s}$ (1)

and

$v \neq k + k(k - 1)/s$ (2)

then there exists a K_k-factorization of $^\lambda K_v$ (= resolvable (v, k, λ)-design).

Due to Theorems 12.43—12.45 the conjecture is true for $k = 2, 3, 4$. The most significant step towards settling the conjecture was done by Ray-Chaudhuri and Wilson (1973): In the case $\lambda = 1$ (and hence $s = 1$), for each $k \geq 2$ there exists a $C(k)$ such that for each $v > C(k)$ satisfying (1) there exists a resolvable $(v, k, 1)$-design.

12.48 Comparing the preceding conjecture with Theorem 12.5 we see that the only possible exception is admitted in the case $v = k + k(k - 1)/s$.

Problem. *If $v \geq 2$, $k \geq 2$, $s = $ g.c.d. $\{k - 1, \lambda\}$, find a necessary and sufficient condition for the existence of a K_k-factorization of the graph $^\lambda K_v$ for* $v = k + k(k - 1)/s$.

Note that this problem is a very difficult one since it includes, as a special case, the problem of the existence of finite affine planes of a given order (cf. 8.23—8.25).

We add that, according to Bose (1963), a resolvable $(v, k, 1)$-design can also be constructed in the case when $k - 1$ is a prime power and $v - 1 = = (k - 1)^3$.

12.49 For further results on the existence of resolvable (v, k, λ)-designs the reader is referred to Bose (1942), Raghavarao (1971), Kageyama (1972, 1973, 1978), Vanstone (1979), Hedayat and Kageyama (1980), Rokowska (1980), Rumov (1982), Wilczyńska (1982), Baker (1983), Kageyama and Hedayat (1982) and Mathon and Rosa (1985). A survey of known methods of construction of resolvable $t - (v, k, \lambda)$-designs was given by Mavron (1981).

12.50 The concept of resolvable G-decompositions of a graph H, defined in 12.3, can be modified in several ways. For example, it is possible to replace G-factors by *almost G-factors*, i.e., factors in which all but one component are isomorphic to G, the exceptional component being an isolated vertex. Thus, in particular, we can speak about an *almost G-factorization* of $^\lambda K_v$ (or, equivalently, about an *almost resolvable (v, G, λ)-design*). In the case $G = K_k$ we get an *almost resolvable (v, k, λ)-design*; here the blocks of each *replication* cover all but one point of the underlying set. The interested reader is referred to Bennet and Sotteau (1981).

12.51 Theorem. *Let there exist an almost K_k-factorization of the graph $^\lambda K_v$, $v \geq 2$. Then:*

$$k \geq 2, \tag{1}$$

$$k \mid v - 1, \tag{2}$$

$$k - 1 \mid \lambda. \tag{3}$$

Proof. The conditions (1) and (2) are obvious. To prove (3) we first note that, according to Theorem 8.9.1,

$$k - 1 \mid \lambda(v - 1). \tag{4}$$

Clearly, the total number of graphs in a K_k-decomposition of $^\lambda K_v$ is equal to $\lambda v(v-1)/(k(k-1))$. Each replication contains exactly $(v-1)/k$ components, and therefore the number of replications is equal to

$$\frac{\lambda v(v-1)/(k(k-1))}{(v-1)/k} = \frac{\lambda v}{k-1}$$

which implies

$$k-1 \mid \lambda v. \tag{5}$$

Now, (3) is a direct consequence of (4) and (5).

12.52 Corollary. *An almost K_2-factorization of the graph $^\lambda K_v$ exists if and only if $v \equiv 1 \pmod 2$.*

Proof. Necessity follows from 12.51.2. The sufficiency for $\lambda = 1$ (and hence for all λ) is a consequence of Lemma 7.41.

12.53 Corollary. *An almost K_k-factorization of the graph K_v exists if and only if $k = 2$ and $v \equiv 1 \pmod 2$.*

Proof. The necessity is clear from 12.51.3 and 12.51.2; the sufficiency follows from 12.52.

12.54 Corollary. *Let λ be a prime and $v \geq 2$. If there exists an almost K_k-factorization of the graph $^\lambda K_v$ then one of the following cases occurs:*
A. $k = 2$, $v \equiv 1 \pmod 2$.
B. $k - \lambda + 1$, $v \equiv 1 \pmod{\lambda + 1}$.

Proof. If λ is a prime number, we get from Theorem 12.51.3 that either $k = 2$ or $k = \lambda + 1$. The rest follows from 12.51.2.

12.55 The case $\lambda = 3$ was investigated by Pukanow and Wilczyńska (1981), who described a construction of many almost resolvable $(v, 4, 3)$-designs for $v \equiv 1 \pmod 4$. For $\lambda = 2$, however, we have a complete characterization:

Theorem. *An almost K_k-factorization of the graph $^\lambda K_v$ exists if and only if one of the following cases occurs:*
A. $k = 2$, $v \equiv 1 \pmod 2$.
B. $k = 3$, $v \equiv 1 \pmod 3$.

Proof. The necessity can be obtained by inserting $\lambda = 2$ in 12.54. The sufficiency of A follows from 12.52, and that of B was proved by Hanani (1974).

12.56 Similar questions have been studied also for directed graphs, see e.g. Bennet et al. (1980). To illustrate this we present here a result of Bennet and Sotteau (1981).

Theorem. *The following three assertions are equivalent:*
A. *There exists an almost $K_3^\rightarrow$-factorization of the graph K_r^*.*

B. *There exists an almost C_3^o-factorization of K_r^*.*

C. $v \equiv 1 \pmod 3$.

Proof. For the implications $A \Rightarrow C$ and $B \Rightarrow C$ see Theorem 12.55.B (cancelling the orientations in K_r^* yields the graph ^2K_v). The reverse implications are proved in Bennet and Sotteau (1981).

12.57 Another modification of the concept of resolvable (or resolved) G-decomposition of a graph H can be obtained by considering packing instead of factorizing. For example, if $k \geq 2$ and $(v-1)/(k-1)$ is not an integer then no K_k-factorization of the graph $^\lambda K_v$ exists (12.4.1). However, assuming that $k|v$ it is reasonable to ask whether there is a packing of the graph $^\lambda K_v$ by the maximum possible number (namely, $\lfloor \lambda(v-1)/(k-1) \rfloor$) of K_k-factors. The case $\lambda = 1$ has attracted the greatest interest so far, and leads directly to the concept of a so-called *nearly Kirkman k-tuple system* of order v (cf. 12.2; see also Petrenyuk 1970, Kotzig and Rosa 1974, Smith 1977, Huang et al. 1979, Vanstone 1980). In particular, for $k = 3$ we get a nearly Kirkman triple system of order v, which corresponds to a K_3-factorization of the *cocktail-party graph* of (an even) order v ($= K_v$ minus a 1-factor). It is known (Huang et al. 1979, 1982) that such a factorization exists if and only if $v \equiv 0 \pmod 6$ and $v \geq 18$. This result can be derived from the results of Kotzig and Rosa (1974), Baker and Wilson (1977), Brouwer (1978) and a construction by R. K. Guy for $v = 84$ (Huang et al. 1982, Exercise 12.62).

Exercises

12.58 For which values of n does there exist a projective plane of order n that is, at the same time, a resolvable design?

12.59 Starting from Example 12.16 (or 12.23) and using Theorems 12.27 and 12.31, give a construction of: A. a projective plane of order 3; B. an orthogonal family of four 3-matrices; C. a code over the alphabet $\{1, 2, 3\}$ with minimum distance 3, consisting of 9 codewords of length 4; D. a group divisible $(4 \times 3, 4, 1)$-design; E. a K_4-decomposition of the graph $K(4 \times 3)$; F. a decomposition of K_9 into four K_3-factors; G. a K_3-decomposition of the graph K_9; H. a K_4-decomposition of the graphs K_{13}.

12.60 What is the largest number of edge-disjoint factors of type K_6 in the graph K_{36}?

12.61 Find a resolvable K_3^--decomposition of the graph K_{12}^* (Bermond et al. 1979b).

12.62 Find a K_3-factorization of the coctail-party graph of order 84 (Huang et al. 1982).

12.63 Using the decomposition of K_{21} into K_3-factors, given in 12.1 ($n = 3$), find a solution to the handcuffed prisoners problem (12.35—12.39) for $v = 21$ (i.e., a P_3-factorization of the graph K_{21}).

12.64 Find a decomposition of the graph K_{16} into Q_3-factors (Q_3 is the graph of the 3-dimensional cube; cf. Kotzig 1979b).

12.65 Find all triples $[v, k, \lambda]$, $k \leq 10$ fulfilling the conditions 12.5.1—2 but not: A. 12.14.1; B. 12.14.2; C. 12.14.3.

12.66 How is it possible to improve Fisher's inequality $b \geq v$ (8.5.F) for resolvable (v, k, λ)-designs such that $v > k$ (equivalently, $r > \lambda$)?

12.67 Prove the equivalence of the conditions 12.4.1—2 with those of 12.5.1—2.

12.68 If there exists a K_k-factorization of the graph ^2K_k ($k \geq 2$) then $v \neq (k^2 + k)/2$.

Chapter 13

Decompositions according to a group, and by using permutations

13.1 Permutations and groups are often used in investigation of graph decompositions, especially of those which satisfy certain extra constraints. In this chapter we present some typical examples of such an approach, accompanied by some results of general nature.

13.2 The automorphisms of graphs, defined in 1.26, will play a key role in subsequent considerations. Above all, we shall be dealing with relations between automorphisms and subgraphs (in particular, factors) of a graph. If α is an automorphism and F is a subgraph of a graph G, the symbol $\alpha(F)$ will stand for the subgraph of G which is determined by images of all vertices and edges of F under the automorphism α.

13.3 As mentioned in 1.30, the set of all automorphisms of a graph G equipped with the operation of composition of mappings constitutes the *automorphism group $A(G)$ of the graph G*. By a *group of automorphisms of G* we shall mean an arbitrary subgroup of $A(G)$.

In what follows we shall investigate the possibility of representation of graph automorphisms (or sometimes even of elements of abstract groups) by factors of graphs. In other words, we shall be interested in mappings which assign factors to automorphisms. Our point of departure will be the following result (here, a *factorization* is to be understood as a set of factors whose edge sets form a decomposition of the edge set of a given graph):

Lemma. *Let R be a factorization of a graph G, and J a group of automorphisms of G. The following three conditions are equivalent:*

A. *There is a mapping g of J onto R such that $\alpha(g(\beta)) = g(\alpha\beta)$ for all $\alpha, \beta \in J$.*

B. *There exists a factor $F \in R$ for which $\{\alpha(F); \alpha \in J\} = R$.*

C. *For each $F \in R$ it holds $\{\alpha(F); \alpha \in J\} = R$.*

Proof. The implication $C \Rightarrow B$ is obvious. To prove $B \Rightarrow A$ put $g(\alpha) = \alpha(F)$ for each $\alpha \in J$; g is clearly a mapping of J onto R. The rest follows from a straightforward computation:

$$\alpha(g(\beta)) = \alpha(\beta(F)) = (\alpha\beta)(F) = g(\alpha\beta).$$

It remains to prove that $A \Rightarrow C$. Obviously, for each $F \in R$ there exists a $\gamma \in J$

166

such that $F = g(\gamma)$. Put $F_0 = g(\varepsilon)$, where ε is the unit element of the group J. Then,

$$\gamma(F_0) = \gamma(g(\varepsilon)) = g(\gamma\varepsilon) = g(\gamma) = F.$$

Again, by a simple computation we get

$$\alpha(F) = \alpha(\gamma(F_0)) = (\alpha\gamma)(F_0) = (\alpha\gamma)(g(\varepsilon)) = g((\alpha\gamma)\varepsilon) = g(\alpha\gamma).$$

Thus,

$$\{\alpha(F); \ \alpha \in J\} = \{g(\alpha\gamma); \ \alpha \in J\} = R,$$

because the set $\{\alpha\gamma; \ \alpha \in J\}$ is equal to J.

13.4 Let R be a factorization of a graph G and H be an arbitrary group. We shall say that R is a *decomposition of the graph G according to the group H* (Zelinka 1967, 1970b) if there is a group J of automorphisms of G, $J \cong H$ which satisfies one of the equivalent conditions 13.3.A—C. As it is evident from 13.3.C, the group J can be formed only by automorphisms preserving the decomposition R (i.e., $F \in R$ implies $\alpha(F) \in R$; cf. 13.2).

In general, if we are given a decomposition R of a graph G into subgraphs, we can define the *automorphism group of the decomposition R* (cf. Robinson 1979) as the collection $A_1(R)$ of all R-preserving automorphisms of G. Thus, we have a chain of subgroups: $J \leq A_1(R) \leq A(G)$. The automorphism groups of 1-factorizations of complete graphs were investigated in Dickson and Safford (1906), Wallis et al. (1972), Gelling and Odeh (1974), Cameron (1976), Lindner et al. (1976), Anderson (1977), Anderson et al. (1977) and Robinson (1979).

Schwenk (1984) has studied decompositions of graphs according to a cyclic group (= *colour cyclic factorization of graphs*, in his terminology). The cyclic decomposition dealt with in 7.1—7.6 is a special case of a decomposition according to a cyclic group in the case when the base system consists of a single graph (see also Exercise 13.24).

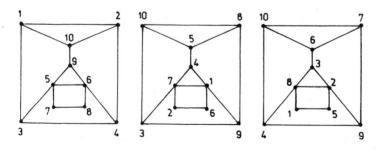

Fig. 13.5.1

13.5 Example [Robinson 1979]. The decomposition $R = \{F_1, F_2, F_3\}$ of K_{10}, depicted in Fig. 13.5.1, cannot be a decomposition according to a group. The reason is that $A_1(R)$ contains only one non-identity permutation $\alpha = (12)(34)(56)(78)(9)(10)$, as it can be easily checked.

13.6 Example. Let G be the complete graph of order 4 with vertices 1, 2, 3, 4. Let R be its factorization into two paths $F_1 = 1243$ and $F_2 = 2314$. Finally, let H be the cyclic group of order 4. Then R is a decomposition of G according to H. To see this, it is sufficient to take $J = A_1(R)$ comprising the identity element ε, the automorphism $\alpha = (1234)$, and its powers α^2 and α^3. Since $\varepsilon(F_1) = \alpha^2(F_1) = F_1$ and $\alpha(F_1) = \alpha^3(F_1) = F_2$, the condition 13.3.B is fulfilled. Note that the mapping g in 13.3.A can be defined by putting $g(\varepsilon) = g(\alpha^2) = F_1$, $g(\alpha) = g(\alpha^3) = F_2$.

13.7 Now we generalize the ideas lying behind our preceding example.

Theorem. *Let b, v be natural numbers, $v \geq 2$. If $v \equiv 0$ or $1 \pmod{2b}$, then there exists a decomposition of the complete graph K_v according to the cyclic group of order $2b$ into b isomorphic factors.*

Proof. First let $v \equiv 0 \pmod{2b}$. Let us denote the elements of the cyclic group H of order $2b$ by a (= the generator), a^2, ..., a^{2b} (= the unit element). Now, we divide vertices of K_v into $s = v/(2b)$ classes $A_1, A_2, ..., A_s$, each containing exactly $2b$ vertices. Let us denote the elements of the i-th class $(i = 1, 2, ..., s)$ by symbols $u_i(a)$, $u_i(a^2)$, ..., $u_i(a^{2b})$. Thus, vertices

$$u_1(a), \ u_1(a^2), \ ..., \ u_1(a^{2b}),$$
$$u_2(a), \ u_2(a^2), \ ..., \ u_2(a^{2b}),$$
$$\cdots\cdots\cdots\cdots\cdots\cdots\cdots\cdots\cdots$$
$$u_s(a), \ u_s(a^2), \ ..., \ u_s(a^{2b})$$

are mutually different and constitute the vertex set $V(K_v)$ of the graph K_v. Define mappings $\alpha_1, \alpha_2, ..., \alpha_{2b}: V(K_v) \to V(K_v)$ in the following way:

$$\alpha_k(u_i(a^j)) = u_i(a^{k+j})$$

for

$$i = 1, 2, ..., s,$$
$$j, k = 1, 2, ..., 2b.$$

It is easily seen that $\alpha_1, \alpha_2, ..., \alpha_{2b}$ are permutations of $V(K_v)$ which, at the same time, form a group J of automorphisms of K_v; obviously $J \cong H$. The mappings α_k can be extended in a straightforward manner to edges and factors. Thus, it is possible to decompose the edge set $E(K_v)$ into classes in such a way that two edges e, f belong to the same class if one of the permutations α_k carries e to f or vice versa. Therefore, if $w \in E(K_v)$, the class containing w consists of edges $\alpha_1(w)$, $\alpha_2(w)$, ..., $\alpha_{2b}(w) = w$. Now let us show that each such class contains either $2b$ or b different edges. To see this, let us realize that, for each vertex $U \in V(K_v)$, the vertices $\alpha_1(U)$, $\alpha_2(U)$, ..., $\alpha_{2b}(U) = U$ are pairwise distinct. Thus, the equality $\alpha_i(w) = \alpha_j(w)$ is possible if and only if $w = u_1 u_2$, where $\alpha_b(u_1) = u_2$ and $\alpha_b(u_2) = u_1$. But then $\alpha_b(w) = w$, and henceforth $\alpha_{b+1}(w) = \alpha_1(w)$, $\alpha_{b+2}(w) = \alpha_2(w)$, ..., $\alpha_{2b-1}(w) = \alpha_{b-1}(w)$.

To complete the proof, it suffices to choose one edge from each class consisting of b edges, and two edges w_1, w_2 such that $\alpha_b(w_1) = w_2$ from each class consisting of $2b$ edges. The chosen edges determine uniquely a factor F of K_v. It is easy to see that the system $R = \{\alpha_1(F); \alpha_2(F), ..., \alpha_b(F) = F\}$ consitutes a decomposition of K_v into b factors. Moreover, the condition 13.3.B is clearly fulfilled; hence R is a decomposition of K_v according to H. Thus, all factors in R are isomorphic.

For $v \equiv 1 \pmod{2b}$ we can proceed analogously, taking into account that vertices of K_v will be partitioned into $(v-1)/(2b)$ sets, each containing b vertices, plus a set consisting of a single vertex u for which we put $\alpha_k(u) = u$ for each $k = 1, 2, ..., b$.

13.8 Example. Set $b = 2$ and $v = 5$. Let us decompose the graph K_5 into two isomorphic factors according to a cyclic group of order 4. For vertices of K_5 we use the notation $u_1(a) = 1$, $u_1(a^2) = 2$, $u_1(a^3) = 3$, $u_1(a^4) = 4$, $u = 0$ (cf. 13.7). Now, $\alpha_1 = (0)(1234)$, $\alpha_2 = \alpha_1^2$, $\alpha_3 = \alpha_1^3$, $\alpha_4 = \alpha_1^4 = \varepsilon$, thus we have the cyclic group of order 4. We get the following classes of the decomposition of $E(K_5)$:

$$T_1 = \{12, 23, 34, 41\},$$

$$T_2 = \{13, 24\},$$

$$T_3 = \{01, 02, 03, 04\}.$$

Let us make the following choice of pairs of edges from T_1, T_3: 12, $34 \in T_1$, 01, $03 \in T_3$. Now, depending on the choice of the edge of T_2, we get two non-isomorphic factors F, F' of K_5 (the fourth and fifth graph in Fig. 14.1.1), and thereby also two non-isomorphic decompositions $\{F, \alpha_1(F)\}$ and $\{F', \alpha_1(F')\}$ of K_5 into two isomorphic factors.

13.9 A simple graph will be called *self-complementary* if it is isomorphic to its complement (1.14).

Corollary [Sachs 1962, Ringel 1963, Harary et al. 1978b]. *For each $v \equiv 0$ or $1 \pmod 4$ there exists a self-complementary graph of order v.*

Proof. See Theorem 13.7 for $b = 2$.

13.10 To introduce a new concept of a simple decomposition of a graph according to a given group, we shall need a result analogous to that of Lemma 13.3.

Lemma. *Let R be a factorization of a graph G, and J be a group of automorphisms of G. The following three assertions are equivalent*:

A. *There exists a bijection* $g: J \to R$ *such that* $\alpha(g(\beta)) = g(\alpha\beta)$ *for all* α, $\beta \in J$.

B. *There exists a factor* $F \in R$ *such that* $\{\alpha(F); \alpha \in J\} = R$, *and for each* α, $\beta \in J$ *the equality* $\alpha(F) = \beta(F)$ *implies* $\alpha = \beta$.

C. *For each* $F \in R$ *it holds* $\{\alpha(F); \alpha \in J\} = R$, *and for each* $F \in R$, α, $\beta \in J$, $\alpha(F) = \beta(F)$ *implies* $\alpha = \beta$.

Proof. Again, obviously C ⇒ B. To prove B ⇒ A it suffices to realize that the mapping g defined by $g(\alpha) = \alpha(F)$ (see 13.3) is a bijection. Finally, we prove A ⇒ C. Let $\alpha(F) = \beta(F)$. By virtue of Lemma 13.3 it is sufficient to prove that $\alpha = \beta$. But, similarly as in 13.3 we have $\alpha(F) = g(\alpha\gamma)$ and $\beta(F) = g(\beta\gamma)$ for a suitable $\gamma \in J$. As g is assumed to be a bijection we conclude that $\alpha = \beta$.

13.11 In our definition 13.4 of decompositions of graphs according to a group we have admitted that one particular factor could be assigned to several automorphisms (by means of the mapping g in 13.3.A). Forbidding this possibility leads to the following concept.

Let R be a decomposition of a graph G according to a group H. If there is a group J of automorphisms of G, $J \cong H$ fulfilling one of the equivalent conditions 13.10.A—C then R is said to be a *simple decomposition of G according to H* (see Zelinka 1967, 1970d). Following the terminology of Porubský (1982), R is a *factorial regular representation of H in G*. In this case we say that G is an R_H-graph (Zelinka 1970d). If H is the cyclic group of order 2 we get the so-called R_2-graphs which we shall investigate in 14.24.

13.12 Example [Robinson 1979]. The factorization of K_6, $V(K_6) = \{1, 2, 3, 4, 5, 6\}$, into three paths $F_1 = 123456$, $F_2 = 246135$ and $F_3 = 415263$ is a simple decomposition of K_6 according to the cyclic group of order 3. To see this it is sufficient to realize that the automorphism group $A_1(R)$ comprises exactly 3 automorphisms: $\alpha = (124)(365)$, α^2, and $\varepsilon = \alpha^3$; then we have $\varepsilon(F_1) = F_1$, $\alpha(F_1) = F_2$, $\alpha^2(F_1) = F_3$ (cf. 13.10.B).

13.13 Example [Robinson 1979]. Employing the notation of the preceding example, the factorization of K_6 into another three paths $F_1 = 123456$, $F_2 = 253164$ and $F_3 = 362415$ cannot be a decomposition of K_6 according to a group. The reason is that this decomposition has only a trivial automorphism group.

13.14 Now we come to the question of the existence of a simple decomposition of the graph K_v according to a group of a given order b. It may sound surprising that, in many cases, the existence of such a decomposition depends only on b and v, and not on the structure of the group H.

Theorem. *Let* $v \geq 2$, *and* H *be a group of order* b. *Then*:

A. *For* $b \equiv 0$ (mod 2) *there is no simple decomposition of* K_v *according to* H.

B. *For* $b \equiv 1$ (mod 2) *and* $v^2 \not\equiv v$ (mod b) *again there is no simple decomposition of* K_v *according to* H.

C. *If* $b \equiv 1$ (mod 2) *and* $v \equiv 0$ *or* 1 (mod b) *then there is a simple decomposition of* K_v *according to* H (*into* b *isomorphic factors*).

Note that this result is implicitly contained in Zelinka (1967).

Proof. A. Suppose there exists a simple decomposition of K_v according to a group of order $b \equiv 0$ (mod 2). Then there is a group J of automorphisms of K_v such that $J \cong H$ and J satisfies 13.10.C. Since J has even order, it contains an element α of order 2, i.e., $\alpha^2 = \varepsilon$. Take a vertex u for which $\alpha(u) \neq u$ and put $\alpha(u) = U$. The edge uU is contained in a unique factor $F \in R$. But relations $\alpha(u) = U$ and $\alpha(U) = \alpha^2(u) = u$ imply that both F and $\alpha(F)$ contain the edge uU. Thus, $\alpha(F) = F = \varepsilon(F)$, contradicting 13.10.C.

B. Assume $b \equiv 1$ (mod 2), $v^2 \not\equiv v$ (mod b), and let a simple decomposition of K_v exist according to a group of order b. Obviously, the number $\binom{v}{2}$ of edges has to be divisible by the number of factors b. For b odd, this is equivalent to the fact that $b|v(v-1)$. But then $v^2 \equiv v$ (mod b) — a contradiction.

C. Let $b \equiv 1$ (mod 2). Consider first the case $v \equiv 0$ (mod b). Denote the elements of the group H by $a_1, a_2, \ldots, a_b$, where a_1 is the unit element. Similarly as in 13.7 we partition vertices of K_v into $t = v/b$ sets $A_1, A_2, \ldots, A_t$, each containing b elements. The elements of the i-th set will be denoted $u_i(a_1)$, $u_i(a_2), \ldots, u_i(a_b)$. Let us define mappings $\alpha_1, \alpha_2, \ldots, \alpha_b \colon V(K_v) \to V(K_v)$ by putting $\alpha_k(u_i(a_j)) = u_i(a_k a_j)$ for $i = 1, 2, \ldots, t; j, k = 1, 2, \ldots, b$. It is easy to see that these mappings are, in fact, permutations of $V(K_v)$ constituting a group J of automorphisms of K_v (which is clearly isomorphic to H). Extending the action of the group J to edges of K_v we may decompose the set $E(K_v)$ into classes corresponding to orbits of the action of J on $E(K_v)$. Since b is odd, the group J cannot have an element of order 2. Thus, each orbit contains exactly b edges. Now, choose an edge from each orbit, thereby creating a factor denoted by $g(a_1)$. We proceed by forming further $b - 1$ factors $g(a_k) = \alpha_k(g(a_1))$ for $k = 2, 3, \ldots, b$. It is a matter of routine to check that the factors $g(a_1), g(a_2), \ldots, g(a_b)$ form a decomposition R of the graph K_v and that the map $g \colon J \to R$ fulfils 13.10.A. Therefore R is a simple decomposition of K_v according to H.

In the case $v \equiv 1$ (mod b) we can modify preceding arguments using the same trick as in the proof of Theorem 13.7.

13.15 Analysing the proof of 13.14 it is evident that changing the group

H or making another choice of the first factor $g(a_1)$ can have a considerable effect on the resulting factorization of K_v.

Example [Harary et al. 1978b]. Set $b = 6$, $v = 3$. Let $H = \{a_1, a_2, a_3\}$ be the cyclic group of order 3, whereby $a_3^2 = a_2$, $a_3^3 = a_1$. Let us denote vertices of the graph K_6 by symbols $1 = u_1(a_1)$, $2 = u_1(a_2)$, $3 = u_1(a_3)$, $4 = u_2(a_1)$, $5 = u_2(a_2)$, $6 = u_2(a_3)$. The permutations a_k then have the form

$$a_1 = (1)(2)(3)(4)(5)(6),$$
$$a_2 = (123)(456),$$
$$a_3 = (132)(465).$$

Thus, we get 5 orbits of $E(K_6)$:

$$\{12, \ 23, \ 31\},$$
$$\{14, \ 25, \ 36\},$$
$$\{15, \ 26, \ 34\}, \qquad\qquad (1)$$
$$\{16, \ 24, \ 35\},$$
$$\{45, \ 56, \ 64\}.$$

Considering all possible choices of one edge from each orbit we get nine non-isomorphic factors depicted in Fig. 13.15.1. For each of them, there is a decomposition of K_6 (into 3 isomorphic factors) according to the cyclic group of order 3 (cf. Harary et al. 1978b). We shall return to this example later in 15.4.

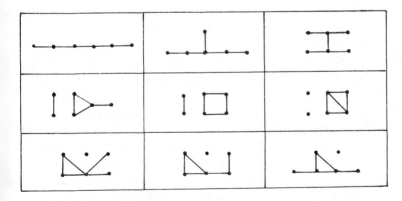

Fig. 13.15.1

13.16 In order to be able to state a general result on the existence of a simple decomposition of a complete graph according to a finite group, we need to introduce some new notation and terminology. The order of a group Γ will be denoted by $|\Gamma|$, its unit element by ε. A subgroup Γ_1 of Γ will be called *Frobenius subgroup* of Γ if $\{\varepsilon\} \neq \Gamma_1 \neq \Gamma$ and $\Gamma_1 \cap a\Gamma_1 a^{-1} = \{\varepsilon\}$ for every $a \in \Gamma - \Gamma_1$.

Theorem [Porubský 1982]. *Let $v \geq 2$, and Γ be a group of order $|\Gamma| = b$. A simple decomposition of the graph K_v according to the group Γ (into b isomorphic factors) exists if and only if $v \equiv 1$ (mod 2) and one of the following cases occurs:*

A. *Either $b|v$ or $b|v - 1$.*

B. *There exists a Frobenius subgroup Γ_1 of Γ such that $|\Gamma|/|\Gamma_1| \equiv v$ (mod b).*

P r o o f. Omitted.

13.17 Corollary [Zelinka 1967]. *Let Γ be an Abelian group of order b. A simple decomposition of K_v, $v \geq 2$ according to Γ exists if and only if b is odd and either $b|v$ or $b|v - 1$.*

P r o o f: An Abelian group cannot have any Frobenius subgroups.

13.18 In 13.4 we have introduced the concept of the *automorphism group* $A_1(R)$ for an arbitrary *decomposition* R of a graph G, comprising all the automorphisms of G which preserve the factorization R. Thus, $A_1(R)$ has an action on R which can be viewed as a group of certain permutations of factors in R. This permutation group is the so-called *symmetry group of the decomposition* R (Robinson 1979), denoted by $A_2(R)$. For example, the factorization R in 13.5 has a symmetry group consisting of two permutations $(F_1)(F_2)$ (F_3) and $(F_1)(F_2 F_3)$; therefore in this case $A_2(R) \cong Z_2$.

As regards the possibility of representation of groups as symmetry groups of decompositions, Robinson (1979) has proposed the following.

13.19 Conjecture. *Every finite group is isomorphic to a symmetry group of a decomposition of a complete graph into isomorphic factors.*

13.20 There is a great number of papers dealing with relations between graph factorization and groups (or permutations). The interested reader is referred to Zelinka (1965, 1967, 1968, 1970a—d, 1971,1973), Camion (1975), Bermond and Sotteau (1976), Cockayne and Hartnell (1976), Tomasta (1977), Harary et al. (1978b), Robinson (1979) and Schwenk (1984).

It is hardly possible to mention all papers considering permutations and groups in connection with graph decompositions. Apart from those listed above let us draw the reader's attention to Sachs (1962), Ringel (1963), Graver and Yackel (1968), Bosák et al. (1971), Clapham (1979), or Gardner (1980).

Similar methods as explained here will also be used in other parts of this book (as in 14.5—14.7 where we shall bring an application of groups in the enumeration of graph decompositions).

Exercises

13.21 Does there exist a group for which the decomposition in 13.6 would be a simple one according to this group?

13.22 [Robinson 1979]. Find a permutation group which does not correspond to any symmetry group of a decomposition of a complete graph into isomorphic factors.

13.23 [Robinson 1979]. Each permutation group on a set of 3 elements corresponds to the symmetry group of a suitable decomposition of a complete graph into isomorphic factors.

13.24. Give an example of a decomposition of a graph according to a cyclic group, which is not a cyclic decomposition in the sense of 7.1.

Chapter 14

Self-complementary graphs and the enumeration of decompositions

14.1 In 13.9, we have introduced the concept of a self-complementary graph as a simple graph G being isomorphic to its complement $\bar{G}$. Thus, for each self-complementary graph there exists a vertex isomorphism $G \rightarrow \bar{G}$, sometimes called *complementing permutation* for G. Our first aim will be to determine the structure of such permutations, focusing on their cycle decomposition. In this sense we will speak about cycles in *complementing permutations*.

In Fig. 14.1.1 there are 5 self-complementary graphs (the first one being the empty graph). As we shall see later, these are (up to isomorphism) all self-complementary graphs of order <8. It is easy to check that each cycle in a complementing permutation of a graph in Fig. 14.1.1 has length 1 or 4.

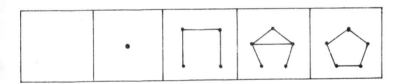

Fig. 14.1.1

For example, one of the complementing permutations of the pentagon can be written as $(1)(2345)$. In general, we have the following theorem:

14.2 Theorem [Sachs 1962, Ringel 1963]. *Let α be a complementing permutation of a self-complementary graph of order v. Then, each cycle of α has length divisible by 4, with a possible exception of one cycle of length 1.*

Proof. The action of α on vertices of a self-complementary graph G can be extended to $E(G)$ in an obvious manner, putting $\alpha(uU) = \alpha(u)\,\alpha(U)$ for $u,\, U \in V(G)$. Since $(G, \bar{G})$ is a decomposition of K_r, α cannot contain two cycles (i), (j) of length 1. Otherwise we would have $\alpha(ij) = \alpha(i)\,\alpha(j) = (ij)$, and the edge ij would have to belong either to both G and $\bar{G}$ or to none of them.

Suppose that $(a_1 a_2 \ldots a_n)$ is a cycle of α of odd length $n \geq 3$, and that $a_1 a_2 \in E(G)$. Then clearly $\alpha(a_1 a_2) = a_2 a_3 \in E(\bar{G})$, $\alpha(a_2 a_3) = a_3 a_4 \in E(G)$, and so on. Because n is odd we get $a_n a_1 \in E(G)$, $a_1 a_2 \in E(\bar{G})$ — a contradiction.

Finally, let $(a_1 a_2 \ldots a_n)$ be a cycle of α of length $n = 4m - 2$ and let $a_1 a_{2m} \in E(G)$. Similarly as above we successively obtain $a_2 a_{2m+1} \in E(\bar{G})$, $a_3 a_{2m+2} \in E(G)$, ..., $a_{2m-1} a_{4m-2} \in E(G)$, $a_{2m} a_1 \in E(\bar{G})$, a contradiction.

14.3 It is possible to obtain a reversed version of the preceding theorem in the following sense: Given an arbitrary permutation α of v elements such that each cycle of α has length divisible by 4 (except, possibly, one trivial cycle), there exists a self-complementary graph G of order v whose complementing permutation is α. The construction goes as follows. Choose an arbitrary edge w (i.e., a pair of elements of a fixed set of size v) and place it into $E(G)$ together with edges $\alpha^2(w)$, $\alpha^4(w)$, $\alpha^6(w)$, ...; the edges $\alpha(w)$, $\alpha^3(w)$, $\alpha^5(w)$, ... will be placed into $E(\bar{G})$. If still there are edges which have not been examined yet, choose one of them and repeat the above. The whole procedure is to be carried out until there is no remaining edge (i.e., each edge is placed either into $E(G)$ or $E(\bar{G})$). It is straightforward to see that the graph G thus defined is a self-complementary graph with complementing permutation α. We have just proved:

Theorem [Sachs 1962, Ringel 1963]. *Let α be a permutation of v elements such that each cycle of α has length divisible by 4 (with a possible exception of one trivial cycle). Then there exists a self-complementary graph of order v with complementing permutation α.*

14.4 Now it is easy to determine all the values of v for which there are self-complementary graphs of order v.

Theorem [Sachs 1962, Ringel 1963]. *A self-complementary graph of finite order v exists if and only if $v \equiv 0$ or 1 (mod 4).*

Proof. The cases $v \leq 1$ are trivial. For $v \geq 2$ we have two different supporting arguments:

I. The necessity is a consequence of Theorem 14.2, the sufficiency follows from 14.3.

II. The condition $v \equiv 0$ or 1 (mod 4) is implied by 6.3.D putting $b = 2$ and $e(H) = \binom{v}{2}$. The reverse implication follows from the Corollary 13.9.

14.5 Harary (1960) posed the problem of counting non-isomorphic self-complementary graphs of a given order. A complete solution was given by Read (1963, 1967). His method is based on enumeration theory originated by Redfield (1927) and Pólya (1937), developed further by de Bruijn (1959, 1964) and Harary and Palmer (1973). Here we shall confine ourselves with a list of the numbers s_v of non-isomorphic self-complementary graphs of order $v < 20$

Table 14.5.1

v	0	1	4	5	8	9	12	13	16	17
s_v	1	1	1	2	10	36	720	5 600	703 760	11 220 000

(Table 14.5.1). Note that an asymptotic formula for s_v, $v \to \infty$, is derived in Palmer (1970); see also Robinson (1978), Sridharan (1978), Palmer (1981) and Schwenk (1984).

14.6 Sometimes it is convenient to have a catalogue of self-complementary graphs of a given order v. For small values of v it is possible to employ the method used in the proof of Theorem 14.3; such a catalogue for $v < 8$ is depicted in Fig. 14.1.1. As regards further values of v, the reader is referred to Morris (1973), Alter (1975), Venkatachalam (1976), Faradzhev (1978) for $v = 8$, Morris (1973) and Faradzhev (1978) for $v = 9$, and finally Faradzhev (1978) and Kropar and Read (1979) for $v = 12$. Moreover, in the paper by Faradzhev (1978) there are depicted 35 self-complementary graphs as a sample of the total of 5600 such graphs.

14.7 The enumeration methods for self-complementary graphs, due to Read (1963), have been generalized by Schwenk (1984) to the case of enumerating simple decompositions of a given graph G according to a cyclic subgroup of an arbitrary order b (i.e., into b factors).

Further results and methods concerning the enumeration of self-complementary graphs, as well as other types of decompositions, can be found in Gallai (1959), Kopylova (1965), Read (1967, 1981), Parthasarathy and Sridharan (1969), Palmer (1970), Sridharan (1970, 1978), Sridharan and Parthasarathy (1972), Frucht and Harary (1974), Gibbs (1974), Wille (1974, 1978), Harary et al. (1977a), Quinn (1979), Mullin and Wallis (1980), Akiyama and Harary (1981), Hedge et al. (1983).

A particular attention has been focused on the enumeration of 1-factorizations of (mostly complete) graphs; see e.g. Dickson and Safford (1906), Wallis et al. (1972), Gelling and Oden (1974), Cameron (1976), Lindner et al. (1976), Anderson et al. (1977), Korovina (1980), Phelps (1980), Petrenyuk and Petrenyuk (1980), Petrenyuk and Shniter (1981), Schrijver (1983), Hartman and Rosa (1985).

The enumeration of decompositions of the complete λ-fold graphs of order v into complete subgraphs of order k has already been mentioned in 7.38.A, 8.16.D, 8.34—8.36, and 12.8; for similar problems concerning other classes of graphs see 2.9, 2.10, 3.9, 3.12, 3.13, 8.43, 8.52—8.54, and 10.11.C.

14.8 In order to be able to count self-complementary blocks of order v, let us start with some preliminary results. First of all, notice that (according to Theorem 5.2) each self-complementary graph is connected.

Let b_v denote the number of pairwise non-isomorphic self-complementary blocks of order v. For a non-negative integer i let $s_v^{(i)}$ be the number of mutually non-isomorphic self-complementary graphs of order v with exactly i end-vertices. The symbol s_v has been defined in 14.5.

Lemma [Akiyama and Harary 1981]. *For $v \geq 4$ the following holds*:

$$s_v^{(i)} = \begin{cases} b_v & \text{for } i = 0, \\ 0 & \text{for } i = 1, \\ s_{v-4} & \text{for } i = 2, \\ 0 & \text{for } i \geq 3. \end{cases}$$

Proof. Lemma 5.7 implies that $s_v^{(i)} = 0$ for $i \geq 3$; the similar was shown to be true by Akiyama and Harary (1981) for $i = 1$. It is clear that for each self-complementary graph G it holds: If G is a block then G has no endvertices. The converse is also true — see Akiyama and Harary (1981). Thus, $s_v^{(0)} = b_v$. Finally, according to Theorem 5.8 there is a 1-1 correspondence between the simple graphs H of order $v - 4$ and the graphs G of the form $G = H + K_2 \circ K_1$. It is a matter of routine to check that G is self-complementary if and only if H is. Therefore $s_v^{(2)} = s_{v-4}$.

14.9 Theorem [Akiyama and Harary 1981]. *For $v \geq 4$ the number b_v of self-complementary graphs of order v is given by*

$$b_v = s_v - s_{v-4}.$$

Proof. By virtue of Lemma 14.8 we have $s_v = s_v^{(0)} + s_v^{(1)} + \ldots + s_v^{(v)} = b_v + s_{v-4}$.

14.10 On the basis of Lemma 14.8 and Theorem 14.9 it is possible to compute the values of b_v given in Table 14.10.1.

Table 14.10.1

v	4	5	8	9	12	13	16	17
$s_v^{(0)} = b_v$	0	1	9	34	710	5 564	703 040	11 214 400
$s_v^{(2)} = s_{v-4}$	1	1	1	2	10	36	720	5 600

Colbourn et al. (1978) dealt with the computational complexity of determining whether two self-complementary graphs are isomorphic. It has turned out that this problem (even when restricted to regular self-complementary graphs) is polynomially equivalent to the graph isomorphism problem in general.

14.11 The property of being a self-complementary graph is quite a restrictive one, as it can be seen from part A of the following theorem.

Theorem [Ringel 1963]. A. *Each self-complementary graph of order $v > 1$ has diameter 2 or 3.*

B. *For each $v \in \{5, 8, 9, 12, 13, 16, 17, \ldots\}$ there exists a self-complementary graph of order v and diameter 2.*

C. *For each $v \in \{4, 5, 8, 9, 12, 13, \ldots\}$ there exists a self-complementary graph of order v and diameter 3.*

Proof is easy and therefore omitted.

A generalization of the claim A to arbitrary complementary graphs can be found in Bosák et al. (1968).

14.12 There are several graph invariants related to complementary graphs. Probably the most famous is the *Ramsey number* $R(n, n)$ (see 4.24), defined as the least order v of a complete graph K_v such that, for every decomposition of K_v into two factors, at least one of them contains an n-clique. Clapham (1979) found an example of a self-complementary graph of order 113 that contained no 7-clique, implying $R(7, 7) \geq 114$. This bound has been improved by Hill and Irving (1982) to $R(7, 7) \geq 126$. Using a slightly more sophisticated method, Guldan and Tomasta (1983) obtained the bounds $R(10, 10) \geq 458$ and $R(11, 11) \geq 542$. For some other results related to this topic see e.g. Chvátal et al. (1972).

14.13 With the aid of two invariants introduced by Araoz and Tannenbaum (1979) it is possible to express, in a way, the measure of *non-self-complementarity* of simple graphs. The first, so-called *self-complement index* $s(G)$, is defined as the largest order of a graph H such that both H and $\bar{H}$ are induced subgraphs of G. The second, the *induced number* $m(G)$, is the smallest order of a graph H that contains both G and $\bar{G}$ as induced subgraphs. Obviously we have $s(G) \leq v(G) \leq m(G)$, with equalities if and only if G is a self-complementary graph. For further results on these invariants we refer the reader to the original paper of Araoz and Tannenbaum (1979).

14.14 Let G be an undirected graph with $V(G) = \{u_1, u_2, ..., u_v\}$, and let d_i be the degree of the vertex u_i, $i = 1, 2, ..., v$. The sequence $[d_1, d_2, ..., d_v]$ is said to be *degree sequence* of the graph G.

In general, the only constraint for a degree sequence of a finite undirected graph is that the sum of all degrees be even. However, the situation becomes more complicated if we consider simple graphs only (cf. Havel 1955, Erdős and Gallai 1960, Hakimi 1962; see also Harary 1969, Theorems 6.1 and 6.2). As regards the self-complementary graphs, we have the following characterization.

Let $d = [d_1, d_2, ..., d_v]$ be a sequence of non-negative integers, $d_1 \geq d_2 \geq \geq ... \geq d_v$. Clapham and Kleitman (1976) proved that d is the degree sequence of a self-complementary graph if and only if:

A. d is a graphical sequence, which means that there exists a simple graph whose degree sequence is d.

B. $v \equiv 0$ or $1 \pmod 4$.

C. $d_i + d_{v+1-i} = v - 1$ for $i = 1, 2, ..., \lfloor (v + 1)/2 \rfloor$.

D. $d_{2j-1} = d_j$ for $j = 1, 2, ..., \lfloor v/4 \rfloor$.

Later on, Clapham (1976a) proved that the condition A may be replaced by a much simpler one, namely

E. $d_2 + d_4 + ... + d_{2k} \leq k(v - 1 - k)$ for $k = 1, 2, ..., \lfloor v/4 \rfloor$.

In Chernyak (1983) and Das Prabir (1983) similar questions are studied for *edge degree* sequences (the edge degree is understood to be the unordered pair of degrees of endvertices of the edge). Rao (Preprint) characterized the degree sequences which are realizable only by self-complementary graphs (forcibly self-complementary sequences).

14.15 The regular self-complementary graphs have attracted the interest of many authors, e.g. Sachs (1962), Colbourn and Colbourn (1979), Kotzig (1979b), Lovász (1979b), Zelinka (1979), Ruiz (1981) and Rosenberg (1982). Clearly, for each such graph of finite order v we have $v \equiv 1 \pmod 4$, the degree of each vertex being $(v - 1)/2$. It is possible to impose further restriction here and to study, for example, vertex transitive self-complementary graphs (cf. Lovász 1979b, Zelinka 1979, Ruiz 1984, for such graphs having a *cyclic* automorphism of order v see Sachs 1962).

Almost regular self-complementary graphs of order $v \equiv 0 \pmod 4$ (i.e., every vertex is of degree $v/2$ or $v/2 - 1$) have been studied by Sachs (1962), in connection with regular self-complementary graphs and their characteristic polynomials.

14.16 The existence of paths or cycles of prescribed length in self-complementary graphs was investigated by several authors: Clapham (1974, 1975, 1976b), Camion (1975) and Rao (1979a—d). For instance, in Clapham (1974, 1976b) and Camion (1975) it was proved that each non-empty self-complementary graph has a Hamiltonian path. If the order of the graph is $v \le 5$ then it has an odd number of Hamiltonian paths (1 or 5, see Fig. 14.1.1); otherwise the number of Hamiltonian paths in self-complementary graphs is always even (Camion 1975; see also Rao 1979c). According to Rao (1977b), each self-complementary graph of order $v > 5$ has a cycle of length i for each $i = 3, 4, ..., v - 2$, but need not have cycles of length $v - 1$ or v. However, if it does have a Hamiltonian cycle then it contains cycles of all lengths $i = 3, 4, ..., v$, i.e. it is *pancyclic* (Rao 1977b). The problem of the existence of a Hamiltonian cycle in self-complementary graphs was solved by Rao (1979b). The minimum number of triangles in a self-complementary graph of a given order was determined by Clapham (1973); for the maximum number of triangles there see Rao (1979d).

14.17 It is hardly possible to give a list of all results, generalizations and applications related to self-complementary graphs. Let us discuss very briefly at least some of them, all the details being left to the reader.

A self-complementary graph can be defined as a solution of the graph equation $G = \bar{G}$. Several modifications of such an approach can be found in Capobianco (1979).

Chao et al. (1979) studied self-complementary graphs with a given chromatic number. Among others they proved that, for each n, the maximum order of an n-chromatic self-complementary graph is n^2. Further, for each $n \geq 3$ they constructed n-chromatic self-complementary graphs of diameter 2 and 3.

The so-called Shannon capacity of self-complementary graphs was investigated by Lovász (1979b).

Blass et al. (1981) focused the attention to a logical analysis of some graph properties. Using Paley graphs the authors showed that, for instance, neither regularity nor self-complementarity is a first-order property. (The *first-order language* contains predicate symbols for equality and adjacency, logical connectives "non", "and", "or", the existence quantifier, and constants "true" and "false".)

Various constructions of self-complementary graphs can be found in Ruiz (1980). For further results on the topic the reader is referred to Sedláček (1966), Cvetković (1970), Grant (1976), Rao (1977a) and Pinto (1981); for a survey of problems and results we recommend Rao (1979a).

14.18 The attractivity of self-complementary graphs led to the investigation of many other self-complementary structures. A self-complementary graph arises as a result of a decomposition of a complete graph into two isomorphic factors. Replacing the complete graph by another graph or by a hypergraph and retaining the conditions on the decomposition we get several new concepts which are easily understood on the basis of the following scheme:

complete graph — self-complementary graph,

λ-fold complete graph — λ-fold self-complementary graph,

complete m-partite graph — self-complementary m-partite graph,

complete digraph — self-complementary digraph,

complete m-partite digraph — self-complementary m-partite digraph,

complete t-uniform hypergraph — self-complementary t-uniform hypergraph,

complete m-partite t-uniform hypergraph — self-complementary m-partite t-uniform hypergraph.

14.19 Let us bring now some references concerning the concepts mentioned above. For λ-fold self-complementary graphs see Salvi Zagaglia (1977) and Wille (1978). Self-complementary bipartite graphs were studied by Zelinka (1970c), Gupta (1978), Harary et al. (1978a), Quinn (1979), and Gangopadhyay (1981, 1982); for the multipartite case see Zelinka (1970c), Gupta (1978), Harary et al. (1978a) and Gangopadhyay and Hebbare (1980b, c, 1982). Self-complementary digraphs were dealt with in Read (1963), Sachs (1965), Harary and Palmer (1966), Parthasarathy and Sridharan (1969), Sridharan (1970, 1976, 1978), Zelinka (1970a), Harary and Palmer (1973), Das Prabir (1981), Pinto (1982) and Hedge et al. (1983). In particular, for self-complementary tournaments we refer to Zelinka (1970a), Reid and

Thomassen (1976), Reid and Beineke (1978), Eplett (1979), Salvi Zagaglia (1979, 1980) and Das Prabir (1981). More general self-complementary structures were investigated by Wille (1974, 1978). As regards the self-complementary multipartite digraphs we recommend Zelinka (1970c) and Gupta (1978). Self-complementary t-uniform hypergraphs were studied in Zelinka (1968) and Palmer (1973) and the m-partite such graphs (for $m = t$) in Zelinka (1970c).

14.20 The papers listed above cover all the topics considered in studying standard self-complementary graphs: cycles in complementing permutations, existence and enumeration of self-complementary structures with given parameters, degree sequences, invariants, paths and cycles as well as other substructures, characterizations and constructions of particular classes of self-complementary structures, decompositions, etc. For example, Zelinka (1970a) proved that for each cardinal number v there exists a self-complementary digraph of order v. As another example, let us mention that complementing permutations for digraphs are characterized by the fact that their cycles are either infinite or of even length, with one possible exceptional cycle of length 1. A similar result holds for self-complementary tournaments (the even cycle lengths are to be replaced by lengths congruent to 2 mod 4).

14.21 The numbers s_r^* $[s_r^{**}]$ of mutually non-isomorphic self-complementary digraphs [self-complementary directed graphs] of order $v \leq 8$ are in Tables 14.21.1 and 14.21.2 (cf. Read 1963, Harary and Palmer 1973).

Table 14.21.1

v	0	1	2	3	4	5	6	7	8
s_v^*	1	1	1	4	10	136	720	44 224	703 760

Table 14.21.2

v	0	1	2	3	4	5	6	7	8
s_v^{**}	1	0	2	0	36	0	5 600	0	11 220 000

14.22 There are several curious formulae involving self-complementary structures. In Read (1963) it is proved that $s_{2n}^* = s_{4n}$ for each $n \geq 0$. As another example we present here the formula $s_{2n}^{**} = s_{4n+1}$, found by Wille (1974, 1978). (Since the graph K_v^{**} (1.18) has an odd number of loops for v odd, we have $s_v^{**} = 0$ in this case.) Note that neither of these formulae has been given a proof by means of a 1-1 correspondence defined in a natural way (cf. also Morris 1974).

14.23 Quinn (1979) investigated the number $s(n_1, n_2)$ of self-complementary bipartite graphs with parts of order n_1 and n_2. The values of $s(n_1, n_2)$ for $n_1 \leq 3$ and $n_2 \leq 4$ are in Table 14.23.1.

Table 14.23.1

$s(n_1, n_2)$		n_2			
		1	2	3	4
n_1	1	0	1	0	1
	2	1	2	3	6
	3	0	3	0	7

14.24 The self-complementary structures dealt with in 14.18—14.23 arose by decomposing graphs (or hypergraphs) into two isomorphic subgraphs (subhypergraphs). In a more general setting, one can ask about graphs admitting at least one decomposition into two isomorphic subgraphs. Such graphs have been called R_2-graphs in Zelinka (1965, 1970c); for R_2-digraphs see Zelinka (1970a).

It is possible to generalize the latter approach to an arbitrary decomposition into b isomorphic subgraphs, focusing only on the existence of such a decomposition, and not on the structure of the factors. This is going to be the central topic of Chapter 15.

14.25 A related concept to that of self-complementary graph is the concept of self-converse graph, i.e., a (partially directed) graph that is isomorphic to its converse graph (6.19). This means that reversing the orientation of all directed edges yields a graph isomorphic to the original one. Such graphs were studied in Harary and Palmer (1966), Harary et al. (1967), Sridharan (1970, 1976), Cavalieri D'Oro (1971), Salvi Zagaglia (1977, 1978), Robinson (1978), Das Prabir (1981); see also Palmer (1981). In the case when the self-converse graph is a tournament we obtain the self-comlementary tournament as in 14.19. Even in a more general case of self-converse oriented graph, this can be viewed as a factor of a decomposition of some symmetric (1.22) digraph into two isomorphic factors.

Exercises

14.26 [Sachs 1962]. Each self-complementary graph of odd order v has at least one vertex of degree $(v - 1)/2$.

14.27 [Ringel 1963]. For each $v \in \{5, 8, 9, 12, 13, ...\}$ there exists a self-complementary graph of order v and diameter 2.

14.28 [Ringel 1963]. For each $v \in \{4, 5, 8, 9, 12, 13, ...\}$ there exists a self-complementary graph of order v and diameter 3.

14.29 Find all types of A. self-complementary tournaments of order 4; B. self-complementary digraphs of order 4 that are not tournaments.

14.30 Determine the number $s(v)$ of self-complementary bipartite graphs of order v, $2 \le v \le 6$.

14.31 [Ruiz 1984]. For each $n \ge 2$ there exists a regular self-complementary graph of order $4n + 1$ which is not vertex transitive.

Chapter 15

The sufficiency of the divisibility conditions

15.1 In Theorem 6.3 we have derived a number of necessary conditions for the existence of a G-decomposition of a graph H. Among them there were also two divisibility conditions, namely 6.3.C—D. Simple examples show that there two conditions are not always sufficient: Take, for example, $b = 2$, $G = P_3$, and for H the graph $K_3 \vee K_2$ of order 5 consisting of two components K_3 and K_2.

Nevertheless, in many cases the conditions 6.3.C—D are of principal importance in the process of deciding whether there exists a G-decomposition of a graph H or not. In this sense we shall be dealing with the sufficiency of divisibility conditions for the existence of decompositions.

When applying the condition 6.3.C ($e(G)|e(H)$) to the question of the existence of a G-decomposition of a graph H, we often consider the number of factors b as unimportant (although, of course, $b = e(H)/e(G)$ for $e(G) \neq 0$). Such approach was successfully used several times, especially in Chapters 8—10, to establish the sufficiency of 6.3.C for some kinds of decompositions.

15.2 From now on we shall concentrate only on the second *divisibility condition* 6.3.D ($b|e(H)$); however, its sufficiency will be understood in a slightly modified way. Throughout, the only object given will be a (finite and non-empty) graph H. We shall say that the (second) divisibility condition is *sufficient with respect to the graph H* if for each divisor b of the number $e(H)$ there exists a graph G such that H admits a G-decomposition into b subgraphs. Thus, only the existence will be essential here (not the structure of the graph G). In the case there exists a decomposition of H into b isomorphic factors we shall write $b|H$. On the other hand, the set of all graphs G for which there exists a G-decomposition of H into b isomorphic factors will be denoted by H/b. Obviously, H/b is non-empty if and only if $b|H$.

For instance, if H is the graph of the example given in 15.1 then we have $1|H$ and $4|H$, but $b \nmid H$ for each $b \neq 1, 4$. Thus, in our terms, the divisibility condition 6.3.D is not sufficient with respect to H, for $2|e(H)$ but $2 \nmid H$ ($H/2$ is empty).

15.3 One of the most significant results on the sufficiency of the (second)

186

divisibility condition is that it is sufficient with respect to complete graphs.

Theorem [Harary et al. 1977b, 1978b, Schönheim and Bialostocki 1978].
Let b and v be positive integers. Then $b\,|K_v$ if and only if $b\,|e(K_v)$ (or, equivalently, $2b\,|v(v-1)$).

Proof. The necessity is obvious (see 6.3.D and 15.2). The sufficiency of the condition $b\,|e(K_v)$ is trivial for $b = 1$, and for $b = 2$ it was proved in Theorem 14.4. If b is a power of 2 [a prime power], the sufficiency is a consequence of Theorem 13.7 [Theorem 13.14.C]; for $b = 4$ see also Zelinka (1965). In the general case, the proof can be found in Harary et al. (1978b); for a more elementary proof see Schönheim and Bialostocki (1978). A significant particular case is included in the Exercise 15.14. A number of other particular cases of the theorem have already been handled in Chapters 9 and 10. For example, the sufficiency of the condition $b\,|e(K_v)$ follows from Theorem 9.10 in the case $b \geq v/2$, putting there $\lambda = 1$ and $k = v(v-1)/(2b) + 1$.

15.4 As mentioned above, trivially $1\,|K_v$, and the relation $2\,|K_v$ was dealt with in Chapter 14. The next interesting case is $b = 3$. By virtue of Theorem 15.3, $3\,|K_v$ if and only if $v \equiv 0$ or $1 \pmod 3$. It can be easily checked that

$$K_1/3 = \{K_1\},$$

$$K_3/3 = \{K_2 \vee K_1\},$$

$$K_4/3 = \{P_3 \vee K_1,\, 2K_2\}.$$

In 13.15 we found 9 graphs belonging to $K_6/3$. By a systematical examination (Harary et al. 1978b) it is possible to check that none of the remaining 6 simple graphs of order 6 and size 5 belongs to $K_6/3$. According to Harary et al. (1978b) and Robinson (1979), the method of constructing a simple decomposition of K_v according to a cyclic group of order 3 might prove useful in determining completely the sets $K_v/3$ for $v = 7$ and 9, but certainly not for $v \geq 10$ (see 13.5).

15.5 Our next result shows that the divisibility condition 6.3.D is sufficient with respect to complete bipartite graphs.

Theorem [Harary et al. 1978a]. *Let b, n_1, n_2 be positive integers. If $b|n_1 n_2$ then $b|K(n_1 n_2)$.*

Proof. If $b|n_1 n_2$ there exist positive integers r and s such that $b = rs$, $r|n_1$, and $s|n_2$. But then the graph $K(n_1, n_2)$ obviously admits a $K(n_1/r, n_2/s)$-decomposition into b factors.

15.6 The situation with the sufficiency of the divisibility condition 6.3.D with respect to complete tripartite graphs is more complicated, as it is clear from the following result:

Theorem [Harary et al. 1978a]. *If $b \equiv 1 \pmod 2$, $b \geq 3$ and $n \geq b(b+1)$ then the set $K(1, 1, n)/b$ is empty.*

Proof — see Harary et al. (1978a).

15.7 Corollary [Harary et al. 1980]. *For $b \equiv 1 \pmod 2$, $b \geq 3$ the divisibility condition 6.3.D is not sufficient with respect to the graph $K(1, 1, b(b + 1) + (b - 1)/2)$.*

Proof. Put $n = b(b + 1) + (b - 1)/2$ in 15.6. Now, the number of edges $2b^2 + 3b$ is divisible by b, in contrast to the fact that $K(1, 1, n)/b$ is empty.

15.8 The preceding example leads naturally to the following question: For which positive integers b the relation $b \mid e(K(n_1, n_2, n_3))$ implies $b \mid K(n_1, n_2, n_3)$? Obviously this is true for $b = 1$, and false for each odd $b \geq 3$ (see 15.7). As regards the remaining values of b, we have:

Tripartite conjecture [Harary et al. 1978a]. *If b is even then $b \mid (n_1 n_2 + n_1 n_3 + n_2 n_3)$ implies $b \mid K(n_1, n_2, n_3)$.*

The correct version of this conjecture was the subject of a discussion between the authors of the announcements (Walikar 1980 and Harary et al. 1980).

15.9 The validity of the tripartite conjecture was proved for $b = 2$ and $b = 4$ by Harary et al. (1978a), and for $b = 6$ by S. Quinn (cf. Harary et al. 1978a, Quinn 1983). For b a power of 2 the conjecture was settled by Yang Shi-hui (1983). In the case $n_1 = n_2$ and b is arbitrary, the tripartite conjecture was shown to be true by Beka (1985).

15.10 Analogous questions for the remaining complete m-partite graphs are even more difficult. For instance (Harary et al. 1978a), if $4 \mid m$ and $m < s$, the graph $K_m(1, 1, \ldots, 1, s)$ (1.15) does not admit a decomposition into two isomorphic subgraphs despite the fact that its size is even if both m and s are even. Similarly, for $m = 5$ the graph $K(1, 1, 1, 2, 5)$ is not decomposable into two isomorphic factors (Harary et al. 1978a).

15.11 Harary et al. (1978a) raised the so-called *equipartite conjecture* which says that the divisibility condition 6.3.D is sufficient with respect to all complete equipartite graphs $K(m \times n)$. In the case $n = 1$ [$m = 2$] the equipartite conjecture is true by virtue of Theorem 15.3 [15.5]. For some other particular cases we refer to Tomasta and Zelinka (1981). Surprisingly enough, the equipartite conjecture has been settled lately:

Theorem [Wang-Jiangfang 1982, Quinn 1983]. *The complete equipartite graph $K(m \times n)$ is decomposable into b isomorphic factors if and only if $b \mid n^2 \binom{m}{2}$.*

Proof is lengthy and therefore omitted.

15.12 The situation with the divisibility condition 6.3.D for digraphs is similar to that of undirected graphs. In Harary et al. (1978c) it is proved that the divisibility condition is sufficient with respect to all complete digraphs. However, there are examples of complete multipartite digraphs for which this is not true. Nevertheless, an analogue to the equipartiteconjecture is valid for all complete equipartite digraphs $K^*(m \times n)$, as announced in Quinn (1983) and proved in Wang-Jiangfang (1983).

15.13 In investigating the sufficiency of the second divisibility condition it is not necessary to confine to complete and complete multipartite graphs or digraphs; examples can be found in Exercises 15.15 and 15.16. One possibility consists in studying other naturally defined classes of graphs, e.g. vertex-transitive graphs:

Problem [Alspach 1982]. *Do there exist finite vertex-transitive graphs with respect to which the divisibility condition 6.3.D is not sufficient? If so, find a characterization of such graphs.*

Another possibility is to specify the subgraphs in the decomposition. For example, in Köhler (1974) it is proved that the complete graph K_v can be decomposed into $(v-1)/m$ isomorphic m-regular factors if and only if vm is an even number and $m \,|v-1$.

Exercices

15.14 [Guidotti 1972, Harary et al. 1978b]. Let g.c.d. $\{b, v\} = 1$ or g.c.d. $\{b, v-1\} = 1$. Then, $2b \,|v(v-1)$ implies $b \,|K_v$.

15.15 Find all the simple graphs of order ≤ 5 with respect to which the divisibility condition 6.3.D is not sufficient.

15.16 Solve an analogous problem as in the previous exercise for digraphs of order ≤ 3.

RESULTS OF EXERCISES

R 1.53 No. Counterexample: Take a graph G with an undirected link w, and let G' be obtained from G by replacing the link w by a loop incident with an end of the link w.

R 1.54 No. Counterexample: Let G be a graph containing a directed edge w, and let G' be obtained from G by "forgetting" the orientation of w.

R 2.8 A, B. No; C. Yes.

R 2.9 There are exactly $|L|^{|E|} = l^e$ different edge L-colourings, i.e., mappings of the edge set E of the graph into the set L.

R 2.10 To every L-decomposition of a graph (into factors) there exist exactly $l!$ edge L-colourings associated with the L-decomposition. Thus, the claim is a consequence of R 2.9.

R 3.11 A, D, F. No; B, C, E, G. Yes.

R 3.12 It suffices to realize that to every partial edge l-colouring of a graph there corresponds its edge $(l + 1)$-colouring obtained from the original one by colouring the so far uncoloured edges with an additional colour. Now, apply R 2.9.

R 3.13 To every L-packing of a graph by factors there exist exactly $l!$ partial edge L-colourings associated with the L-packing. The result follows from 3.12.

R 3.14 $(2^{|L|})^{|E|} = 2^{le}$.

R 4.41 A. None; B. 4.39.2. (In all cases a self-complementary graph of diameter 3 can serve as a counterexample; see Fig. 14.1.1.)

R 4.42 Proof: If there is a vertex u that does not belong to any G_i then delete u and obtain the result by induction on the number of vertices. Otherwise, partition the vertices of K_v into classes U_i and U_{ij} ($i, j = 1, 2, \ldots, l$; $i \neq j$) in the following way: Let U_i consist of all vertices in G_i which do not belong to any other graph of the packing P. Further put $U_{ij} = V(G_i) \cap V(G_j)$ for $i \neq j$. Making use of Lemma 4.11 we then get

$$\chi(G_1) + \chi(G_2) + \ldots + \chi(G_{l-1}) + \psi(G_l) \leq$$

$$\leq \sum_{i=1}^{l-1} \chi(G_i[U_i]) + \sum_{\substack{i=1 \\ i \neq j}}^{l-1} \sum_{j=1}^{l} \chi(G_i[U_{ij}]) + \psi(G_l[U_l]) + \sum_{j=1}^{l-1} \psi(G_l[U_{lj}]) \leq$$

$$\leq \sum_{i=1}^{l} |U_i| + \frac{1}{2} \sum_{\substack{i=1 \\ i \neq j}}^{l} \sum_{j=1}^{l} (|U_{ij}| + 1) = v + \binom{l}{2}.$$

R 4.43 It is sufficient to use the preceding exercise, the inequality 4.5.1 and the fact that every decomposition is a packing.

R 4.44 As $\lfloor ((v-1)/2)^2 \rfloor = \lfloor (v-1)/2 \rfloor \lceil (v-1)/2 \rceil$, in view of the upper bound of Theorem 4.29.2 it suffices to consider the cases $v = 4$ and $v \equiv 3 \pmod 4$. The first one is trivial (every decomposition of K_4 into two connected factors consists of two paths of order 4). In the second case it is sufficient to prove the impossibility of the equation

$$\varkappa'(G)\varkappa'(\bar{G}) = ((v-3)/2)\,((v+1)/2) + 1 = ((v-1)/2)^2.$$

Assuming the contrary we would get that both G and $\bar{G}$ are regular graphs of degree $(v-1)/2$ of an odd order v, which is clearly absurd.

R 4.45 Take into account that $\alpha_1(S_v) = v - 1$, $\alpha_1(\bar{S}_v) = \lceil (v-1)/2 \rceil$.

R 4.46 For each $n \neq 2$. (Such a graph does not exist for $n = 2$, since each graph G containing at least one non-loop edge satisfies $\chi''(G) \geq 3$.)

R 4.47 Two distinct vertices of a graph G at distance at least 2 in G have distance 1 in $\bar{G}$. Thus, their sum is at least 3.

R 4.48 We shall construct a graph G of order $v = 5$ for which equality in 4.47 is attained. Let G be a simple graph with vertices $u_1, u_2, \ldots, u_v$ and edges $u_1u_2, u_2u_3, u_3u_4, u_4u_5, u_4u_6, u_4u_7, \ldots, u_4u_v, u_5u_1, u_6u_1, \ldots, u_vu_1$. It is a matter of routine to check that the sum of the distances of arbitrary two vertices in G and $\bar{G}$ is always 3.

R 4.49 Take an example of a graph of order v in which $\lceil 2v/3 \rceil$ vertices form a clique and the remaining $\lfloor v/3 \rfloor$ vertices are null vertices.

R 4.50 No; even the condition that both G and $\bar{G}$ be connected is not sufficient. Namely, for each $v \geq 4$ there exists a connected graph G having a connected complement $\bar{G}$ of order v for which $\beta_1(G) = \lfloor (v+3)/3 \rfloor$ and $\beta_1(\bar{G}) = \lfloor (v+1)/3 \rfloor$. For instance, let G be a graph with vertex set $V(G) = A \cup B \cup \{u, U\}$, where $|A| = \lfloor (v-2)/3 \rfloor$, $|B| = \lfloor (2v-2)/3 \rfloor$, the vertex u be adjacent to U as well as to all vertices in A, the vertex U be adjacent to all vertices in B, and moreover let B induce a clique in G.

R 5.15 The claim is a consequence of Theorem 5.8, since pendant vertices in G are at distance 3 and no greater distance can occur in G.

R 5.16 Let G_1 be the graph of order v whose one component is the star $K(1, \lfloor v/2 \rfloor)$ and the remaining components are isolated vertices. Further, let G_2 be the graph of order v in which all components are isomorphic to K_2

except, possibly, of one isolated vertex (if v is odd). Both G_1 and G_2 are simple bipartite graphs of order v and size $\lfloor v/2 \rfloor$. Clearly, for the pair (G_1, G_2) we have $U = \lfloor v/2 \rfloor \geq v/2 - 1/2$, and the result follows.

R 5.17 From the fact that the graphs $G_1, G_2, \ldots, G_n$ appearing in the part III of the proof of Theorem 5.13 are all bipartite we obtain for each $n = 3$ the inequalities

$$\frac{3}{4} v - \frac{7}{2} \sqrt{2v} - \frac{21}{4} \leq U_n^*(v) \leq U_n(v) \leq \frac{3}{4} v + o_n(v).$$

(The second inequality is obvious, and the third follows from Theorem 5.13.) Thus it is sufficient to put

$$o_n^*(v) = U_n^*(v) - 3v/4.$$

R 6.19 A holds obviously, but B does not (e.g. there is a C_3^o-decomposition of K_4^* but — by virtue of 6.14.3 — there is no C_3-decomposition of K_4). Similarly, C holds but D does not (e.g. there exists a K_3-decomposition of the graph $^2K_6$ — cf. 8.3 — but K_6^* admits no C_3^o-decomposition, as seen from 10.9 and R 10.22).

R 6.20 The union of all G-decompositions of the graphs $F \in R$ forms a G-decomposition of the graph H.

R 6.21 Take an F-decomposition of H for the set R in 6.20.

R 6.22 In 6.20 put $H = {}^{\lambda_1 + \lambda_2 + \cdots + \lambda_n}K$, $R = ({}^{\lambda_1}K, {}^{\lambda_2}K, \ldots, {}^{\lambda_n}K)$.

R 6.23 A. No. (A simple counterexample: $H = K_2 \vee K_3$, $R = (K_2, K_3)$, $G = K_2$.) B, C. Yes.

R 7.43 All of these graphs are graceful (grac $G = e(G)$) except for three graphs of Fig. R 7.43.1 for which grac $G = e(G) + 1$. Graceful labellings of the remaining graphs of order ≤ 5 can be found in Golomb (1972). It is easily verified that the graphs in Fig. R 7.43.1 do not admit any graceful labelling.

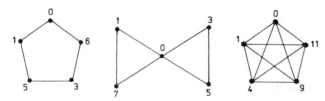

Fig. R 7.43.1

(For the first one this is a consequence of 7.16.B, for the third one it follows from 7.12.)

R 7.44 The vertices of one base of the prism will have labels 0, 2, 13, 1, and 14; the corresponding vertices of the other base will be labelled 15, 12, 4, 8, and 9.

R 7.45 For $n \equiv 2$ or 3 (mod 4), the number $e = 3n$ of edges of the Dutch n-windmill satisfies the congruence $e \equiv 2$ or 1 (mod 4) which contradicts Lemma 7.15. In the case $n \equiv 0$ or 1 (mod 4), according to Lemma 7.35, the set $\{1, 2, ..., 2n\}$ is a Skolem set, i.e. a set of the form $\{a_1, b_1, ..., a_n, b_n\}$ where $b_i = a_i + i$ for $i = 1, 2, ..., n$. But then it is sufficient to label the vertices of the i-th triangle of the Dutch n-windmill by 0, $a_i + n$, $b_i + n$, the common vertex of these triangles being labelled 0. It is a matter of routine to check that the labelling thus obtained is graceful (cf. Bermond et al. 1978a, d, Bermond 1979). (This provides a solution to a problem posed by C. Hoede — see Proc. 5th Hung. Colloq. on Combinatorics, Keszthely, 1976).

R 7.46 It is graceful for $n = 1$ and 4 (Bermond 1979) and not for $n = 2$ and 3 (Bermond et al. 1978d, Bermond 1979). There is a conjecture due to J. C. Bermond that for each $n \geq 4$ the French n-windmill is graceful.

7.47 It is if and only if $m = 1$ and $n \leq 4$ (Kotzig and Turgeon 1978), for $m = 1$ see also 7.12. However, the general problem is still open: For which values of m, n and p is the graph consisting of m copies of K_n having a common p-clique graceful? Confer also Bermond et al. (1978d) and Kotzig and Turgeon (1980).

R 7.48 See Fig. R 7.48.1 obtained by means of the labelling in Fig. 7.8.1.

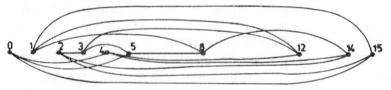

Fig. R 7.48.1

R 7.49 Let us denote vertices of the graph $K(9 \times 5)$ by the numbers 0, 1, 2, ..., 44, the edges being of lengths 1, 2, ..., 8, 10, 11, ..., 17, 19, 20, 21, 22. The base system of the decomposition will consist of the cycles with vertices 0, 4, 1, 2; 0, 8, 1, 6; 0, 13, 1, 11; 0, 17, 1, 15; and 0, 22, 1, 20.

R 7.50 For $v = 4$, the base system of the decomposition comprises two 1-factors $\{01, 23\}$ and $\{02, 13\}$; for $v = 6$ it consists of two 1-factors $\{01, 23, 45\}$ and $\{03, 15, 24\}$; for $v = 10$ we need three 1-factors $\{01, 23, 45, 67, 89\}$, $\{03, 25, 47, 69, 81\}$ and $\{05, 19, 28, 37, 46\}$. Suppose there were a cyclic decomposition of the graph K_8 into 1-factors. Then, the number of edges of length 2 in each factor of the base system should divide 4 (the number of edges in the 1-factor). However, the number of edges of length 2 in one factor cannot be 1 (because rotating it we would obtain 8 factors whereas only 7 are at disposal); it cannot be 2 (since the remaining two edges would have to be of length 4 and they would coincide after performing the rotation twice), and it cannot be 4 (some of the edges of length 2 would coincide after a rotation, and some not). Thus, such a decomposition does not exist.

R 7.51 It is the tree of order 7 depicted in Fig. R 7.51.1. All trees of smaller order are either paths or caterpillars which, according to Theorem 7.23, admit an α-labelling.

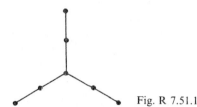

Fig. R 7.51.1

R 7.52 A. $h(00) = 0$, $h(01) = 2$, $h(10) = 4$, $h(11) = 1$. B. $h(000) = 0$, $h(001) = 5$, $h(010) = 10$, $h(011) = 3$, $h(100) = 12$, $h(101) = 4$, $h(110) = 1$, $h(111) = 7$.

R 7.53 If and only if $n \geq 2$. Namely, according to 7.18.5 and Lemma 7.9.B there exists an α-labelling h' of the cube Q_n with $\alpha(h') = 2^{n-2}(n+1) - 1$, which is not equal to $(n-1)2^{n-2}$ for $n \geq 2$. Clearly, for $n = 1$ each α-labelling of Q_1 has $\alpha = 0$.

R 7.54 From Theorems 7.11 and 7.12 it follows that $\operatorname{grac} K_5 \geq 11$. The reverse inequality follows from the existence of a labelling of vertices of K_5 by the numbers 0, 1, 4, 9 and 11 (Fig. R 7.43.1).

R 7.55 See Theorem 7.12 for $k \leq 4$. For $k = 5$ [or 6; 8; 9; 10; 12] consider a labelling of K_k by the numbers 0, 2, 7, 8, 11 [0, 1, 3, 8, 12, 18; 0, 4, 11, 20, 25, 26, 28, 38; 0, 5, 21, 29, 33, 35, 36, 46, 55; 0, 7, 12, 28, 32, 42, 43, 45, 51, 69; 0, 4, 18, 26, 35, 37, 38, 62, 67, 77, 83, 90].

R 7.56 It is obvious for $k = 1$. For other values of k it is a consequence of 7.55 and Theorem 7.10 putting there $G = K_k$ and $e = \binom{k}{2}$.

From some of the known results on the existence of cyclic (v, k, λ)-designs (8.16.C) (Singer 1938, Hall 1967, Theorem 11.3.1, Dembowski 1968, Kárteszi 1976) it follows that if n is a prime power, $k = n + 1$ and $v = n^2 + n + 1$ then there exists a cyclic K_k-decomposition of the graph K_v (and hence a ϱ-labelling of K_k).

R 7.57 The base system will consist of a single graph — the complete graph with vertices 0, 1, 2, 4, 5, 8 and 10 [or 3, 6, 7, 9, 11, 12, 13, 14].

R 8.44 It is a consequence of 8.20 and 8.22.

R 8.45 A. Follows from the definition of a (v, k, λ)-design. B. If two lines had more than one point in common then we would have a contradiction with A. According to 8.44, each projective plane is a symmetric $(v, k, 1)$-design in which $k \geq 3$. From 8.7.1—2 we get

$$v = b = v(v - 1)/(k(k - 1)), \qquad r = (v - 1)/(k - 1),$$

which implies $v = 1 + k(k - 1) = b$, $k = (v - 1)/(k - 1) = r$, i.e., each point lies on k lines. Thus, the number of lines which have a point in common with a fixed line is exactly $1 + k(k - 1)$, and this is at the same time, the total number of lines. C. Let u_1, u_2 be two distinct points. As $k \geq 3$, there are at least two distinct lines H_1, H_2 passing through u_1 and not containing u_2, and similarly two lines $H_3 \neq H_4$ passing through u_2 and avoiding u_1. Let u_3 [or u_4] be the common point of the lines H_1 and H_3 [H_2 and H_4]. It can be easily checked that the points u_1, u_2, u_3, u_4 have the required properties.

R 8.46 A. If and only if $k \in \{2, 3, 6, 7, 10, 91\}$. B. If and only if $k \in \{2, 3, 129\}$. This is a consequence of Theorems 8.29 and 8.30.

R 8.47 If and only if one of the following cases occur: A. $m = 1$, $n_1 = v$, $1 \leq \lambda_1 = \lambda$. B. $m = v$, $n_1 = n_2 = \ldots = n_m = 1$, $1 \leq \lambda_2 = \lambda$. C. $m = n_1 = v = 1$. D. $n_1 + n_2 + \ldots + n_m = v$, $1 \leq \lambda_1 = \lambda_2 = \lambda$.

R 8.48 In the case C we should add $v = 141, 151, 156, 166, 171, 181, 186$ and 196; in the case D $v = 169$ and 175; in the case E $v = 169$ and 176, in the case F $r = 145$ and 153; in the case G $v = 145, 181$ and 190; in the case I $r = 144$ and 177. Further, we should add the cases J ($k = 13$, $v \in \{157, 169\}$) and K ($k = 14$, $v = 183$).

R 8.49 In the case C, one should add $v = 151, 156, 181$ and 186; in the case D $v = 169$. Further one should consider two more cases: J ($k = 13$, $v = 169$) and K ($k = 14$, $v = 183$).

R 8.50 26.

R 8.51 $[15, 5], [21, 6], [36, 8], [55, 10], [78, 12], [91, 13]$.

R 8.52 A. $k!^{\binom{k}{2} - 1}$. B. 1.

R 8.53 Exactly one.

R 8.54 For $v = k = \lambda$ the existence follows from 8.52. For $v = k + 1$, $\lambda = k - 1$ see 8.53.

R 8.55 To a given (v, k, λ)-design it is sufficient to construct a complementary design on the same set of points whose blocks are the complements of the blocks of the original design. If the original one had parameters v, b, r, k, λ. then the new one has $v' = v$, $b' = b$, $r' = b - r$, $k' = v - k$, $\lambda' = b - 2r + \lambda$, as it can be obtained easily using incidence matrices (8.3). The rest is a consequence of 8.7.

R 8.56 All ordered triples $[k, k, k]$ and $[k + 1, k, k - 1]$, $2 \leq k \leq 8$, as well as the triples $[7, 3, 1], [7, 4, 2], [13, 4, 1], [11, 5, 2], [21, 5, 1], [11, 6, 3], [16, 6, 2], [31, 6, 1], [15, 7, 3], [15, 8, 4], [57, 8, 1]$. The necessity follows from 8.20 and 8.24; the sufficiency from 8.20, R 8.54, 8.14, 8.26, 8.24, 7.57, and 8.55.

R 8.57 A. (7, 4, 2)-design with blocks 1234, 1256, 1357, 1467, 2367, 2457, 3456. B. Design with blocks 124, 1467, 1257, 1236, 2347, 1345, 2456.

R 8.58 A. The $(7, 4, 2)$-design of R 8.57. A. B. $(7, \{3, 5\}, 2)$-design with blocks 12345, 12567, 13467, 236, 247, 357, 456.

R 8.59 A design with block 12, 13, ..., $1v$, $23\ldots v$, for each $v \geq 2$. If $v = n^2 + n + 1$, $n \geq 2$, the conditions are satisfied also by each projective plane of order n (if it exists).

R 9.32 The necessity follows from 6.14.2, putting $e(G) = 2$, $\lambda = 1$. The sufficiency can be proved by means of a construction. Let us denote the vertices of the graph K_v by $u_0, u_1, ..., u_{v-1}$. If $v \equiv 0 \pmod 4$, the subgraphs of the L_4-decomposition of K_v will be induced by pairs of edges $\{u_n u_{n+1}, u_{n+2} u_{n+3}\}$ for $n \equiv 0$ or $1 \pmod 4$, $0 \leq n \leq v - 3$ and $\{u_n u_{n+i}, u_{n+1} u_{n+i+1}\}$ for $n = 0, 1, 2, ..., v - 1$; $i = 2, 3, ..., v/2$. (All subscripts should be taken mod v.) In the case $v \equiv 1 \pmod 4$ we first decompose the edges of K_v into classes A_n ($n = 0, 1, 2, ..., v - 1$) comprising the edges $u_i u_j$ for which $i + j \equiv n$ $\pmod v$. Each class A_n contains an even number of edges and these are pairwise non-adjacent. Partitioning each class A_n into pairs we get the required decomposition.

R 9.33 If $G |^\lambda K_v$ we have $2e|\lambda v(v - 1)$ by virtue of 6.14.2. Since e and λ are coprime we get $e|v(v - 1)$ or $2e|v(v - 1)$ depending on whether e is odd or even.

R 9.34 A consequence of 9.33 for $e = k - 1$, $\lambda = 1$, $G = T$.

R 9.35 For $V(K_{16}) = \{0, 1, 2, ..., 15\}$ and $V(Q_3) = \{000, 001, ..., 111\}$ the decomposition is given in Table R 9.35.1 which represents isomorphisms between Q_3 and the subgraphs H_i of K_{16} ($i = 1, 2, ..., 10$).

Table R 9.35.1

Q_3	H_1	H_2	H_3	H_4	H_5	H_6	H_7	H_8	H_9	H_{10}
000	0	8	0	2	0	4	0	1	0	1
001	1	9	4	6	8	12	2	3	11	10
010	3	11	7	5	9	13	6	7	13	12
011	2	10	3	1	1	5	4	5	6	7
100	5	13	12	14	14	10	10	11	15	14
101	4	12	8	10	6	2	8	9	4	5
110	6	14	11	9	7	3	12	13	2	3
111	7	15	15	13	15	11	14	15	9	8

R 9.36 Let H be a simple cubic graph containing a 1-factor L. Deleting the edges of L from H yields a 2-regular graph Q which is obviously decomposable into cycles and 2-way infinite paths. Thus, it is possible to direct edges of Q in such a way that each vertex is an initial vertex of exactly one arc. Assign to each edge w of L the two arcs of Q terminating at endvertices of w. These three edges (after cancelling the orientation) create a path of order 4, and all paths thus obtained determine a P_4-decomposition of the graph H.

R 10.21 The necessary condition is a consequence of Theorem 11.9.B: Let us denote by a [b, c, respectively,] the number of cycles of length 4 of the form 12341 [13421, 14231] occurring in a decomposition of $^\lambda K_4^*$ with vertices 1, 2, 3, 4 into cycles of length 4. Since edges of the form 23, 34 and 42 can occur only in the above 4-cycles we get $a + c = a + b = b + c = \lambda$, which yields $a = b = c = \lambda/2$, and thus λ must be even. Conversely, if λ is even we easily construct a C_4°-decomposition of $^\lambda K_4^*$ using $\lambda/2$ copies of the cycles 12341, 13421, 14231, 12431, 13241, and 14321.

R 10.22 If and only if $k = 2$ or $k = 5$. A sketch of the proof: From 10.7 for $v = 6$, $\lambda = 1$ and from 11.9.B we get the necessary condition $k \in \{2, 3, 5\}$. An examination of all possible cases excludes the value $k = 3$ (cf. Bermond 1974). For example, for $k = 5$ we have the following C_5°-decomposition of the graph K_6^* with vertices 1, 2, ..., 6: (123451, 136421, 152631, 162541, 143561, 246532). The assertion is obvious for $k = 2$.

R 10.23 The assertions follow from the facts listed below: A. The number of vertices in a cycle cannot exceed the order of the graph $K(m \times n)$. B. Each vertex of $K(m \times n)$ has an even degree. C. The size of $K(m \times n)$ has to be divisible by the order of the cycle.

(Note that, in some cases, the conditions A—C are also sufficient, e.g. if $k = 4$ or if n is odd and k even ≥ 4 (cf. Cockayne and Hartnell 1976, 1977.)

R 10.24 Let $v \equiv 12 \pmod{24}$ and let G be a loopless undirected graph of order v with vertex set $\{1, 2, ..., v\}$, two vertices i and j being adjacent if and only if $i \neq j$ and $i - j \not\equiv 2 \pmod 4$; every pair of adjacent vertices is supposed to be joined by λ edges. Clearly, G is a regular graph of even order equal to $3\lambda v/4 - \lambda$, its order being v, size $e \equiv 0 \pmod 3$, and multiplicity λ. Call an edge ij *even* if $i \equiv j \pmod 4$; otherwise ij will be *odd*. Obviously, each triangle contains an even edge. If there were a K_3-decomposition of G, the number of odd edges would not exceed twice the number of even edges — but it is a matter of routine to check that it does not hold in G. Thus, replacing the ratio 3/4 in the Problem 10.17 by a number $c < 3/4$ we get (for $v \geq (3/4 - c)^{-1}$, $v \equiv 12 \pmod{24}$) a statement which cannot be true in general.

R 10.25 (123, 214, 341, 432), (123, 214, 342, 431), (123, 241, 314, 432).

R 10.26 Only one, e.g., (123, 134, 142, 243).

R 11.56 If $(G_0, G_1, ..., G_{v-1})$ is a decomposition of K_v into almost 1-factors, constructed on the basis of 7.10 and 7.41, then $(G_0 \cup G_1, G_2 \cup G_3, ..., G_{v-3} \cup G_{v-2}, G_{v-1})$ is a decomposition with all the required properties.

R 11.57 Based on 7.41 and 7.42 let us construct a decomposition $(G_0, G_1, ..., G_{v-2})$ of K_v into $v - 1$ 1-factors. The required decomposition is then $(G_0 \cup G_1, G_2 \cup G_3, ..., G_{v-4} \cup G_{v-3}, G_{v-2})$.

R 11.58 It is sufficient to replace each Hamiltonian path in the decomposition of 11.56 by two directed Hamiltonian paths and each edge in the factor G_{v-1} by a directed cycle of order 2.

R 11.59 Apply a similar method as in R 11.58.

R 11.60 If v is even then use a similar method as in the proof of Lemma 11.1.A to obtain a P_v-decomposition of K_v. Replacing each path of the decomposition by two antipaths (oppositely directed) we get the required decomposition of the graph K_v^*. If v is odd we employ an analogous method using the proof of Lemma 11.1.B.

R 11.61 If the vertices are 1, 2, ..., 7, then one of the decompositions can be determined by the following 7 Hamiltonian paths: 1234567, 2135476, 3146275, 4157263, 5361742, 6524371, 7325164 (cf. Bermond and Faber 1976).

R 11.62 (Thomassen 1979) A. No. B. Yes, see Fig. R 11.62.1. C. Yes, see Fig. 11.62.2.

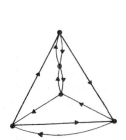

Fig. R 11.62.1

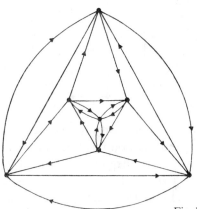

Fig. R 11.62.2

R 11.63 A follows from 11.22.1. Let 1, 2, ..., 5 be vertices of the graph K_5. There are exactly 6 decompositions of K_5 into Hamiltonian cycles: (123451, 135241), (123541, 134251), (124351, 145231), (124531, 143251), (125341, 154231), (125431, 153241). It is easy to check that for each of these decompositions there exists exactly one orthogonal decomposition. Thus, each maximal set of 1-orthogonal decompositions of K_5 into Hamiltonian cycles has exactly 2 elements, which implies B.

R 11.64 No. For instance, let F be a simple graph of order 4 and size 4 with a triangle. Put $G = H = F$. Clearly, both G and H have an F-decomposition but $G \times H$ has not any. The reason is that such a decomposition would have to comprise 8 graphs isomorphic to F, i.e., 8 graphs containing triangles; however, $G \times H$ contains only 6 triangles.

R 11.65 Let 0, 1, ..., $v-1$ be the vertices of the graph in question. Let the vertices u and U be adjacent in the graph if and only if $|u - U| \in \{1, 2, v - 2, v - 1\}$. We thus obtain a simple 4-regular graph of order v. The fact that each of its decompositions into 2-factors contains a Hamiltonian cycle is proved in Doutre and Kotzig (1980).

R 11.66 For $n = 0$ see Fig. 11.40.1, for $n = 4$ Fig. R 11.66.1. If n is odd, $n \geq 5$, then consider the graph of the n-prism, and for even $n \geq 6$ the graph of the $(n - 2)$-prism.

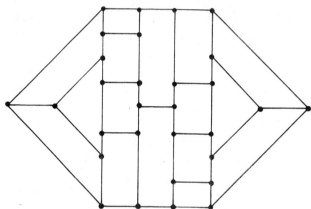

Fig. R 11.66.1

R 11.67 The necessity is a consequence of the fact that ${}^\lambda K_v$ must contain a 1-factor and be of odd degree. If v is even we can construct a 1-factorization $(F_1, F_2, ..., F_{v-1})$ of the graph K_v with vertices $0, 1, ..., v - 2$ and ∞ as follows: The factor F_i ($i = 0, 1, ..., v - 2$) consists of the edges uU for which $u + U \equiv i$ (mod $v - 1$) and the edge $j\infty$ such that $2j \equiv i$ (mod $v - 1$); compare with 7.42. Then, $(F_0 \cup F_1, F_2 \cup F_3, ..., F_{v-4} \cup F_{v-3}, F_{v-2})$ is a quasi-Hamiltonian decomposition of K_v. Since the graph ${}^{\lambda - 1}K_v$ has a Hamiltonian decomposition (cf. Corollary 11.4.B) the graph ${}^\lambda K_v$ admits a quasi-Hamiltonian decomposition.

R 11.68 Never, since it is of even degree.

R 12.58 It does not exist for any n. This is a consequence of the fact that any two lines in a projective plane intersect (see Exercise 8.45.B).

R 12.59 A. The projective plane comprises the points $1, 2, 3, ..., 9, a, b, c, d$ and the lines $123a, 456a, 789a, 147b, 258b, 369b, 159c, 267c, 348c, 168d, 249d, 357d, abcd$.

B. The following four 3-matrices form an orthogonal set:

$$\begin{bmatrix} 1 & 2 & 3 \\ 2 & 3 & 1 \\ 3 & 1 & 2 \end{bmatrix} \begin{bmatrix} 1 & 2 & 3 \\ 3 & 1 & 2 \\ 2 & 3 & 1 \end{bmatrix} \begin{bmatrix} 1 & 1 & 1 \\ 2 & 2 & 2 \\ 3 & 3 & 3 \end{bmatrix} \begin{bmatrix} 1 & 2 & 3 \\ 1 & 2 & 3 \\ 1 & 2 & 3 \end{bmatrix}.$$

C. The required code consists of the words 1111, 2212, 3313, 2321, 3122, 1223, 3231, 1332, 2133.

D. The group divisible $(4 \times 3, 4, 1)$-design has points 11, 12, 13, 14, 21, 22, 23, 24, 31, 32, 33, 34; groups (11, 21, 31), (12, 22, 32), (13, 23, 33) and blocks {11, 12, 13, 14}, {21, 22, 13, 24}, {31, 32, 13, 34}, {21, 32, 23, 14}, {31, 12, 23, 24}, {11, 22, 23, 34}, {31, 22, 33, 14}, {11, 32, 33, 24}, {21, 12, 33, 34}.

E. We can obtain a K_4-decomposition of the graph $K(4 \times 3)$ from D by considering the points to be vertices, groups to be partites, and blocks to be the subgraphs of the decomposition.

F. Let us denote the vertices of K_9 by 1, 2, ..., 9. The first factor consists of the triangles 123, 456 and 789, the second of 147, 258 and 369, the third of 159, 267 and 348, and the fourth of 168, 249 and 357.

G. It is a consequence of F.

H. Follows from A, if the points of the projective plane are considered to be the vertices, and the lines to be the complete subgraphs of the graph K_{13}.

R 12.60 Three, since $M(6) = N(6) + 2 = 3$ (cf. 12.27.F, 12.29, 13.30).

R 12.61 Let us denote the vertices of the graph K_{12} by 0, 1, 2, ..., 10, ∞. The first replication will consist of four transitive tournaments [0, 2, 10], [7, 1, 5], [8, 9, 4], [3, ∞, 6] (every ordered triple represents the vertices of one transitive tournament of order 3 whose edges are directed "from the left to the right"). Further classes are obtained from the above by rotating cyclically the subscripts modulo 11, the vertex ∞ being left unchanged.

R 12.62 Let the vertices of the graph be 0, 1, ..., 81, u, U; the omitted 1-factor has the edges (0,41), (1,42), ..., (40, 81), (u, U). The first factor of the decomposition consists of the triangles (u, 35, 75), (U, 34, 76), (0, 38, 70), (1, 3, 64), (2, 6, 18), (4, 10, 24), (5, 12, 39), (7, 50, 67), (8, 32, 77), (9, 26, 48), (11, 57, 68), (13, 60, 69), (14, 37, 81), (15, 25, 33), (16, 27, 52), (17, 22, 71), (19, 28, 54), (20, 21, 72), (23, 42, 44), (29, 41, 79), (30, 58, 63), (31, 61, 62), (36, 49, 73), (40, 55, 78), (43, 47, 59), (45, 51, 65), (46, 53, 80), (56, 66, 74). The remaining 40 factors are obtained by shifting cyclically the numbers assigned to vertices, respectively by 1, 2, ..., 40 modulo 82, leaving the vertices u and U unchanged.

R 12.63 Let us take the union of two consecutive K_3-factors (step by step) so as to form 5 new factors, and let us decompose each of them into three P_3-factors, obtaining thereby a decomposition of the graph K_{21} into 15 P_3-factors. An arrangement for the handcuffed prisoners problem for $v = 21$ can then have the following form:

1st day: (1 2 5,   8 11 10, 18 15 6,   3 17 9, 12 19 21, 13 20 4, 14 7 16).
2nd day: (2 4 1, 11 9 8, 15 16 18, 17 10 3, 19 5 12, 20 6 13,   7 21 14).
3rd day: (2 3 5, 11 13 10, 15 14 6, 17 1 9, 19 18 21, 20 12 4,   7 8 16).
4th day: (3 4 5, 12 8 10, 21 17 18, 19 1 11, 20 2 15, 14 16 13,   7 9 6).
5th day: (4 6 3,   8 13 12, 17 15 21,   1 10 19,   2 11 20, 16 5 14,   9 18 7).

6th day: (4 7 5,   8 14 10,   17 20 18,   1 21 11,   2 12 15,   16 3 13,   9 19 6).
7th day: (1 6 2,   10 12 11,   21 16 19,   17 13 18,   14 3 20,   8 4 9,   7 15 5).
8th day: (6 5 1,   12 9 10,   16 20 21,   13 2 17,   3 18 14,   4 19 8,   15 11 7).
9th day: (6 7 2,   12 14 11,   16 17 19,   13 1 18,   3 8 20,   4 21 9,   15 10 5).
10th day: (3 1 15,   9 13 7,   19 20 5,   8 2 10,   16 4 14,   17 11 6,   18 12 21).
11th day: (1 7 3,   13 14 9,   20 15 19,   2 21 8,   4 10 16,   11 5 17,   12 6 18).
12th day: (1 8 15,   13 19 7,   20 9 5,   2 18 10,   4 17 14,   11 16 6,   12 3 21).
13th day: (9 2 19,   11 3 15,   13 4 18,   8 5 21,   10 6 17,   12 7 20,   14 1 16).
14th day: (2 16 9,   3 19 11,   4 15 13,   5 18 8,   6 21 10,   7 17 12,   1 20 14).
15th day: (2 14 19,   3 9 15,   4 11 18,   5 13 21,   6 8 17,   7 10 20,   1 12 16).

R 12.64 If H_1, H_2, ..., H_{10} are as in R 9.35 then $(H_1 \cup H_2, H_3 \cup H_4, ..., ..., H_9 \cup H_{10})$ is a suitable decomposition.

R 12.65 A. If 12.14.1 does not hold then clearly $\lambda \neq 1$, and according to the proof of the Corollary 12.11 we have $\lambda | k - 1$. This leaves 14 ordered triples $[v, k, \lambda]$ to be considered. However, five of them, namely [8, 4, 3], [12, 6, 5], [16, 8, 7], [27, 9, 4] and [20, 10, 9] satisfy the condition 12.14.1. Thus there are nine ordered triples fulfilling 12.5.1—2 but not 12.14.1: [6, 3, 2], [10, 5, 4], [15, 5, 2] (in this case see also 8.14.4), [14, 7, 6], [21, 7, 3], [28, 7, 2], [18, 9, 8], [45, 9, 2], [40, 10, 3].

B. $[2k, k, \lambda]$ for $k \in \{3, 5, 7, 9\}$, $\lambda \equiv k - 1 \pmod{2k - 2}$.

C. [36, 6, 1].

R 12.66 $v \geq k + k(k - 1)/\lambda$, thus $b \geq v + r - 1$. Namely, by virtue of the proof of Theorem 12.10 we obtain $\lambda v - \lambda k - k^2 + k \geq 0$ which implies the first inequality. The second one is a consequence of the first one, using 8.7.1—2. (The equality corresponds to the case of the affine plane dealt with in 12.9—12.12.) Cf. Bose (1942), Raghavarao (1971, pp. 60—61) and Vanstone (1979).

R 12.67 Assume that the conditions 12.5.1—2 hold. Then the similar is true for 12.5.3—4. From 12.5.3 it follows that $k - 1|s(v - k) + s(k - 1)$, i.e. $s(v - 1)|\lambda(v - 1)$, thus we have 12.4.1. From 12.5.4 we get $k|s(v - k) + sk = sv$, and since g.c.d. $\{k, s\} = 1$ we have 12.4.2. The reverse implication is a consequence of the proof of Theorem 12.5.

R 12.68 Assuming that $v = (k^2 + k)/2$ we would have $(k^2 + k)/2|k^2$ according to 12.14.1, which is true only for $k = 1$.

R 13.21 No.

R 13.22 A group consisting of the identity permutation of two elements.

R 13.23 Let $I_n [S_n, Z_n,$ respectively] denote the identity [symmetric, cyclic] permutation group of order n (cf. Harary 1969, Tab. 14.2.1). For the group I_3 it is sufficient to consider the Example 13.13, for $I_1 \times S_2$ the Example 13.5 (or 13.18), for the group S_3 — the decomposition of K_3 into three factors of size 1, and finally for the group Z_3 the decomposition of the graph K_4 into 3 factors of size 2 containing one null vertex.

R 13.24 The graph with vertices 1, 2, ..., 6 and edges 12, 23, 45, 56 has a decomposition ({12, 45}, {23, 56}) according to the cyclic group of order two generated by the permutation (13)(2)(46)(5). This decomposition is not cyclic because the number of edges of the graph is not divisible by 6. As another example can serve the graph with vertices 1, 2, ..., 8, edges 12, 23, 34, 41, 56, 67, 78, 85 and its decomposition ({12, 56}, {23, 67}, {34, 78}, {41, 85}) according to the cyclic group of order 4 generated by the permutation (1234)(5678); it can be easily checked that it is not a cyclic decomposition.

R 14.26 From Theorem 14.2 it follows that any complementing permutation of a self-complementary graph of odd order has a cycle of length 1. The corresponding vertex has to be incident with the same number of edges in G and $\bar{G}$, having thus the degree as required.

R 14.27 Let G be a self-complementary graph of order $v - 4$ (such a graph exists by Theorem 14.4). Let us add to G the new vertices u_1, u_2, u_3, u_4, edges u_1u_2, u_2u_3, u_3u_4, and all the edges joining the vertices of G with u_1 or u_4. It is easy to see that we obtain a self-complementary graph of order v and diameter 2.

R 14.28 We proceed similarly as in R 14.27 (the edge u_2u_3 is to be replaced by u_1u_4).

R 14.29 Denote the vertices by 1, 2, 3, 4. A. The self-complementary tournaments have edges 12, 13, 23, 24, 34, and one of the edges 14, 41. B. The remaining self-complementary digraphs of order 4 have edges 12, 21 and four other edges for which we have the following possibilities: {23, 32, 34, 43}, {13, 31, 32, 41}, {13, 14, 31, 32}, {13, 14, 23, 24}, {13, 23, 24, 41}, {13, 24, 32, 41}, {31, 32, 41, 42}, {13, 32, 41, 42}.

R 14.30 $s(2) = 0, s(3) = 1, s(4) = 2, s(5) = 4, s(6) = 6$.

R 14.31 It is sufficient to take a simple graph with vertices u and u_{ij}, where $i = 1, 2, 3, 4; j = 1, 2, ..., n$, the adjacency being defined as follows: u is adjacent to u_{ij} if and only if $i \in \{1, 4\}$; u_{ij} and u_{IJ} are adjacent if and only if $|i - I| = 1$, or $i = I \in \{1, 4\}$ and $j \neq J$.

R 15.14 Assume first that g.c.d.$\{b, v\} = 1$. Then we obtain successively implies $2b|v - 1$. According to 13.7, $b|K_v$. In the case when g.c.d.$\{2b, b$ is even then v must be odd. But then $2b$ and v are coprime, and $2b|v(v - 1)$ implies $2b|v - 1$. According to 13.7, $b|K_v$. In the case when g.c.d. $\{2b, v - 1\} = 1$ we proceed analogously.

R 15.15 Up to isomorphism, only the graph $K_2 \vee K_3$.

R 15.16 The digraph of order 3 and size 4, with vertices 1, 2, 3 and edges 12, 21, 23, 31 (and all digraphs isomorphic to it).

REFERENCES

Abramson, H. D.: 'An Algorithm for Finding Euler Circuits in Complete Graphs of Prime Order', in *Théorie des graphes — Theory of Graphs* (Journées int. d'étude, Rome, 1966), ed. Rosenstiehl, P., Dunod, Paris — Gordon and Breach, New York, 1967, pp. 1—7, MR 36: 5015. Cit. 11.5.

Achuthan, N.: 'On the n-Clique Chromatic Number of Complementary Graphs, in *Combinatorics '79*, Part II (Proc. Colloquium Univ. Montreal, Montreal, 1979), eds. Deza, M. and Rosenberg, I. G., Annals of Discrete Mathematics 8—9, North-Holland, Amsterdam, 1980, pp. 285—290, MR 82c: 05046. Cit. 4.18, 4.24, 4.25.

Ahrens, W.: *Mathematische Unterhaltungen und Spiele,* Teubner, Leipzig, 1901, 1918. Cit. 12.8.

Ajtai, M., Komlós, J., Rödl, V. and Szemerédi E.: 'On Coverings of Random Graphs', *Comment. Math. Univ. Carolin.* **23** (1982) 193—198, MR 83f: 05056. Cit. 10.17.

Akiyama, J. and Harary, F.: 'A Graph and Its Complement with Specified Properties, I, Connectivity', *Int. J. Math. Math. Sci.* **2** (1979a) 223—228, MR 82i: 05062a. Cit. 5.5, 5.10.

Akiyama, J. and Harary, F.: 'A Graph and Its Complement with Specified Properties, II, Unary Operations', *Nanta Math.* **12** (1979b) 110—116, MR 82i: 05062b. Cit 5.10.

Akiyama, J. and Harary, F.: 'A Graph and Its Complement with Specified Properties, III, Girth and Circumference', *Int. J. Math. Math. Sci.* **2** (1979c) 685—692, MR 82i: 05062c. Cit. 5.5, 5.10.

Akiyama, J., Exoo, G. and Harary, F.: 'A Graph and Its Complement with Specified Properties, V, The Self-Complement Index', Mathematika **27** (1980) 64—68, MR 82i: 05062e. Cit. 5.10, 11.54.

Akiyama, J. and Harary, F.: 'A Graph and Its Complement with Specified Properties, IV, Counting Self-Complementary Blocks', *J. Graph Theory* **5** (1981) 103—107, MR 82i: 05062d. Cit. 5.7—5.9, 5.15, 14.7—14.9.

Akiyama, J., Harary, F. and Ostrand, P.: 'A Graph and Its Complement with Specified Properties, VI, Chromatic and Achromatic Numbers', *Pac. J. Math.* **104** (1983) 15—27. Cit. 5.10.

Alavi, Y. and Behzad, M.: 'Complementary Graphs and Edge Chromatic Numbers', *SIAM J. Appl. Math.* **20** (1971) 161—163, MR 44: 3923. Cit 4.26.

Alavi, Y. and Mitchem, J.: 'The Connectivity and Line-Connectivity of Complementary Graphs', in *Recent Trends in Graph Theory* (Proc. 1st New York City Graph Theory Conference, New York, 1970), eds. Capobianco, M., Frechen, J. B. and Krolik, M., Lecture Notes in Mathematics 186, Springer-Verlag, Berlin, 1971, pp. 1—3, MR 43: 3142. Cit. 4.18, 4.28, 4.30, 4.40.

Alekseev, V. E.: 'On the Number of Steiner Triple Systems', *Mat. Zametki* **15** (1974) 767—774 (in Russian); *Math. Notes* **15** (1974) 461—464 (Engl. transl.), MR 50: 12754. Cit 8.34.

Alon, N.: 'A Note on the Decomposition of Trees into Isomorphic Subtrees', *Ars Combin.* **12** (1981) 117—121, MR 83g: 05006. Cit. 9.17, 9.18.

204

Alon, N. and Caro, A.: 'More on the Decomposition of Trees into Isomorphic Subtrees', *Ars Combin.* **14** (1982) 123—130. Cit. 9.17, 9.18.

Alon, N.: 'A Note on the Decomposition of Graphs into Isomorphic Matchings', *Acta Math. Hung.* **42** (1983) 221—223, MR 84h: 05101a. Cit. 9.28.

Alspach, B., Mason, D. W. and Pullman, N. J.: 'Path Numbers of Tournaments', *J. Combin. Theory* **B20** (1976) 222—228, MR 54: 10065. Cit. 11.54.

Alspach, B., Heinrich, K. and Varma, B. N.: 'Decompositions of Complete Symmetric Digraphs into the Oriented Pentagons', *J. Aust. Math. Soc.* **A28** (1979) 353—361, MR 81g: 05087. Cit. 10.11.

Alspach, B.: 'Hamiltonian Partitions of Vertex-Transitive Graphs of Order $2p$', in *Proc. 11th Southeastern Conference on Combinatorics, Graph Theory and Computing* (Florida Atlantic Univ., Boca Raton, 1980), eds. Hadlock, F. O., Hoffman, F., Mullin, R. C., Stanton, R. G. and van Rees, G. H. J., Congressus Numerantium 28—29, Utilitas Math., Winnipeg, 1980, Vol. I, pp. 217—221, MR 82f: 05066. Cit. 11.15, 11.52.

Alspach, B. and Varma, B. N.: 'Decomposing Complete Graphs into Cycles of Length $2p^{e\perp}$', in *Combinatorics '79*, Part II (Proc. Colloquium University Montreal, Montreal, 1979), eds. Deza, M. and Rosenberg, I. G., Annals of Discrete Mathematics 8—9, North-Holland, Amsterdam, 1980, pp. 155—162, MR 82b: 05087. Cit. 10.5, 10.6.

Alspach, B.: 'The Search for Long Paths and Cycles in Vertex-Transitive Graphs and Digraphs', in *Combinatorial Mathematics VIII* (Proc. 11th Australian Conference on Combinatorial Mathematics, Deakin University, Geelong, 1980), ed. McAvaney, K. L., Lecture Notes in Mathematics 884, Springer-Verlag, Berlin—Heidelberg—New York, 1981, pp. 14—22, MR 83b: 05080. Cit. 11.52.

Alspach, B., Heinrich, K. and Rosenfeld, M.: 'Edge Partitions of the Complete Symmetric Directed Graph and Related Designs', *Israel J. Math.* **40** (1981) 118—128, MR 82k: 05051. Cit 10.16.

Alspach, B.: 'Problem 19', *Discrete Math.* **40** (1982) 321—322. Cit 15.13.

Alter, R.: 'A Characterization of Self-Complementary Graphs of Order 8', *Portug. Math.* **34** (1975) 157—161, MR 53: 10643. Cit 14.6.

Andersen, L. D., Hilton, A. J. W. and Mendelsohn E.: 'Embedding Partial Steiner Triple Systems', *Proc. London Math. Soc.* **41** (1980) 557—576, MR 82a: 05010. Cit. 8.16.

Anderson, B. A.: 'Symmetry Groups of Some Perfect 1-Factorizations of Complete Graphs', *Discrete Math.* **18** (1977) 227—234, MR 57: 2979. Cit 13.4.

Anderson, B. A., Barge, M. M. and Morse, D.: 'A Recursive Construction of Asymmetric 1-Factorizations', *Aeqs. Math.* **15** (1977) 201—211, MR 57: 9605. Cit. 13.4, 14.7.

Araoz, J. and Tannenbaum, P.: 'Minimal Sets of Conditions for Balanced Incomplete Block Designs', in *Proc. 2nd Int. Conference on Combinatorial Mathematics* (New York, 1978), eds. Gewirtz, A. and Quintas, L. V., Annals of the New York Academy of Sciences 319, New York Academy of Sciences, New York, 1979, pp. 28—32, MR 81h: 05021. Cit. 8.36.

Arnautov, V. I.: 'On the External Stability Number of a Graph', *Diskret. Analiz* **20** (1972) 3—8 (in Russian), MR 47: 3246. Cit. 4.18.

Aubert, J. and Schneider, B.: 'Décomposition de $K_m + K_n$ en cycles hamiltoniens', *Discrete Math.* **37** (1981) 19—27. Cit 11.33.

Aubert, J. and Schneider, B.: 'Graphes orientés indécomposables en circuits hamiltoniens', *J. Combin. Theory* **B32** (1982a) 347—349. Cit. 11.12.

Aubert, J. and Schneider, B.: 'Décomposition de la somme cartésienne d'un cycle et de l'union de deux cycles hamiltoniens en cycles hamiltoniens', *Discrete Math.* **38** (1982b) 7—16. Cit. 11.30, 11.34.

Ayel, J. and Favaron, O.: 'The Helms Are Graceful', in *Progress in Graph Theory* (Proc.

Conference Univ. Waterloo, Waterloo, 1982), eds. Bondy, J. A. and Murty, U. S. R., Academic Press, Toronto, 1984, pp. 89—92. Cit. 7.18.

Babai, L.: 'Almost All Steiner Triple Systems Are Asymmetric', in *Topics on Steiner Systems*, eds. Lindner, C. C. and Rosa, A., Annals of Discrete Mathematics 7, North-Holland, Amsterdam, 1980, pp. 37—39. Cit 8.34.

Baker, R. D.: 'Whist Tournaments', in *Proc. 6th Southeastern Conference on Combinatorics, Graph Theory and Computing* (Florida Atlantic Univ., Boca Raton, 1975), eds. Hoffman, F., Mullin, R. C., Levow, R. B., Roselle, D., Stanton, R. G. and Thomas, R. S. D., Congressus Numerantium 14, Utilitas Math., Winnipeg, 1975, pp. 89—100, MR 54: 2513. Cit. 12.45.

Baker, R. D. and Wilson, R. M.: 'The Whist Tournament Problem of E. H. Moore', *Proc. 6th Southeastern Conference on Combinatorics, Graph Theory and Computing* (Florida Atlantic Univ., Boca Raton, 1975), eds. Hoffman, F., Mullin, R. C., Levow, R. B., Roselle, D., Cit. 12.45, 12.49.

Baker, R. D. and Wilson, R. M.: 'The Whist Tournament Problem of E. H. Moore', *Proc. 6th Southeastern Conference on Combinatorics, Graph Theory and Computing* (Florida Atlantic Univ., Boca Raton, 1975), eds Hoffman, F., Mullin, R. C., Levow, R. B., Roselle, D., Stanton, R. G. and Thomas, R.S.D., Congressus Numerantium 14, Utilitas Math., Winnipeg, 1975, pp. 101—110. Cit. 12.45.

Bange, D. W., Barkauskas, A. E. and Slater, P. J.: 'Simply Sequential and Graceful Graphs', in *Proc. 10th Southeastern Conference on Combinatorics, Graph Theory and Computing*, Vols. I, II (Florida Atlantic Univ., Boca Raton, 1979), eds. Hoffman, F., McCarthy, D., Mullin, R. C. and Stanton, R. G., Congressus Numerantium 23—24, Utilitas Math., Winnipeg, 1979, pp. 155—162, MR 82a: 05072. Cit 7.19, 7.21.

Bange, D. W., Barkauskas, A. E. and Slater, P. J.: 'Conservative Graphs', *J. Graph Theory* 4 (1980) 81—91, MR 81b: 05089. Cit. 7.19.

Baranyai, Z.: 'On the Factorization of the Complete Uniform Hypergraph', in *Infinite and Finite Sets* (Proc. Colloquium, Keszthely, 1973), eds. Hajnal, A., Rado, R. and Sós, V. T., Colloquia Mathematica Societatis János Bolyai 10, Bolyai János matematikai társulat, Budapest — North-Holland, Amsterdam, 1975, pp. 91—108, MR 54: 5047. Cit. 12.33.

Baranyai, Z. and Szász, G. R.: 'Hamiltonian Decomposition of Lexicographic Product', *J. Combin. Theory* **B31** (1981) 253—261, MR 83c: 05085. Cit 11.38.

Bays, S.: 'Une question de Cayley relative au problème des triades de Steiner', *Enseignement Math.* **19** (1917) 57—67. Cit. 12.33.

Bays, S.: 'Recherche des systèmes cycliques de triples de Steiner différents pour *N* premier (ou puissance de nombre premier)', *J. Math. Pures Appl.* **2** (1923) No. 9, 73—98. Cit 8.35.

Behzad, M. and Chartrand, G.: *Introduction to the Theory of Graphs*, Allyn and Bacon, Boston, 1971, MR 55: 5449. Cit. 1.1.

Beineke, L. W.: 'Derived Graphs with Derived Complements', in *Recent Trends in Graph Theory* (Proc. 1st New York City Graph Theory Conference, New York, 1970), eds. Capobianco, M., Frechen, J. B. and Krolik, M., Lecture Notes in Mathematics 186, Springer-Verlag, Berlin, 1971, pp. 15—24, MR 44: 3901. Cit. 5.10.

Beka, J.: 'On the Tripartite Conjecture', *Math. Slov.* **35** (1985) 239—241. Cit. 15.9.

Bennet, F. E., Mendelsohn, E. and Mendelsohn, N. S.: 'Resolvable Perfect Cyclic Designs', *J. Combin. Theory* **A29** (1980) 142—150, MR 82b: 05030. Cit. 12.28, 12.42, 12.56.

Bennet, F. E. and Sotteau, D.: 'Almost Resolvable Decomposition of K_n^*', *J. Combin. Theory* **B30** (1981) 228—232. Cit. 12.50, 12.56.

Berge, C.: *Théorie des graphes et ses applications*, Dunod, Paris, 1958, MR 21: 1608 (Engl. transl.: *The Theory of Graphs and Its Applications*, Methuen, London, 1962, MR 24: A 2381). Cit 11.6, 11.15.

Berge, C.: *Graphes et hypergraphes*, Dunod, Paris, 1970, 1973, MR 50: 9641, 9639 (Engl. transl.: *Graphs and Hypergraphs*, North-Holland, Amsterdam—London — American Elsevier, New York, 1973, 1976, MR 50: 9640, 52: 5453). Cit. 1.1, 11.5, 11.6, 11.15.

Bermond, J.-C.: 'An Application of the Solution of Kirkman's Schoolgirl Problem: The Decomposition of the Symmetric Oriented Complete Graph into 3-Circuits', *Discrete Math.* **8** (1974) 301—304, MR 49: 4850. Cit. 10.9, 10.12, R 10.22.

Bermond, J.-C.: 'Cycles dans les graphes et G-configurations', Thesis, University of Paris XI (Orsay), Paris, 1975a. Cit. 9.11, 9.15, 10.9, 10.10, 10.12, 10.14, 10.16, 11.11.

Bermond, J.-C.: 'Decomposition of K_n^* into k-Circuits and Balanced G-Designs', in *Recent Advances in Graph Theory* (Proc. 2nd Czechoslovak Symposium, Prague, 1974), ed. Fielder, M., Academia, Prague, 1975b, pp. 57—68, MR 53: 13036. Cit. 9.2, 10.9, 10.10, 10.12, 10.14, 11.25.

Bermond, J.-C. and Faber, V.: 'Decomposition of the Complete Directed Graph into k-Circuits', *J. Combin. Theory* **B21** (1976) 146—155, MR 55: 2638. Cit. 10.3, 10.9, 10.12, 10.16, 11.7, 11.8, 11.10, 11.11, 11.18, R 11.61.

Bermond, J.-C. and Sotteau, D.: 'Graph Decompositions and G-Designs', in Proc. *5th British Combinatorial Conference* (Aberdeen, 1975), eds. Nash-Williams, C. St. J. A. and Sheehan J., Congressus Numerantium 15, Utilitas Math., Winnipeg, 1976, pp. 53—72, MR 56: 8401. Cit. 6.5, 6.14, 6.16—6.19, 8.2, 9.2, 9.10, 9.20, 10.16, 11.7, 13.20.

Bermond, J.-C. and Sotteau, D.: 'Cycle and Circuit Designs, Odd Case', in *Beiträge zur Graphentheorie und deren Anwendungen* (Colloquium, Oberhof, 1977), Technische Hochschule Ilmenau, Ilmenau, 1977, pp. 11—32, MR 83b: 05039. Cit. 9.2, 10.3—10.5, 10.7, 10.12, 10.16.

Bermond, J.-C. and Schönheim, J.: 'G-Decomposition of K_n, Where G Has Four Vertices or Less', *Discrete Math.* **19** (1977) 113—120, MR 58: 371. Cit. 9.4

Bermond, J.-C., Germa, A. and Sotteau, D.: 'Hypergraph-Designs', *Ars Combin.* **3** (1977) 47—66. Cit. 8.33, 9.2, 9.4.

Bermond, J.-C.: 'Hamiltonian Decompositions of Graphs, Directed Graphs and Hypergraphs', in *Advances in Graph Theory* (Cambridge Combinatorial Conference, Trinity College, Cambridge, 1977), ed. Bollobás, B., Annals of Discrete Mathematics 3, North-Holland, Amsterdam, 1978, pp. 21—28, MR 58: 21803. Cit. 11.16, 11.30, 11.32, 11.36, 11.38, 11.55.

Bermond, J.-C., Brouwer, A. E. and Germa, A.: 'Systèmes de triplets et différences associées', in *Problèmes combinatoires et théorie des graphes* (Colloquium, Orsay, 1976), eds. Bermond, J. C., Fournier, J.-C., Las Vergnas, M. and Sotteau, D., Colloq. Int. de C.N.R.S. 260, Centre national de la recherche scientifique, Paris, 1978a, pp. 35—38, MR 80j: 05020. Cit. 7.33, R 7.45.

Bermond, J.-C., Germa, A., Heydemann, M. C. and Sotteau, D.: 'Hypergraphes hamiltoniens', in *Problèmes combinatoires et théorie des graphes* (Colloquium, Orsay, 1976), eds. Bermond, J.-C., Fournier, J.-C., Las Vergnas, M. and Sotteau, D., Colloq. Int. de C.N.R.S. 260, Centre national de la recherche scientifique, Paris, 1978b, pp. 39—43, MR 80j: 05093. Cit. 11.55.

Bermond, J.-C., Huang, C. and Sotteau, D.: 'Balanced Cycle and Circuit Designs: Even Cases', *Ars Combin.* **5** (1978c) 293—318, MR 80c: 05050. Cit. 9.2, 10.2—10.5, 10.7, 10.12, 10.15.

Bermond, J.-C., Kotzig, A. and Turgeon, J.: 'On a Combinatorial Problem of Antennas in Radioastronomy', in *Combinatorics* (Proc. 5th Hungarian Colloquium on Combinatorics, Keszthely, 1976), eds. Hajnal, A. and Sós, V. T., Colloquia Mathematica Societatis János Bolyai 18, Bolyai János matematikai társulat, Budapest — North-Holland, Amsterdam, 1978d, pp. 135—149, MR 80h: 05014. Cit. 7.14, 7.19, R 7.45—7.47.

Bermond, J.-C.: 'Graceful Graphs, Radio Antennas and French Windmills', in *Graph Theory and Combinatorics* (Proc. Conference Open University Milton Keynes, 1978), ed. Wilson, R. J., Pitman, London, 1979, pp. 18—37, MR 81i: 05109. Cit. 7.14, 7.17, 7.19, 7.21.

Bermond, J.-C., Germa, A. and Heydemann, M. C.: 'Hamiltonian Cycles in Strong Products of Graphs', *Can. Math. Bull.* **22** (1979a) 305—309, MR 81h: 05091. Cit. 11.25.

Bermond, J.-C., Germa, A. and Sotteau, D.: 'Resolvable Decompositions of K_n^*', *J. Combin. Theory* **A26** (1979b) 179—185, MR 80k: 05083. Cit. 12.42, 12.61.

Bermond, J.-C., Huang, C., Rosa, A. and Sotteau, D.: 'Decomposition of Complete Graphs into Isomorphic Subgraphs with Five Vertices', *Ars Combin.* **10** (1980) 211—254, MR 82c: 05079. Cit. 6.5, 9.4.

Bermond, J.-C. and Thomassen, C.: 'Cycles in Digraphs — a Survey', *J. Graph Theory* **5** (1981) 1—43, MR 82k: 05053. Cit. 10.1, 11.3, 11.14.

Bermond, J.-C.: 'Sur un problème combinatoire d'antennes en radioastronomie', in *Surveys in Combinatorics*, ed. Lloyd, K., London Math. Soc. Lecture Note Series 82, Cambridge University Press, Cambridge, 1983, pp. 49—53. Cit. 7.14, 7.19.

Beth, T. and Spraque, A. P.: 'Trees of British Number Systems Are Graceful', *Arch. Math.* **33** (1979/80) 383—391, MR 81f: 05063. Cit. 7.21.

Bialostocki, A. and Roditty, Y.: '$3K_2$-Decomposition of a Graph', *Acta Math. Acad. Sci. Hung.* **40** (1982) 201—208, MR 84g: 05117. Cit. 9.28.

Bigalke, A. and Kögler, N.: 'Antigerichtete Hamiltonlinien in vollständig gerichteten Graphen', *Abh. Math. Sem. Univ. Hamburg* **48** (1979) 114—117, MR 80m: 05049. Cit. 11.58, 11.60.

Biggs, N.: 'Some Odd Graphs Theory', in *Proc. 2nd Int. Conference on Combinatorial Mathematics* (New York, 1978), eds. Gewirtz, A. and Quintas, L. V., Annals of the New York Academy of Sciences 319, New York Academy of Sciences, New York, 1979, pp. 71—81, MR 81c: 05026. Cit. 11.54.

Billington, E. J.: 'On Lambda Coverings of Pairs by Triples, Repeated Elements Allowed in Triples', *Utilitas Math.* **C21** (1982) 187—203, MR 84b: 05021. Cit. 8.16.

Billington, E. J. and Stanton, R. G.: 'Bipackings and Lambda Packings of Pairs into Triples, Repeated Elements Allowed in Triples', *J. Combin. Inf. Syst. Sci.* **7** (1982) 83—90, MR 84f: 05034. Cit. 8.16.

Birkhoff, G. and Mac Lane, S.: *A Survey of Modern Algebra* (3rd ed.), Macmillan, New York — Collier — Macmillan, London, 1965, MR 31: 2250. Cit. 8.23.

Blass, A., Exoo, G. and Harary, F.: 'Paley Graphs Satisfy All First-Order Adjacency Axioms', *J. Graph Theory* **5** (1981) 435—439, MR 84i: 05107. Cit. 14.17.

Bloom, G. S. and Golomb, S. W.: 'Applications of Numbered Undirected Graphs', *Proc. IEEE* '65 (1977) 562—570. Cit. 7.14.

Bloom, G. S.: 'A Chronology of the Ringel—Kotzig Conjecture and the Continuing Quest to Call All Trees Graceful', in *Topics in Graph Theory* (New York, 1977), ed. Harary, F., Annals of the New York Academy of Sciences 238, New York Academy of Sciences, New York, 1979, pp. 32—51, MR 82g: 05041. Cit. 7.21.

Bodendiek, R., Schumacher, H. and Wagner, K: 'Über eine spezielle Klasse graziöser Eulerscher Graphen', *Mitt. Math. Gesell. Hamburg* **10** (1976), No. 4, 241—248, MR 58: 21804. Cit. 7.15, 7.19.

Bodendiek, R., Schumacher, H. and Wagner, K.: 'Über graziöse Numerierungen von Graphen', *Elemente Math.* **32** (1977a) 49—58, MR 57: 16139. Cit. 7.16.

Bodendiek, R., Schumacher, H. and Wagner, K.: 'Über graziöse Graphen', *Math.-Phys. Semesterber.* **24** (1977b) 103—126, MR 56: 11853. Cit. 7.19.

Bollobás, B.: *Extremal Graph Theory*, Academic Press, London, 1978, MR 80a: 05120. Cit. 4.12, 11.42.

Bondy, J. A.: 'A Note on the Diameter of a Graph', *Can. Math. Bull.* **11** (1968) 499—501. Cit. 4.39.

Bondy, J. A.: 'Hamilton Cycles in Graphs and Digraphs', in *Proc. 9th Southeastern Conference on Combinatorics, Graph Theory and Computing* (Florida Atlantic Univ., Boca Raton, 1978), eds. Hoffman, F., McCarthy, D., Mullin, R. C. and Stanton, R. G., Congressus Numerantium 21, Utilitas Math., Winnipeg, 1978, pp. 3—28, MR 80k: 05074. Cit. 11.13, 11.54.

Bondy, J. A. and Häggkvist, R.: 'Edge Disjoint Hamilton Cycles in 4-Regular Planar Graphs', *Aeqs. Math.* **22** (1981) 42—45, MR 83a: 05090. Cit. 11.42.

Borowiecki, M., Kwaśnik, M. and Stencel, M.: 'Hereditary Properties and Nordhaus—Gaddum Problem', Congressus Numerantium 37, Utilitas Math., Winnipeg, 1983, pp. 187—190. Cit. 4.18.

Bosák, J.: 'Hamiltonian Lines in Cubic Graphs', in *Théorie des graphes — Theory of Graphs* (Journées int. d'étude, Rome, 1966), ed. Rosenstiehl, P., Dunod, Paris—Gordon and Breach, New York, 1967, pp. 35—46, MR 36: 5022. Cit. 11.40, 11.43, 11.66.

Bosák, J., Rosa A. and Znám, Š.: 'On Decompositions of Complete Graphs into Factors with Given Diameters', in *Theory of Graphs* (Proc. Colloquium, Tihany, 1966), eds. Erdős, P. and Katona, G., Akadémiai Kiadó, Budapest, 1968, pp. 37—56, MR 38: 2041. Cit. 4.39, 5.2, 5.4, 14.11.

Bosák, J., Erdős, P. and Rosa, A.: 'Decompositions of Complete Graphs into Factors with Diameter Two', *Mat. Čas.* **21** (1971) 14—28, MR 48: 165. Cit. 13.20.

Bosák, J. and Nešetřil, J.: 'Complete and Pseudocomplete Colourings of a Graph', *Math. Slov.* **26** (1976) 171—184, MR 55: 12558. Cit. 4.5.

Bose, R. C.: 'On the Construction of Balanced Incomplete Block Designs', *Ann. Eugenics* **9** (1939) 353—399, MR 1: 199. Cit. 8.20.

Bose, R. C.: 'A Note on the Resolvability of Incomplete Block Designs', *Sankhyā* **6** (1942) 105—110, MR 4: 237. Cit. 12.4, 12.10, 12.49, R 12.66.

Bose, R. C.: 'A Note on Fisher's Inequality for Balanced Incomplete Block Designs', *Ann. Math. Statist.* **20** (1949) 619—620, MR 11: 306. Cit. 8.5.

Bose, R. C. and Connor, W. S.: 'Combinatorial Properties of Group Divisible Incomplete Block Designs', *Ann. Math. Statist.* **23** (1952) 367—383, MR 14: 124. Cit. 8.40, 8.42.

Bose, R. C., Shrikhande, S. S. and Bhattacharya, K. N.: 'On the Construction of Group Divisible Incomplete Block Designs', *Ann. Math. Statist.* **24** (1953) 167—195, MR 15: 3. Cit. 8.40, 8.42.

Bose, R. C. and Shrikhande, S. S.: 'On the Falsity of Euler's Conjecture About the Nonexistence of Two Orthogonal Latin Squares of Order $4t + 2$', *Proc. Nat. Acad. Sci. U.S.A.* **45** (1959) 734—737, MR 21: 3343. Cit. 12.30.

Bose, R. C. and Shrikhande, S. S.: 'On the Composition of Balanced Incomplete Block Designs', *Can J. Math.* **12** (1960a) 177—188, MR 22: 1046. Cit. 12.28.

Bose, R. C. and Shrikhande, S. S.: 'On the Construction of Sets of Mutually Orthogonal Latin Squares and the Falsity of a Conjecture of Euler', *Trans. Amer. Math. Soc.* **95** (1960b) 191 —209, MR 22: 2557. Cit. 12.30.

Bose, R. C. Shrikhande, S. S. and Parker, E. T.: 'Further Results on the Construction of Mutually Orthogonal Latin Squares and the Falsity of Euler's Conjecture', *Can. J. Math.* **12** (1960) 189—203, MR 23: A69. Cit. 12.30.

Bose, R. C.: 'On the Application of Finite Projective Geometry for Deriving a Certain Series of Balanced Kirkman Arrangements', in *Calcutta Math. Soc. Golden Jubilee Commemoration*, Vol. 1958/59, Part II, 1963, 341—354, MR 27: 4769. Cit. 8.27, 12.46, 12.49.

Bouchet, A. and Fouquet, J. L.: 'Trois types de décomposition en chaînes dans les graphes', in *Combinatorial Mathematics* (Proc. Int. Colloquium on Graph Theory and Combinatorics C.N.R.S., Marseille-Luminy, 1981), eds. Berge, C., Bresson, D., Maurros, J. F. and Sterboul, F., North-Holland Mathematics Studies 75, Annals of Discrete Mathematics 17, North-Holland, Amsterdam, 1983. Cit. 9.9.

Brouwer A. E.: 'Two New Nearly Kirkman Triple Systems', *Utilitas Math.* **13** (1978) 311—314, MR 58: 10527. Cit. 12.57.

Bruck, R. H. and Ryser, H. J.: 'The Nonexistence of Certain Finite Projective Planes', *Can. J. Math.* **1** (1949) 88—93, MR 10: 319. Cit. 8.23.

Bruck, R. H.: 'Existence of the A-Maps of A. K. Dewney and N. Robertson', Preprint. Cit. 10.9, 10.12.

de Bruijn, N. G. and Erdős, P.: 'On a Combinatorial Problem', Nederl. Akad. Wctcsch. Proc. 51, *Indagationes Math.* **10** (1948) 421—423, MR 10: 424. Cit. 8.5.

de Bruijn, N. G.: 'Generalization of Pólya's Fundamental Theorem in Enumeration Combinatorial Analysis', *Indagationes Math.* **21** (1959) 59—69, MR 21: 4112. Cit. 14.5.

de Bruijn, N. G.: 'Pólya's Theory of Counting', in *Applied Combinatorial Mathematics*, ed. Beckenbach, E. F., J. Wiley, New York—London—Sydney, 1964, pp. 144—184. Cit. 14.5.

Burr, S. A.: 'A Ramsey-Theoretic Result Involving Chromatic Numbers', *J. Graph Theory* **4** (1980) 241—242, MR 81f: 05078. Cit. 4.7.

Burr, S. A. and Grossman, J. W.: 'Ramsey Numbers of Graphs with Long Tails', *Discrete Math.* **41** (1982) 223—227. Cit. 8.40.

Bush, K. A.: 'Families of Regular Designs Based on Geometries', *J. Statist. Plann. Inference* **5** (1981) 391—398, MR 83a: 05030. Cit. 8.42.

Cahit, I. and Cahit, R.: 'On the Graceful Numbering of Spanning Trees', *Inf. Process. Lett.* **3** (1974/75) 115—118, MR 52: 5455. Cit. 7.21, 7.23.

Cahit, I.: 'Realization of Graceful Permutation by a Shuffle-Exchange Network', *Inf. Process. Lett.* **6** (1977) 171—173, MR 56: 10120. Cit. 7.21.

Cain, P.: 'Decomposition of Complete Graphs into Stars', *Bull. Aust. Math. Soc.* **10** (1974) 23—30, MR 49: 10601. Cit. 9.20.

Cameron, P. J.: 'Biplanes', *Math. Z.* **131** (1973) 85—101, MR 50: 12757. Cit. 8.5.

Cameron, P. J.: 'Parallelisms of Complete Designs', London Math. Soc. Lecture Note Series 23, Cambridge Univ. Press, Cambridge, 1976, MR 54: 7269. Cit. 8.4, 12.15, 12.33, 13.4, 14.7.

Camion P.: 'Hamiltonian Chains in Self-Complementary Graphs', in *Colloque sur la théorie des graphes* (Paris, 1974), eds. Gillis, P. P. and Huyberechts, S., Cahiers du Centre d'Etudes de Recherche Opérationelle 17, 1975, Nos. 2—4, Inst. Stat. U. L. B., Bruxelles, 1975, pp. 173—183, MR 53: 5372. Cit. 13.20, 14.16.

Camion, P.: 'Une classe de graphes non planaires sommet-transitifs représentables sur le tore', in *Problèmes combinatoires et théorie des graphes* (Colloquium, Orsay, 1976), eds. Bermond, J.-C., Fournier, J.-C., Las Vergnas, M. and Sotteau, D., Colloq. Int. de C.N.R.S. 260, Centre national de la recherche scientifique, Paris, 1978, pp. 73—77, MR 81b: 05035. Cit. 11.52.

Capobianco, M. and Molluzzo, J. C.: *Examples and Counterexamples in Graph Theory*, North-Holland, New York, 1978, MR 58: 10536. Cit. 4.13, 4.26, 11.40, 11.41.

Capobianco, M. F.: 'Graph Equations', in *Proc. 2nd Int. Conference on Combinatorial Mathematics* (New York, 1978), eds. Gewirtz, A. and Quintas, L. V., Annals of the New York Academy of Sciences 319, New York Academy of Sciences, New York, 1979, pp. 114—118, MR 80m: 05089. Cit. 14.17.

Caro, Y. and Schönheim, J.: 'Decomposition of Trees into Isomorphic Subtrees', *Ars Combin.* **9** (1980) 119—130. Cit. 9.6, 9.17—9.20.

Caro, Y.: 'The Decomposition of Graphs into Graphs Having Two Edges', Preprint. Cit. 9.28.

Cavalieri D'Oro, L.: 'Sui grafi orientati finiti isomorfi con l'opposto', *Boll. Un. Mat. Ital.* **4** (1971) 859—869, MR 45: 8577. Cit. 14.25.

Chao, Chong Yun and Whitehead, E. G., Jr.: 'Chromaticity of Self-Complementary Graphs' *Arch. Math.* **32** (1979) 295—304, MR 81c: 05039. Cit. 14.17.

Chartrand, G. and Mitchem, J.: 'Graphical Theorems of the Nordhaus—Gaddum Class', in *Recent Trends in Graph Theory* (Proc. 1st New York City Graph Theory Conference, 1970), eds. Capobianco, M., Frechen, J. B. and Krolik, M., Lecture Notes in Mathematics 186, Springer-Verlag, Berlin, 1971, pp. 55—61, MR 44: 6545. Cit. 4.1, 4.30.

210

Chartrand, G., Polimeni, A. D. and Stewart, M. J.: 'The Existence of 1-Factors in Line Graphs, Squares, and Total Graphs', *Indagationes Math.* **35** (Proc. Konikl. Nied. Akad. Wetensch. Math. Sci. A76) (1973), 228—232, MR 48: 176. Cit. 9.7.

Chartrand, G. and Polimeni, A. D.: 'Ramsey Theory and Chromatic Numbers', *Pac. J. Math.* **55** (1974) 39—43, MR 51: 7924. Cit. 4.6.

Chartrand, G. and Schuster, S.: 'On the Independence Numbers of Complementary Graphs', *Trans. New York Acad. Sci.* **36** (1974) 247—451, MR 50: 4390. Cit. 4.33, 4.36, 4.37.

Chen, C. C.: 'On the Enumeration of Certain Graceful Graphs', in *Combinatorial Mathematics* (Proc. Int. Conference on Combinatorial Theory, Australian Nat. Univ., Canberra, 1977), eds. Holton, D. A. and Seberry, J., Lecture Notes in Mathematics 686, Springer-Verlag, Berlin—New York — Australian Academy of Sciences, Canberra, 1978, pp. 111—115, MR 81i: 05112. Cit. 7.19.

Chen, Ching-Shui: 'Construction of Optimal Balanced Incomplete Block Designs for Correlated Observations', *Ann. Statist.* **11** (1983) 240—246. Cit. 8.42, 11.3.

Cherepanov, M. G.: 'On the Number of Central and Peripheral Vertices of a Graph and Its Complement (in Russian), in *Heuristic Algorithms of Optimization*, ed. Mamatov, Yu. A., Yaroslav. Gosud. Univ., Yaroslavl, 1978, pp. 154--160, MR 81h: 05086. Cit. 4.39, 5.3.

Chernyak, Zh. A.: 'Edge Degree Sequences in Self-Complementary Graphs', *Mat. Zametki* **34** (1983) 297—308 (in Russian). Cit. !4.14.

Cho, C. J.: 'Rotational Triple Systems', *Ars Combin.* **13** (19ˇ2) 203—209, MR 83i: 05014. Cit. 8.16.

Chowla, S. and Ryser, H. J.: 'Combinatorial Problems', *Can. J. Math.* **2** (1950) 93—99, MR 11: 306. Cit. 8.20, 8.23.

Chung, F. R. K., Erdős, P., Graham, R. L., Ulam, S. M. and Yao F. F.: 'Minimal Decompositions of Two Graphs into Pairwise Isomorphic Subgraphs', in *Proc. 10th Southeastern Conference on Combinatorics, Graph Theory and Computing* (Florida Atlantic Univ., Boca Raton, 1979), eds. Hoffman, F., McCarthy, D., Mullin, R. C. and Stanton, R. G., Congressus Numerantium 23—24, Utilitas Math., Winnipeg, 1979, pp. 3—18, MR 82b: 05080. Cit. 5.11, 5.13, 5.14, 5.16.

Chung, F. R. K., Erdős, P. and Graham, R. L.: 'Minimal Decompositions of Graphs into Mutually Isomorphic Subgraphs', *Combinatorica* **1** (1981) 13—24, MR 82j: 05071. Cit. 5.11, 5.13, 5.14, 5.17.

Chung, F. R. K. and Graham, R. L.: 'Recent Results in Graph Decompositions', in *Combinatorics* (Proc. 8th British Combinatorial Conference, Univ. College, Swansea, 1981), ed. Temperley, H. N. V., London Math. Soc. Lecture Note Series 52, Cambridge University Press, Cambridge—New York, 1981, pp. 103—123, MR 83a: 05118. Cit. Preface, 8.2, 9.20, 11.30, 11.54.

Chung, F. R. K. and Hwang, F. K.: 'Rotatable Graceful Graphs', *Ars Combin.* **11** (1981) 239—250, MR 83c: 05110. Cit. 7.19, 7.21.

Chung, F. R. K., Erdős, P. and Graham, R. L.: 'Minimal Decompositions of Hypergraphs into Mutually Isomorphic Subhypergraphs', *J. Combin. Theory* **A32** (1982) 241—251, MR 83i: 05057. Cit. 5.11.

Chvátal, V., Erdős, P. and Hedrlín, Z.: 'Ramsey's Theorem and Self-Complementary Graphs', *Discrete Math.* **3** (1972) 301—304, MR 47: 1674. Cit. 14.12.

Clapham, C. R. J.: 'Triangles in Self-Complementary Graphs', *J. Combin. Theory* **B15** (1973) 74—76, MR 48: 1975. Cit. 14.16.

Clapham, C. R. J.: 'Hamiltonian Arcs in Self-Complementary Graphs', *Discrete Math.* **8** (1974) 251—255, MR 49: 133. Cit. 14.16.

Clapham, C. R. J.: 'Hamiltonian Arcs in Infinite Self-Complementary Graphs', *Discrete Math.* **13** (1975) 307—314, MR 52: 13486. Cit. 14.16.

Clapham, C. R. J.: 'Potentially Self-Complementary Degree Sequences', *J. Combin. Theory* **B20** (1976a) 75—79, MR 54: 7315. Cit. 14.14.

Clapham, C. R. J.: 'Piecing Together Paths in Self-Complementary Graphs', in *Proc. 5th British Combinatorial Conference* (Aberdeen, 1975), eds. Nash-Williams, C. St. J. A. and Sheehan, J., Congressus Numerantium 15, Utilitas Math., Winnipeg, 1976b, pp. 125—130, MR 53: 7855. Cit. 14.16.

Clapham, C. R. J. and Kleitman, D. J.: 'The Degree Sequences of Self-Complementary Graphs', *J. Combin. Theory* **B20** (1976) 67—74, MR 54: 7314. Cit. 14.14.

Clapham, C. R. J.: 'A Class of Self-Complementary Graphs and Lower Bounds of Some Ramsey Numbers', *J. Graph Theory* **3** (1979) 287—289, MR 81d: 05054. Cit. 13.20, 14.12.

Clatworthy, W. H.: 'Tables of Two-Associate-Class Partially Balanced Designs', Nat. Bur. Standards, Applied Math. Ser. 63, U.S. Dept. of Commerce, Washington, 1973, MR 54: 4029. Cit. 8.42.

Cockayne, E. J. and Lorimer, P. J.: 'Ramsey Numbers for Stripes', *J. Aust. Math. Soc.* **A19** (1975) 252—256, MR 51: 7950. Cit. 4.33.

Cockayne, E. J. and Hartnell, B. L.: 'Edge Partitions of Complete Multipartite Graphs into Equal Length Circuits — a Summary, in *Proc. 5th British Combinatorial Conference* (Aberdeen, 1975), eds. Nash-Williams, C. St. J. A. and Sheehan, J., Congressus Numerantium 15, Utilitas Math., Winnipeg, 1976, pp. 131—134, MR 52: 13487. Cit. 7.49, 8.41, 10.17, 10.23, 13.20, R. 10.23.

Cockayne, E. J. and Hartnell, B. C.: 'Edge Partitions of Complete Multipartite Graphs into Equal Length Circuits', *J. Combin. Theory* **B23** (1977) 174—183. Cit. 10.17, R 10.23.

Cockayne, E. J. and Thomason, A. G.: 'Ordered Colourings of Graphs', *J. Combin. Theory* **B32** (1982) 286—292, MR 83k: 05046. Cit. 4.18.

Colbourn, M. J. and Colbourn, C. J.: 'Graph Isomorphism and Self-Complementary Graphs', *SIGACT News* **10** (1978) 25—29. Cit. 14.10.

Colbourn, M. J.: 'An Analysis Technique for Steiner Triple Systems', in *Proc. 10th Southeastern Conference on Combinatorics, Graph Theory and Computing* (Florida Atlantic Univ., Boca Raton, 1979), eds. Hoffman, F., McCarthy, D., Mullin, R. C. and Stanton, R. G., Congressus Numerantium 23—24,. Utilitas Math., Winnipeg, 1979, pp. 289—303. Cit. 8.35.

Colbourn, C. J. and Colbourn, M. J.: 'Isomorphism Problems Involving Self-Complementary Graphs and Tournaments, in *Proc. 8th Manitoba Conference on Numerical Mathematics and Computing* (Univ. Manitoba, Winnipeg, 1978), eds. McCarthy, D. and Williams, H. C., Congressus Numerantium 22, Utilitas Math., Winnipeg, 1979, pp. 153—164, MR 81c: 05082. Cit. 14.15.

Colbourn, M. J.: 'Cyclic Block Designs: Computational Aspects of Their Construction and Analysis', Technical Report 146/80, Dept. of Computer Science, Univ. of Toronto, 1980. Cit. 12.46.

Colbourn, M. J. and Mathon, R. A.: 'On Cyclic Steiner 2-Designs', in *Topics on Steiner Systems*, eds. Lindner, C. C. and Rosa, A., Annals of Discrete Mathematics 7, North-Holland, Amsterdam, 1980, pp. 215—253. Cit. 7.31—7.37, 8.35.

Colbourn, M. J. and Colbourn, C. J.: 'Cyclic Block Designs with Block Size 3', *Eur. J. Combin.* **2** (1981a) 21—26, MR 82g: 05018. Cit. 8.16.

Colbourn, M. J. and Colbourn, C. J.: 'Some Small Directed Triples', in *Proc. 10th Manitoba Conference on Numerical Mathematics and Computing* (Winnipeg, 1980), Congressus Numerantium 30, Utilitas Math., Winnipeg, 1981b, pp. 247—255, MR 83b: 05041. Cit. 10.11.

Colbourn, C. J.: 'Distinct Cyclic Steiner Triple Systems', *Utilitas Math.* **22** (1982a) 103—126, MR 84d: 05041. Cit. 7.38, 8.35.

Colbourn, C. J.: 'Hamiltonian Decompositions of Complete Graphs', *Ars Combin.* **14** (1982b)

212

261—269, MR 84d: 05095. Cit. 11.11.

Colbourn, C. J. and Mendelsohn, E.: 'Kotzig Factorizations: Existence and Computational Results', in *Theory and Practice of Combinatorics*, eds. Rosa, A., Sabidussi G. and Turgeon, J., North-Holland Mathematics Studies 60, Annals of Discrete Mathematics 12, North-Holland, Amsterdam, 1982, pp. 65—78. Cit. 11.19.

Colbourn, C. J., Colbourn, M. J., Harms, J. J. and Rosa, A.: 'A Complete Census of (10, 3, 2)-Block Designs and of Mendelsohn Triple Systems of Order Ten, III, (10, 3, 2)-Block Designs Without Repeated Blocks', Congressus Numerantium 37, Utilitas Math., Winnipeg, 1983a, pp. 211—234. Cit. 8.35.

Colbourn, C. J., Colbourn, M. J. and Rosa, A.: 'Completing Small Partial Triple Systems', *Discrete Math.* **45** (1983b) 165—179, MR 84g: 05028. Cit. 7.32, 8.16.

Colbourn, C. J. and Harms, J. J.: 'Directing Triple Systems', *Ars Combin.* **15** (1983) 261—266. Cit. 8.37, 10.11, 12.42.

Colbourn, C. J.: 'Embedding Partial Steiner Triple Systems Is NP-Complete', *J. Combin. Theory* **A** (in press). Cit. 8.16.

Cole, F. N.: 'Kirkman Parades', *Bull. Amer. Math. Soc.* **28** (1922) 435—437. Cit. 12.8.

Connor, W. S., Jr: 'On the Structure of Balanced Incomplete Block Designs', *Ann. Math. Statist.* **23** (1952) 57—71, MR 13: 617. Cit. 8.14, 8.26.

Cook, R. J.: 'Complementary Graphs and Total Chromatic Numbers', *SIAM J. Appl. Math.* **27** (1974) 626—628, MR 50: 4365. Cit. 4.27.

Cook, R. J.: 'Some Theorems of the Nordhaus—Gaddum Class', in *Finite and Infinite Sets* (Eger, 1981), Colloquia Mathematica Societatis János Bolyai 37, North-Holland, Amsterdam—New York, 1984a, pp. 223—229. Cit. 4.17, 4.18, 4.42.

Cook, R. J.: 'Graph Factorization and Theorems of the Nordhaus—Gaddum Class', *Period. Math. Hung.* **15** (1984b) 109—120. Cit. 4.17, 4.18, 4.21, 4.22, 4.29.

Crampin, D. J. and Hilton, A. J.: 'Remarks on Sade's Disproof of the Euler Conjecture with an Application to Latin Squares Orthogonal to Their Transpose', *J. Combin. Theory* **A18** (1975) 47—59, MR 51: 197. Cit. 12.30.

Cvetković, D. M.: 'The Generating Function for Variations with Restrictions and Paths of the Graph and Self-Complementary Graphs', *Univ. Beograd Publ. Elektroteh. Fak. (Mat. Fiz.)*, (1970) Nos. 320—328, 27—34, MR 43: 7356. Cit. 5.10, 14.17.

Das Prabir: 'Characterization of Potentially Self-Complementary, Self-Converse Degree-Pair Sequences for Digraphs', in *Combinatorics and Graph Theory* (Proc. 2nd Symposium Indian Statist. Inst., Calcutta, 1980), ed. Rao, S. B., Lecture Notes in Mathematics 885, Springer-Verlag, Berlin—Heidelberg—New York, 1981, pp. 212—226, MR 83i: 05033. Cit. 14.19, 14.25.

Das Prabir: 'Integer-Pair Sequences with Self-Complementary Realizations', *Discrete Math.* **45** (1983) 189—198. Cit. 14.14.

Dehon, M.: 'On the Existence of 2-Designs $S_\lambda(2, 3, v)$ without Repeated Blocks', *Discrete Math.* **43** (1983) 155—171, MR 84d: 05030. Cit. 8.16.

Delorme, Ch., Maheo, M. Thuillier, H., Koh, K. M. and Teo, H. K.: 'Cycles with a Chord Are Graceful', *J. Graph Theory* **4** (1980) 409—415, MR 82g: 05078. Cit. 7.16.

Dembowski, P.: *Finite Geometries*, Springer-Verlag, Berlin, 1968, MR 38: 1597. Cit. 8.1, 8.4, 8.5, 8.20, 8.22, 8.23, R 7.56.

Dénes, J. and Keedwell, A. D.: *Latin Squares and Their Applications,* Academic Press, New York, 1974, MR 50: 4338. Cit. 11.11, 12.22, 12.25, 12.30.

Dénes, J. and Gergely, E.: 'Groupoids and Codes', in *Topics in Information Theory* (Proc. Colloquium, Keszthely, 1975), eds. Csiszár, I. and Elias, P., Colloquia Mathematica Societatis János Bolyai 16, North-Holland, Amsterdam, 1977, pp. 155—162, MR 56: 18100. Cit. 12.3, 12.27, 12.31.

Denniston, R. H. F.: 'Double Resolvability of Some Complete 3-Designs', *Manuscripta Math.* **12** (1974a) 105—112, MR 50: 1950. Cit. 12.33, 12.34.

Denniston, R. H. F.: 'Sylvester's Problem of the 15 Schoolgirls', *Discrete Math.* **9** (1974b) 229—233, MR 51: 5322. Cit. 12.33.

Denniston, R. H. F.: 'On the Number of Non-Isomorphic Reverse Steiner Triple Systems of Order 19', in *Topics on Steiner Systems*, eds. Lindner, C. C. and Rosa, A., Annals of Discrete Mathematics 7, North-Holland, Amsterdam, 1980a, pp. 255—264. Cit. 8.35.

Denniston, R. H. F.: 'Steiner System with a Maximal Arc', *Ars Combin.* **9** (1980b) 247—248, MR 81f: 05024. Cit. 8.27.

Deza, M., Mullin, R. C. and Vanstone, S. A.: 'Recent Results on (r, λ)-Designs and Related Configurations', *Rev. Técn. Fac. Ingr. Univ. Zulia* **4** (1981) 139—158, MR 83c: 05016. Cit. 8.1.

Dickson, L. E. and Safford, F. H. (1906): Solution to Problem 8 (Group Theory)', *Amer. Math. Monthly* **13** (1906) 150—151. Cit 13.4, 14.7.

Di Paola, J. W. and Nemeth, E. (1979): 'Applications of Parallelism', in *Proc. 2nd Int. Conference on Combinatorial Mathematics* (New York, 1978), eds. Gewirtz, A. and Quintas, L. V., Annals of the New York Academy of Sciences 319, New York Academy of Sciences, New York, 1979, pp. 153—163, MR 81g: 05091. Cit. 7.21, 12.15, 12.33.

Dirac, G. A.: 'Graph Union and Chromatic Number', *J. London Math. Soc.* **39** (1964) 451—454, MR 33: 63. Cit. 4.18.

Dirac, G. A.: 'On Hamilton Circuits and Hamilton Paths', *Math. Ann.* **197** (1972) 57—70, MR 46: 1644. Cit. 11.16.

Doutre, P. and Kotzig, A.: 'Les décompositions des graphes Q_n en facteurs quadratiques', *Ann. Soc. Math. Québec* **4** (1980) 17—26, MR 81h: 05087. Cit. 11.65, R 11.65.

Doyen, J. and Rosa, A.: 'An Extended Bibliography and Survey of Steiner Systems', in *Proc. 7th Manitoba Conference on Numerical Mathematics and Computing* (Univ. Manitoba, Winnipeg, 1977), eds. Hartnell, B. L. and Williams, H. C., Congressus Numerantium 20, Utilitas Math., Winnipeg, 1978, pp. 297—361, MR 80g: 51009. Cit. 8.34, 8.35.

Duchet, P.: 'Sur les hypergraphes invariants', *Discrete Math.* **8** (1974) 269—280, MR 49: 4786. Cit. 5.10.

Dudeney, H. E.: *Amusements in Mathematics*, Nelson, London—New York, 1917 (reprinted by Dover Publications, New York, 1959), MR 21: 4087. Cit. 11.20, 12.35.

Dudeney, H. E.: *536 Puzzles and Curious Problems*, Charles Scribner's Sons, New York, 1967. Cit. 12.35.

Engel, K.: 'Über die Anzahl elementarer, teilweise balancierter, unvollständiger Blockpläne', *Rostock. Math. Kolloq.* (1980), No. 13, 19—41, MR 82e: 05027. Cit. 8.40, 8.43.

Entringer, R. C., Jackson, D. E. and Snyder, P. A.: 'Distance in Graphs', *Czechoslov. Math. J.* **26** (1976) 283—296, MR 58: 2783. Cit. 4.47, 4.48.

Entringer, R. C. and Swart, H.: 'Spanning Cycles of Nearly Cubic Graphs', *J. Combin. Theory* **B29** (1980) 303—309, MR 82e: 05093. Cit. 11.43.

Eplett, W. J. R.: 'Self-Converse Tournaments', *Can. Math. Bull.* **22** (1979) 23—27, MR 80j: 05068. Cit. 14.19.

Erdős, P. and Gallai, T.: 'Graphs with Vertices of Prescribed Degrees', *Mat. Lapok* **11** (1960) 264—274 (in Hungarian). Cit. 14.14.

Erdős, P.: 'On a Lemma of Hajnal-Folkman', in *Proc. Colloquium on Combinatorial Theory and Its Applications* (Balatonfüred, 1969), eds. Erdős, P., Rényi, A. and Sós, V. T., Colloquia Mathematica Societatis János Bolyai 4, North-Holland, Amsterdam — Bolyai János matematikai társulat, Budapest, 1970, pp. 311—316, MR 45: 6655. Cit. 8.39.

Erdős, P. and Schönheim, J.: 'Edge Decompositions of the Complete Graph into Copies of a Connected Subgraph', in *Proc. Conference on Algebraic Aspects of Combinatorics* (Univ.

214

Toronto, Toronto, 1975), eds. Corneil, D. and Mendelsohn, E., Congressus Numeratium 13, Utilitas Math., Winnipeg, 1975, pp. 271—278, MR 52: 10501. Cit. 6.11, 9.4.

Erdős, P. and Meir, A.: 'On Total Matching Numbers and Total Covering Numbers of Complementary Graphs', *Discrete Math.* **19** (1977) 229—233, MR 57: 2985. Cit. 4.37.

Erdős, P.: 'Old and New Problems in Combinatorial Analysis and Graph Theory', in *Proc. 2nd Int. Conference on Combinatorial Mathematics* (New York, 1978), eds. Gewirtz, A. and Quintas, L. V., Annals of the New York Academy of Sciences 319, New York Academy of Sciences, New York, 1979, pp. 177—187, MR 81a: 05033. Cit. 4.17.

Erdős, P. and Schuster, S.: 'Existence of Complementary Graphs with Specified Independence Numbers', in *The Theory and Applications of Graphs* (Proc. 4th Int. Conference Western Michigan Univ., Kalamazoo, 1980), eds. Chartrand, G., Alavi, Y., Goldsmith, D. L., Lesniak-Foster, L. and Lick, D. R., J. Wiley, New York, 1981, pp. 343—349, MR 82m: 05080. Cit. 4.33, 4.36, 4.37, 4.49.

Escalante, F. and Simões-Pereira, J. M. S.: 'Just Two Total Graphs Are Complementary', *Monatsh. Math.* **81** (1976) 5—13, MR 53: 7860. Cit. 5.10.

Faradzhev, I. A.: 'A Complete List of Self-Complementary Graphs with 12 or Less Vertices', in *Algorithmic Investigations in Combinatorics*, ed. Faradzhev, I. A., Nauka, Moscow, 1978, pp. 69—75 (in Russian), MR 80b: 05035. Cit. 14.6.

Finck, H.-J.: 'Über die chromatischen Zahlen eines Graphen und seines Komplements, I, II', *Wiss. Z. Tech. Hochschule Ilmenau* **12** (1966) 243—246, MR 35: 5358. Cit. 4.13.

Finck, H.-J.: 'On the Chromatic Number of a Graph and Its Complement', in *Theory of Graphs* (Proc. Colloquium Tihany, 1966), eds. Erdős, P. and Katona, G., Akadémiai Kiadó, Budapest, 1968, pp. 99—113, MR 38: 1027. Cit. 4.13.

Finck, H.-J. and Sachs, H.: 'Über eine von H. S. Wilf angegebene Schranke für die chromatische Zahl endlicher Graphen', *Math. Nachr.* **39** (1969) 373—386, MR 40: 60. Cit. 4.18.

Fiorini, S. and Wilson, R. J.: *Edge Colourings of Graphs*, Research Notes in Mathematics 16, Pitman, London, 1977. Cit. Introduction.

Fisher, R. A.: 'An Examination of the Different Possible Solutions of a Problem in Incomplete Blocks' *Ann. Eugenics* **10** (1940) 52—75, MR 1: 348. Cit. 8.4, 8.34.

Foregger, M. F.: 'Hamiltonian Decompositions of Product of Cycles', *Discrete Math.* **24** (1978) 251—260, MR 80a: 05138. Cit. 11.27, 11.29.

Foregger, M. F. and Foregger, T. H.: 'The Tree-Covering Number of a Graph', *Czechoslov. Math. J.* **30** (1980) 633—639, MR 82d: 05086. Cit. 4.18.

Fort, M. K. and Hedlund, G. A.: 'Minimal Coverings of Pairs by Triples', *Pac. J. Math.* **8** (1958) 709—719. Cit. 8.16.

Frucht, R. and Harary, F.: 'Self-Complementary Generalized Orbits of a Permutation Group', *Can. Math. Bull.* **17** (1974) 203—208, MR 51: 5381. Cit. 14.7.

Frucht, R. W.: 'Graceful Numbering of Wheels and Related Graphs', in *Proc. 2nd Int. Conference on Combinatorial Mathematics* (New York, 1978), eds. Gewirtz, A. and Quintas, L. V., Annals of the New York Academy of Sciences 319, New York Academy of Sciences, New York, 1979, pp. 219—229, MR 82e: 05113. Cit. 7.17, 7.19.

Fuji-Hara, R. and Vanstone, S. A.: 'On the Spectrum of Doubly Resolvable Kirkman Systems', in *Proc. 11th Southeastern Conference on Combinatorics, Graph Theory and Computing* (Florida Atlantic Univ., Boca Raton, 1980), eds. Hadlock, F. O., Hoffman, F., Mullin, R. C., Stanton, R. G. and van Rees, G. H. J., Congressus Numerantium 28—29, Utilitas Math., Winnipeg, 1980, Vol. I, pp. 399—407, MR 82j: 05039. Cit. 12.34.

Fuji-Hara, R. and Vanstone, S. A.: 'Orthogonal Resolutions of Lines in $AG(n, q)$', *Discrete Math.* **41** (1982) 17—28, MR 83m: 05025. Cit. 12.34.

Gallai, T.: 'Über extreme Punkt- und Kantenmengen', *Ann. Univ. Sci. Budapest L. Eötvös Nom., Sectio Math.* **2** (1959) 133—138, MR 24: A1222. Cit. 4.32, 14.7.

Galvin, F. and Krieger, M. M.: 'The Minimum Number of Cliques in a Graph and Its Complement', in *Proc. 2nd Louisiana Conference on Combinatorics, Graph Theory and Computing*, eds. Mullin, R. C., Reid, K. B., Roselle, D. P. and Thomas, R. S. D., Louisiana State Univ., Baton Rouge, 1971, pp. 345—352, MR 48: 3809. Cit. 4.18.

Gangopadhyay, T. and Hebbare, S. P. Rao: 'Bigraceful graphs', I, *Utilitas Math.* 17 (1980a) 271 —275, MR 81k: 05065. Cit. 7.19.

Gangopadhyay, T. and Hebbare, S. P. Rao: 'Paths in *r*-Partite Self-Complementary Graphs', *Discrete Math.* 32 (1980b) 229—244, MR 82b: 05090a. Cit. 14.19.

Gangopadhyay, T. and Hebbare, S. P. Rao: '*r*-Partite Self-Complementary Graphs — Diameters', *Discrete Math.* 32 (1980c) 245—255, MR 82b: 05090b. Cit. 14.19.

Gangopadhyay, T.: 'Characterization of Forcibly Bipartite Self-Complementary Bipartitioned Sequences', in *Combinatorics and Graph Theory* (Proc. 2nd Symp. Indian Statistical Institute, Calcutta, 1980), ed. Rao, S. B., Lecture Notes in Mathematics 885, Springer-Verlag, Berlin —Heidelberg—New York, 1981, pp. 237—260, MR: 83k: 05092. Cit. 14.19.

Gangopadhyay, T.: 'Characterization of Potentially Bipartite Self-Complementary Bipartitioned Sequences', *Discrete Math.* 38 (1982) 173—184, MR 84e: 05088. Cit. 14.19.

Gangopadhyay, T. and Hebbare, S. P. Rao: 'Multipartite Self-Complementary Graphs', *Ars Combin.* 13 (1982) 87—114, MR 83m: 05120. Cit. 14.19.

Ganter, B., Gülzov, A., Mathon, R. A. and Rosa, A.: 'A Complete Census of (10, 3, 2)-Block Designs and of Mendelsohn Triple Systems of Order Ten, IV, (10, 3, 2)-Designs with Repeated Blocks', Technical Report 5/78, Gesamthochschule, Kassel, 1978a. Cit. 8.35.

Ganter, B., Mathon, R. and Rosa, A.: 'A Complete Census of (10, 3, 2)-Block Designs and of Mendelsohn Triple Systems of Order Ten, I, Mendelsohn Triple Systems without Repeated Blocks', in *Proc. 7th Manitoba Conference on Numerical Mathematics and Computing* (Univ. Manitoba, Winnipeg, 1977), eds. Hartnell, B. L. and Williams, H. C., Congressus Numerantium 20, Utilitas Math., Winnipeg, 1978b, pp. 383—398. Cit. 8.35.

Ganter, B., Mathon, R. and Rosa, A.: 'A Complete Census of (10, 3, 2)-Block Designs and of Mendelsohn Triple Systems of Order Ten, II, Mendelsohn Triple Systems with Repeated Blocks', in *Proc. 8th Manitoba Conference on Numerical Mathematics and Computing* (Univ. Manitoba, Winnipeg, 1978), eds. McCarthy, D. and Williams, H. C., Congressus Numerantium 22, Utilitas Math., Winnipeg, 1979, pp. 181—204. Cit. 8.35.

Gardner, M.: 'The Graceful Graphs of Solomon Golomb, or How to Number a Graph Parsimoniously', *Sci. Amer.* **March** (1972) 108—112. Cit. 7.19.

Gardner, M.: 'Mathematical Games — What Unifies Dinner Guests, Strolling Schoolgirls and Handcuffed Prisoners?', *Sci. Amer.* **May** (1980) 14—21. Cit. 12.1, 13.20.

Gelling, E. N. and Odeh, R. E.: 'On 1-Factorizations of the Complete Graph and the Relationship to Round Robin Schedules, in *Proc. 3rd Manitoba Conference on Numerical Mathematics* (Winnipeg, 1973), eds. Thomas, R. S. D. and Williams, H. C., Congressus Numerantium 9, Utilitas Math., Winnipeg, 1974, pp. 213—221, MR 50: 171. Cit. 13.4, 14.7.

Germa, A.: 'Decomposition of the Edges of a Complete *t*-Uniform Directed Hypergraph', in *Combinatorics* (Proc. 5th Hungarian Colloquium on Combinatorics, Keszthely, 1976), eds. Hajnal, A. and Sós, V. T., Colloquia Mathematica Societatis János Bolyai 18, Bolyai János matematikai társulat, Budapest — North-Holland, Amsterdam, 1978, pp. 393—399, MR 80e: 05086. Cit. 6.18.

Gibbs, R. A.: 'Self-Complementary Graphs', *J. Combin. Theory* **B16** (1974) 106—123, MR 50: 188. Cit. 14.1, 14.7.

Golomb, S. W.: 'How to Number a Graph', in *Graph Theory and Computing* (Proc. Conference Univ. West Indies, Kingston, 1969), ed. Read, R. C., Academic Press, New York—London, 1972, pp. 23—37, MR 49: 4863. Cit. 7.8, 7.11—7.13, 7.15, 7.17, 7.21, 7.43, R 7.43.

Graham, R. L.: 'Comment on Nash-Williams' Problem', in *Combinatorial Theory and Its*

Applications (Proc. Colloquium Balatonfüred, 1969), eds. Erdős, P., Rényi, A. and Sós, V. T., Colloquia Mathematica Societatis János Bolyai 4, North-Holland, Amsterdam — Bolyai János matematikai társulat, Budapest, 1970, p. 1181. Cit. 10.17, 10.24.

Graham, R. L., Rothschild, B. L. and Spencer, J. H.: *Ramsey Theory*, J. Wiley, New York, 1980. Cit. Introduction.

Grant, D. D.: 'Stability of Line Graphs', *J. Aust. Math. Soc.* **A21** (1976) 457—466, MR 55: 5483. Cit. 14.17.

Graver, J. and Yackel, J.: 'Some Graph Theoretic Results Associated with Ramsey's Theorem', *J. Combin. Theory* **4** (1968) 125—175, MR 37: 1278. Cit. 13.20.

Grünbaum, B.: *Convex Polytopes*, J. Wiley, London, 1967, MR 37: 2085. Cit. 11.40.

Grünbaum, B. and Zaks, J.: 'The Existence of Certain Planar Maps', *Discrete Math.* **10** (1974) 93—105, MR 50: 1949. Cit. 11.42, 11.54.

Grünbaum, B. and Malkevitch, J.: 'Pairs of Edge-Disjoint Hamiltonian Circuits', *Aeqs. Math.* **14** (1976) 191—196, MR 54: 2544b. Cit. 11.41, 11.42, 11.54.

Guidotti, L.: 'On Divisibility of Complete Graphs' *Riv. Mat. Univ. Parma* **1** (1972) 231—237, (in Italian), MR 52: 2962. Cit. 15.14.

Guldan, F. and Tomasta, P.: 'New Lower Bounds of Some Diagonal Ramsey Numbers', *J. Graph Theory* **7** (1983) 149—151. Cit. 14.12.

Gupta, R. P.: 'Bounds on the Chromatic and Achromatic Numbers of Complementary Graphs', in *Recent Progress in Combinatorics* (Proc. 3rd Waterloo Conference on Combinatorics, Waterloo, 1968), ed. Tutte, W. T., Academic Press, New York—London, 1969, pp. 229—235, MR 41: 1585. Cit. 4.11, 4.14, 4.15.

Gupta, R. P.: 'An Edge-Coloration Theorem for Bipartite Graphs with Applications', *Discrete Math.* **23** (1978) 229—233, MR 80d: 05023. Cit. 14.19.

Guy, R. K. and Milner, E. C.: 'Graphs Defined by Coverings of a Set', *Acta Math. Acad. Sci. Hung.* **19** (1968) 7—21, MR 36: 6299b. Cit. 5.10.

Guy, R. K. and Klee, V.: 'Monthly Research Problems, 1969—71', *Amer. Math. Monthly* **78** (1971) 1113—1122. Cit. 7.21.

Guy, R. K.: 'Monthly Research Problems, 1969—73', *Amer. Math. Monthly* **80** (1973) 1120—1128. Cit. 7.19, 7.21.

Guy, R. K.: 'Monthly Research Problems, 1969—77', *Amer. Math. Monthly* **84** (1977) 807—815, MR 58: 26657. Cit. 7.19, 7.21.

Gyárfás, A. and Lehel, J.: 'A Method to Generate Graceful Trees', in *Problèmes combinatoires et théorie des graphes* (Colloquium, Orsay, 1976), eds. Bermond, J.-C., Fournier, J.-C., Las Vergnas, M. and Sotteau, D., Colloq. Int. de C.N.R.S. 260, Centre national de la recherche scientifique, Paris, 1978, pp. 207—209, MR 81k: 05041. Cit. 7.21.

Haggard, G. and McWha, P.: 'Decomposition of Complete Graphs into Trees', *Czechoslov. Math. J.* **25** (1975) 31—36, MR 51: 5352. Cit. 7.21.

Hakimi, S.: 'On the Realizability of a Set of Integers as Degrees of the Vertices of a Graph', *SIAM J. Appl. Math.* **10** (1962) 496—506, MR 26: 5558, Cit. 14.14.

Hall, M., Jr. and Connor, W. S.: 'An Embedding Theorem for Balanced Incomplete Block Designs', *Can. J. Math.* **6** (1954) 35—41, MR 15: 494. Cit. 8.14, 8.26, 8.36, 8.51.

Hall, M., Jr.: *Combinatorial Theory*, Blaisdell, Waltham, 1967, MR 37: 80. Cit. 6.11, 7.4, 7.57, 8.4, 8.9, 8.20, 8.23, 8.36, 8.55, 12.15, 12.17, 12.28, 12.31, R 7.56.

Hall, M., Jr.: Combinatorial Constructions', in *Studies in Combinatorics*, ed. Rota, G.-C., MAA Studies in Mathematics 17, Mathematical Association of America, 1978, pp. 218—253, MR 58: 21675. Cit. 8.5.

Hanani, H.: 'The Existence and Construction of Balanced Incomplete Block Designs', *Ann. Math. Statist.* **32** (1961) 361—386, MR 29: 4161. Cit. 8.14, 8.15.

Hanani, H.: 'On Balanced Incomplete Block Designs with Block Having Five Elements',

J. Combin. Theory **A12** (1972) 184—201, MR 45: 6645. Cit. 8.14.

Hanani, H., Ray-Chaudhuri, D. K. and Wilson, R. M.: 'On Resolvable Designs', *Discrete Math.* **3** (1972) 343—357, MR 46: 8857. Cit. 12.2, 12.42.

Hanani, H.: 'On Resolvable Balanced Incomplete Block Designs', *J. Combin. Theory* **A17** (1974) 275—289, MR 50: 6883. Cit. 8.40, 12.2, 12.44, 12.55, 12.57, 12.60.

Hanani, H.: 'Balanced Incomplete Block Designs and Related Designs', *Discrete Math.* **11** (1975) 255—369, MR 52: 2918. Cit. 6.18, 7.4, 8.14, 8.16, 8.18, 8.23, 8.26, 8.27, 8.43.

Hanani, H.: 'Generalization of Kirkman's Schoolgirl Problem', in *Proc. 2nd Int. Conference on Combinatorial Mathematics* (New York, 1978), eds. Gewirtz, A. and Quintas, L. V., Annals of the New York Academy of Sciences 319, New York Academy of Sciences, New York, 1979, A collection of open problems, Problem 35, p. 589. Cit. 12.46.

Harary, F.: 'Unsolved Problems in the Enumeration of Graphs', *Publ. Math. Inst. Hung. Acad. Sci.* **A5** (1960) 63—95, MR 26: 4340. Cit. 14.5.

Harary, F. and Palmer, E. M.: 'Enumeration of Self-Converse Digraphs', *Mathematika* **13** (1966) 151—157, MR 34: 2499. Cit. 14.19, 14.25.

Harary, F., Palmer, E. M. and Smith, C.: 'Which Graphs Have Only Self-Converse Orientations?', *Can. Math. Bull.* **10** (1967) 425—429, MR 35: 2791. Cit. 14.25.

Harary, F.: *Graph Theory*, Addison-Wesley, Reading, 1969, MR 41: 1566. Cit. 1.1, 4.32, 11.1, 11.5, 11.57, 14.5, 14.14, R 13.23.

Harary, F. and Hedetniemi, S.: 'The Achromatic Number of a Graph', *J. Combin. Theory* **8** (1970) 154—161, MR 40: 7143. Cit. 4.14.

Harary, F. and Palmer, E. M.: *Graphical Enumeration*, Academic Press, New York, 1973, MR 50: 9682. Cit. 14.5, 14.19, 14.21.

Harary, F. and Schwenk, A. J.: 'Which Graphs Have Integral Spectra?', in *Graphs and Combinatorics* (Proc. Capital Conference on Graph Theory and Combinatorics, George Washington Univ., Washington, 1973), eds. Bari, R. A. and Harary, F., Lecture Notes in Mathematics 406, Springer-Verlag, Berlin—New York, 1974, pp. 45—51, MR 52: 7970. Cit. 5.10.

Harary, F. and Wallis, W. D.: 'Isomorphic Factorizations, II, Combinatorial Designs', in *Proc. 8th Southeastern Conference on Combinatorics, Graph Theory and Computing* (Louisiana State Univ., Baton Rouge, 1977), eds. Hoffman, F., Lesniak-Foster, L., McCarthy, D., Mullin, R. C., Reid, K. B. and Stanton, R. G., Congressus Numerantium 19, Utilitas Math., Winnipeg, 1977, pp. 13—28, MR 58: 27646b. Cit. 9.2.

Harary, F., Palmer, E. M., Robinson, R. W. and Schwenk, A. J.: 'Enumeration of Graphs with Signed Points and Lines', *J. Graph Theory* **1** (1977a) 295—308, MR 57: 5818. Cit. 14.7.

Harary, F., Robinson, R. W. and Wormald, N. C.: 'The Divisibility Theorem for Isomorphic Factorizations of Complete Graphs', *J. Graph Theory* **1** (1977b) 187—188, MR 56: 183. Cit. 15.3.

Harary, F., Robinson, R. W. and Wormald, N. C., 'Isomorphic Factorizations, III, Complete Multipartite Graphs', in *Combinatorial Mathematics* (Proc. Int. Conference on Combinatorial Theory, Australian Nat. Univ., Canberra, 1977), eds. Holton, D. A. and Seberry, J., Lecture Notes in Mathematics 686, Springer-Verlag, Berlin—New York — Australian Academy of Science, Canberra, 1978a, pp. 47—54. Cit. 14.19, 15.5, 15.6, 15.8—15.11.

Harary, F., Robinson, R. W. and Wormald, N. C.: 'Isomorphic Factorizations, I, Complete Graphs', *Trans. Amer. Math. Soc.* **242** (1978b) 243—260, MR 58: 27646a. Cit. 13.9, 13.15, 13.20, 15.3, 15.4, 15.14.

Harary, F., Robinson, R. W. and Wormald, N. C.: 'Isomorphic Factorizations, V, Directed Graphs', *Mathematika* **25** (1978c) 279—285, MR 80m: 05050. Cit. 15.12.

Harary, F., Wallis, W. D. and Heinrich, K.: 'Decompositions of Complete Symmetric Digraphs into the Four Oriented Quadrilaterals', in *Combinatorial Mathematics* (Proc. Int. Conference on Combinatorial Theory, Australian Nat. Univ., Canberra, 1977), eds. Holton, D. A. and

218

Seberry, J., Lecture Notes in Mathematics 686, Springer-Verlag, Berlin—New York — Australian Academy of Science, Canberra, 1978d, pp. 165—173, MR 82b: 05068. Cit. 10.11.

Harary, F., Robinson, R. W. and Wormald, N. C.: 'Clarification of a Tripartite Conjecture', *Graph Theory Newslett.* **9** (1980), No. 6, 14. Cit. 15.7, 15.8.

Hardy, G. H. and Wright, E. M.: *An Introduction to the Theory of Numbers*, Clarendon Press, Oxford, 1971. Cit. 8.21.

Hartman, A. and Rosa, A.: 'Cyclic One-Factorizations of the Complete Graph', *Eur. J. Combin.* **6** (1985) 45—48. Cit. 7.12, 7.50, 11.25, 14.7.

Hartnell, B.: 'Decomposition of K_{xy}, x and y Odd, into 2-x-Circuits', in *Proc. 4th Manitoba Conference on Numerical Mathematics* (Winnipeg, 1974), eds. Hartnell, B. L. and Williams, H. C., Congressus Numerantium 12, Utilitas Math., Winnipeg, 1975, pp. 265—271, MR 52: 185. Cit. 7.1, 10.16.

Hartnell, B.: 'Decomposition of K_{xy}^*, x and y Odd, into 2-x-Circuits', in *Proc. 4th Manitoba sur la théorie des graphes* (Paris, 1974), eds. Gillis, P. P. and Huyberechts, S., Cahiers du Centre d'Etudes de Recherche Opérationelle 17, 1975, Nos. 2—4, Inst. Stat. U.L.B., Bruxelles, 1975, pp. 221—223, MR 54: 138. Cit. 10.16.

Havel, V.: 'A Note on the Existence of Finite Graphs', *Čas. Pěstov. Mat.* **80** (1955) 477—480, MR 19: 627. Cit. 14.14.

Hebbare, S. P. Rao: 'Graceful Cycles', *Utilitas Math.* **10** (1976) 307—317, MR 55: 7849. Cit. 7.16, 7.19.

Hebbare, S. P. Rao: 'Embedding Graphs into Graceful and Supergraceful Graphs', *Graph Theory Newslett.* **10** (1981), No. 5, 1. Cit. 7.19.

Hedayat, A. and Kageyama, S.: 'The Family of t-Designs — Part I, *J. Statist. Plann. Inference* **4** (1980) 173—212, MR 82b: 05025. Cit. 8.32, 8.36, 12.49.

Hedge, N., Read, R. C. and Sridharan, M. R.: 'The Enumeration of Transitive Self-Complementary Digraphs', *Discrete Math.* **47** (1983) 109—112. Cit. 14.7, 14.19.

Heffter, L.: 'Über Tripelsysteme', *Math. Ann.* **49** (1987) 101—112. Cit. 7.31, 7.34.

Hell, P. and Rosa, A.: 'Handcuffed Prisoners and Balanced P-Designs', in *Proc. 2nd Louisiana Conference on Combinatorics Graph Theory and Computing*, eds. Mullin, R. C., Reid, K. B., Roselle, D. P. and Thomas, R. S. D., Louisiana State Univ., Baton Rouge, 1971, pp. 283—296. Cit. 6.18, 8.39, 9.2, 9.11, 12.40.

Hell, P. and Rosa, A.: 'Graph Decompositions, Handcuffed Prisoners and Balanced P-Designs', *Discrete Math.* **2** (1972) 229—252, MR 46: 1615. Cit. 6.18, 8.5, 8.39, 9.2, 9.10, 9.11, 11.1, 11.3, 11.4, 11.24, 11.25, 12.35, 12.37—12.40.

Hell, P., Kotzig, A. and Rosa, A.: 'Some Results on the Oberwolfach Problem (Decomposition of Complete Graphs into Isomorphic Quadratic Factors)', *Aeqs. Math.* **12** (1975) 1—5, MR 52: 197. Cit. 10.20, 11.25, 12.3, 12.40.

Hell, P. and Kirkpatrick, D. G.: 'Scheduling, Matching, and Colouring', in *Algebraic Methods in Graph Theory*, I—II (Proc. Colloquium, Szeged, 1978), eds. Lovász, L. and Sós, V. T., Colloquia Mathematica Societatis János Bolyai 25, North-Holland, Amsterdam — Bolyai János matematikai társulat, Budapest, 1981, pp. 273—279, MR 83c: 68074. Cit. 6.18.

Hering, F.: 'Block Designs with Cyclic Block Structure', in *Combinatorial Mathematics, Optimal Designs and Their Applications* (Proc. Symposium on Combinatorial Mathematics and Optimal Design, Colorado State Univ., Fort Collins, 1978), ed. Srivastava, J., Annals of Discrete Mathematics 6, North-Holland, Amsterdam, 1980, pp. 201—214, MR 82f: 05021. Cit. 9.2, 10.4.

Hetyei, G.: 'On the 1-Factors and the Hamiltonian Circuits of Complete n-Colorable Graphs', *Acta Acad. Paedagog. Civitate Pécs* (6), *Math. Phys. Chem. Tech.* **19** (1975) 5—10, MR 58: 33. Cit. 11.16.

Hill, R. and Irving, R. W.: 'On Group Partitions Associated with Lower Bounds for Symmetric

Ramsey Numbers', *Eur. J. Combin.* **3** (1982) 35—50, MR 84c: 05065. Cit. 14.12.

Hilton, A. J. W.: 'On Steiner and Similar Triple Systems', *Math. Scand.* **24** (1969) 208—216. Cit. 7.35, 8.16.

Himelwright, P., Wallis, W. D. and Williamson, J. E.: 'On One-Factorizations of Compositions of Graphs', *J. Graph Theory* **6** (1982) 75—80, MR 83d: 05076. Cit. 11.25.

Hoede, C. and Kuiper, H.: 'All Wheels Are Graceful', *Utilitas Math.* **14** (1978) 311, MR 80c: 05114. Cit. 7.19.

Hoffman, D. G. and Lindner, C. C.: 'Mendelsohn Triple Systems Having a Prescribed Number of Triples in Common', *Eur. J. Combin.* **3** (1982) 51—61, MR 83h: 05026. Cit. 8.35.

Hogarth, P. C.: 'Decomposition of Complete Graphs into 6-Stars and into 10-Stars', in *Combinatorial Mathematics III* (Proc. 3rd Australian Conference, Queensland Univ., St. Lucia, 1974), eds. Street, A. P. and Wallis, W. D., Lecture Notes in Mathematics 452, Springer-Verlag, Berlin—New York, 1975, pp. 136—142, MR 52: 198. Cit. 9.20.

Horton, J. D.: 'Room Designs and One-Factorizations', *Aeqs. Math.* **22** (1981) 56—63, MR 83b: 05043. Cit. 12.34.

Horton, J. D.: 'The Construction of Kotzig Factorizations', *Discrete Math.* **43** (1983a) 199—206, MR 84r: 05034. Cit. 11.19.

Horton, J. D.: 'Resolvable Path Designs', Preprint, Univ. New Brunswick, June, 1983b. Cit. 11.64.

Huang, C. and Rosa, A.: 'On Sets of Orthogonal Hamiltonian Circuits', in *Proc. 2nd Manitoba Conference on Numerical Mathematics* (Univ. Manitoba, Winnipeg, 1972), Congressus Numerantium 7, Utilitas Math., Winnipeg, 1973a, pp. 327—332, MR 50: 6920. Cit. 11.20 —11.22.

Huang, C. and Rosa, A.: 'On the Existence of Balanced Bipartite Designs', *Utilitas Math.* **4** (1973b) 55—75, MR 48: 10861. Cit. 9.2, 9.20, 9.25, 9.27.

Huang, C.: 'On the Existence of Balanced Bipartite Designs, II', *Discrete Math.* **9** (1974) 147—159, MR 50: 148. Cit. 7.27, 9.2, 9.20, 9.25, 9.26.

Huang, C.: 'Resolvable Balanced Bipartite Designs', *Discrete Math.* **14** (1976) 319—335, MR 53: 2706. Cit. 12.40.

Huang, C. and Rosa, A.: 'Decomposition of Complete Graphs into Trees', *Ars Combin.* **5** (1978) 23—63, MR 80e: 05091. Cit. 7.21, 7.24, 7.27, 9.3, 9.5, 9.7, 9.10—9.12, 9.25, 9.28, 9.34.

Huang, C., Kotzig, A. and Rosa, A.: 'On a Variation of the Oberwolfach Problem', *Discrete Math.* **27** (1979) 261—277, MR 82b: 05104. Cit. 10.20, 12.57.

Huang, C., Kotzig, A. and Rosa, A.: 'Further Results on Tree Labellings', *Utilitas Math.* **21C** (1982) 31—48, MR 84f: 05036. Cit. 7.21.

Huang, C., Mendelsohn, E. and Rosa, A.: 'On Partially Resolvable *t*-Partitions', in *Theory and Practice of Combinatorics,* eds. Rosa, A., Sabidussi, G. and Turgeon, J., North-Holland Mathematics Studies 60, Annals of Discrete Mathematics 12, North-Holland, Amsterdam, 1982, pp. 169—183. Cit. 12.57, 12.62.

Hung, S. H. Y. and Mendelsohn, N. S.: 'Directed Triple Systems', *J. Combin. Theory* **A14** (1973) 310—318, MR 47: 3190. Cit. 9.20, 10.10.

Hung, S. H. Y. and Mendelsohn, N. S.: 'Handcuffed Designs', *Aeqs. Math.* **11** (1974) 256—266, MR 51: 188. Cit. 9.3, 9.11.

Hung, S. H. Y. and Mendelsohn, N. S.: 'Handcuffed Designs', *Discrete Math.* **18** (1977) 23—33, MR 56: 5318. Cit. 7.27, 9.3, 9.11, 9.12.

Hwang, F. K.: 'Linear Neighbour Designs', in *Proc. 7th Southeastern Conference on Combinatorics, Graph Theory and Computing* (Louisiana State Univ., Baton Rouge, 1976), eds. Hoffman, F., Lesniak, L., Mullin, R., Reid, K. B. and Stanton, R., Congressus Numerantium 17, Utilitas Math., Winnipeg, 1976, pp. 335—340, MR 56: 5332. Cit. 9.3.

Hwang, F. K. and Lin, S.: Construction of 2-balanced (n, k, λ)-arrays. *Pacific J. Math.* **64** (1976)

220

437—454, MR 56: 5331. Cit. 10.4

Indzheyan, S. G., Kareyan, Z. A. and Nikogosyan, Zh. G.: 'Minimum Edge Decomposition of an *m*-Chromatic Graph into *n*-Colorable Graphs', *Trudy Vychisl. Tsentra AN ASSR i Erevan. Gos. Univ., Mat. Voprosy Kibernet. i Vychisl. Tekh.*, No. 9, Teoriya Grafov (1979) 13—15 (in Russian), MR 84d: 05082. Cit. 4.7.

Isbell, J. R.: 'An Inequality for Incidence Matrices', *Proc. Amer. Math. Soc.* **10** (1959) 216—218, MR 21: 3345. Cit. 8.5.

Ivanov, A. V.: *Some Results of a Constructive Enumeration of Block Designs*, Moskovskii Fiz.-Tekh. Inst., Moscow, 1981 (in Russian). Cit. 8.35.

Jackson, B.: 'Hamilton Cycles in Regular Graphs', *J. Graph Theory* **2** (1978) 363—365. Cit. 11.54.

Jackson, B.: 'Edge-Disjoint Hamilton Cycles in Regular Graphs of Large Degree', *J. London Math. Soc.* **19** (1979), No. 2, 13—16, MR 80k: 05078. Cit. 11.54.

Jackson, B.: 'Hamilton Cycles in Regular Two-Connected Graphs', *J. Combin. Theory* **B29** (1980a) 27—46. Cit. 11.12.

Jackson, B.: 'Paths and Cycles in Oriented Graphs', in *Combinatorics '79*, Parts I, II (Proc. Colloquium Univ. Montreal, Montreal, 1979), eds. Deza, M. and Rosenberg, I. G., Annals of Discrete Mathematics 8—9, North-Holland, Amsterdam, 1980b, pp. 275—277, MR 82a: 05051. Cit. 11.54.

Jaeger, F. and Payan, C.: 'Relations du type Nordhaus—Gaddum pour le nombre d'absorption d'un graphe simple', *C. R. Acad. Sci. Paris (A—B)* **274** (1972) A728—A730, MR 45: 3234. Cit. 4.18.

Jaeger, F., Payan, C. and Kouider, M. 'Partition of Odd Regular Graphs into Bistars', *Discrete Math.* **46** (1983) 93—94, MR 84i: 05088. Cit. 9.9.

Janko, Z. and Tran van Trung: 'On Projective Planes of Order 12 with an Automorphism of Order 13, I, Kirkman Designs of Order 27, *Geom. Dedicata* **11** (1981) 257—284, MR 83d: 05022a. Cit. 12.8.

Jimbo, M. and Kuriki, S.: 'On a Composition of Cyclic 2-Designs', *Discrete Math.* **46** (1983) 249—255. Cit. 8.16.

Jucovič, E. and Olejník, F.: 'On Chromatic and Achromatic Numbers of Uniform Hyper-graphs', *Čas. Pěstov. Mat.* **99** (1974) 123—130, MR 52: 2941. Cit. 4.18.

Jungnickel, D.: 'On Automorphism Groups of Divisible Designs', *Can. J. Math.* **34** (1982) 257—297. Cit. 8.42.

Kageyama, S.: 'A Survey of Resolvable Solutions of Balanced Incomplete Block Designs', *Int. Statist. Rev.* **40** (1972) 269—273. Cit. 12.46, 12.49.

Kageyama, S.: 'A Series of 3-Designs', *J. Japan Statist. Soc.* **3** (1973) 67—68, MR 54: 5002. Cit. 12.49.

Kageyama, S.: 'A Condition for the Existence of Certain Resolvable 2-Designs', *J. Japan Statist. Soc.* **8** (1978) 37—38, MR 82a: 62119. Cit. 12.13, 12.49.

Kageyama, S.: 'On Balanced Incomplete Block Designs with $v = 2k$, *J. Japan Statist. Soc.* **10** (1980) 135—137, MR 82g: 62108. Cit. 8.36.

Kageyama, S. and Hedayat, A.: 'The Family of *t*-Designs, II', *J. Statist. Plann. Inference* **7** (1982/83) 257—287, MR 84g: 05024. Cit. 8.20, 8.32, 8.36, 8.57, 12.49.

Kárteszi, F.: *Introduction to Finite Geometries*, Akàdémiai Kiadó, Budapest — North-Holland, Amsterdam—Oxford—New York, 1976, MR 54: 11156. Cit. 7.55, 7.56, 8.22, R 7.56.

Keedwell, A. D.: 'Some Problems Concerning Complete Latin Squares', in *Combinatorics* (Proc. British Combinatorial Conference, Univ. College of Wales, Aberystwyth, 1973), eds. McDonough, T. P. and Mavron, V. C., London Math. Soc. Lecture Note Series 13, Cambridge University Press, London—New York, 1974, pp. 89—96, MR 51: 2943. Cit. 11.11.

Keedwell, A. D.: 'Decompositions of Complete Graphs Defined by Quasigroups', in *Theory and Practice of Combinatorics*, eds. Rosa, A., Sabidussi, G. and Turgeon, J., North-Holland Mathematics Studies 60, Annals of Discrete Mathematics 12, North-Holland, Amsterdam, 1982, pp. 185—192. Cit. 10.6, 10.16.

Kel'mans, A. K.: 'On the Properties of the Characteristic Polynomial of a Graph', in *Let Cybernetics Serve Communism 4*, ed. Berg, A. I., Energiya, Moscow—Leningrad, 1967, pp. 27—41 (in Russian), MR 52: 13450. Cit. 5.10.

Kirkman, T. P.: 'On a Problem in Combinations', *Cambridge and Dublin Math. J.* **2** (1847) 191—204. Cit. 7.39, 12.1.

Kirkman, T. P.: 'Query Lady's and Gentleman's Diary' (1850a) 48. Cit. 12.1.

Kirkman, T. P.: 'Note on an Unanswered Prize Question', *Cambridge and Dublin Math. J.* **5** (1850b) 255—262. Cit. 12.1, 12.33.

Klee, V.: 'Shapes of the Future (Physics, Botany, and Organic Chemistry Provide Illuminating Examples of Some Unsolved Geometric Problems)', *Amer. Scientist* **59** (1971) 84—91 (*The Two-Year College Math. J.* **2** (1971) No. 2, 14—27). Cit. 11.40.

Koh, K. M.: 'Some Problems on Graph Theory', *Southeast Asian Bull. Math.* **1** (1977) 39—43, MR 80h: 05048. Cit. 11.14.

Koh, K. M., Rogers, D. G. and Tan, T.: 'On Graceful Trees', *Nanta Math.* **10** (1977) 207—211, MR 81b: 05033. Cit. 7.21.

Koh, K. M., Rogers, D. G. and Lim, C. K.: 'On Graceful Graphs, I, Sum of Graphs', *Southeast Asian Bull. Math.* **3** (1979a) 58, MR 83h: 05071. Cit. 7.19.

Koh, K. M., Rogers, D. G. and Tan, T.: 'A Graceful Arboretum: a Survey of Graceful Trees', in *Proc. 1st Franco-Southeast Asian Mathematical Conference* (Singapore, 1979), *Southeast Asian Bull. Math.*, 1979b, Special Issue b, pp. 278—287, MR 83e: 05041. Cit. 7.19, 7.21.

Koh, K. M., Tan, T. and Rogers, D. G.: 'Interlaced Trees: A Class of Graceful Trees', in *Combinatorial Mathematics VI* (Proc. 6th Australian Conference, Univ. New England, Armidale, 1978), eds. Horadam, A. F. and Wallis, W. D., Lecture Notes in Mathematics 748, Springer-Verlag, Berlin, 1979c, pp. 65—78, MR 81h: 05054. Cit. 7.21.

Koh, K. M., Tan, T. and Rogers, D. G.: 'Two Theorems on Graceful Trees', *Discrete Math.* **25** (1979d) 141—148, MR 80e: 05050. Cit. 7.21.

Koh, K. M., Rogers, D. G., Lee, P. Y. and Toh, C. W.: 'On Graceful Graphs, V, Unions of Graphs with One Vertex in Common', *Nanta Math.* **12** (1979e) 133—136, MR 83h: 05072b. Cit. 7.19.

Koh, K. M., Rogers, D. G. and Tan, T.: 'Products of Graceful Trees', *Discrete Math.* **31** (1980a) 279—293, MR 81i: 05100. Cit. 7.21.

Koh, K. M., Rogers, D. G., Teo, H. K. and Yap, K. Y.: 'Graceful Graphs: Some Further Results and Problems', in *Proc. 11th Southeastern Conference on Combinatorics, Graph Theory and Computing* (Florida Atlantic Univ., Boca Raton, 1980), eds. Hadlock, F. O., Hoffman, F., Congressus Numerantium 28—29, Utilitas Math., Winnipeg, 1980b, Vol. II, pp. 559—571, MR 83h: 05072a. Cit. 7.16, 7.19.

Koh, K. M., Rogers, D. G. and Tan, R.: 'Another Class of Graceful Trees', *J. Aust. Math. Soc.* A31 (1981) 226—235, MR 83e: 05042. Cit. 7.21.

Koh, K. M. and Punnim, N.: 'On Graceful Graphs: Cycles with 3-Consecutive Chords', *Bull. Malaysian Math. Soc.* **5** (1982), No. 2, 49—64, MR 84b: 05046. Cit. 7.16.

Köhler, E.: 'Das Oberwolfacher Problem', *Mitt. Math. Gesell. Hamburg* **10** (1973) 124—129, MR 49: 4855. Cit. 10.20.

Köhler, E.: 'Reguläre Faktoren in vollständigen Graphen', *Abh. Math. Sem. Univ. Hamburg* **41** (1974) 252—254, MR 50: 4400. Cit. 15.13.

Köhler, N.: 'Zerlegung total gerichteter Graphen in Kreise', *Manuscr. Math.* **19** (1976) 151—164, MR 53: 13021. Cit. 10.16, 10.19.

222

König, D.: 'Über Graphen und ihre Anwendung auf Determinantentheorie und Mengenlehre', *Math. Ann.* **77** (1916) 453—465. Cit. 12.21, 12.22.

Kopylova, A. N.: 'A Problem Concerning Coloured Polygons', *Vestnik Moskov. Univ. (I)* **20** (1965), No. 2, 35—38 (in Russian), MR **32**: 2358. Cit. 14.7.

Korovina, R. P.: 'Systems of Pairs of the Cyclic Type', *Mat. Zametki* **28** (1980) 271—278 (in Russian), MR **82e**: 05108. Cit. 14.7.

Kotzig, A.: 'From the Theory of Finite Regular Graphs of Degree Three and Four', *Čas. Pěstov. Mat.* **82** (1975) 76—92 (in Slovak), MR **19**: 876. Cit. 11.40.

Kotzig, A.: 'Hamiltonian Circuits in Lattice Graphs', *Čas. Pěstov. Mat.* **90** (1956a) 1—11 (in Russian), MR **33**: 2567. Cit. 11.25.

Kotzig, A.: 'On Decompositions of the Complete Graph into 4k-gons', *Mat.-Fyz. Čas.* **15** (1965b) 229—233 (in Russian), MR **33**: 3958. Cit. 7.1, 7.27, 9.31, 10.5, 10.6.

Kotzig, A.: 'The Decomposition of a Directed Graph into Quadratic Factors Consisting of Cycles', *Acta Fac. R. N. Univ. Comenianae, Math.* **22** (1969) 27—29, MR **44**: 117. Cit. 11.14.

Kotzig, A.: 'Every Cartesian Product of Two Circuits is Decomposable into Two Hamiltonian Circuits', Preprint, Univ. de Montréal, Montréal, 1973a. Cit. 11.25—11.28, 11.31.

Kotzig, A.: 'On Certain Vertex-Valuations of Finite Graphs', *Utilitas Math.* **4** (1973b) 261—290, MR **52**: 5490. Cit. 7.19, 7.21.

Kotzig, A. and Rosa, A.: 'Nearly Kirkman Systems', in *Proc. 5th Southeastern Conference on Combinatorics, Graph Theory and Computing* (Florida Atlantic Univ., Boca Raton, 1974), eds. Hoffman, F. et al., Congressus Numerantium 10, Utilitas Math., Winnipeg, 1974, pp. 607—614, MR **51**: 190. Cit. 12.57.

Kotzig, A.: 'β-Valuations of Quadratic Graphs with Isomorphic Components', *Utilitas Math.* **7** (1975) 263—279, MR **52**: 2964. Cit. 7.19.

Kotzig, A.: 'Problems and Recent Results on 1-Factorizations of Cartesian Products of Graphs', in *Proc. 9th Southeastern Conference on Combinatorics, Graph Theory and Computing* (Florida Atlantic Univ., Boca Raton, 1978), eds. Hoffman, F., McCarthy, D., Mullin, R. C. and Stanton, R. G., Congressus Numerantium 21, Utilitas Math., Winnipeg, 1978, pp. 457—460, MR **80i**: 05036. Cit. 11.25.

Kotzig, A. and Turgeon, J.: 'β-Valuations of Regular Graphs with Complete Components', in *Combinatorics* (Proc. 5th Hungarian Colloquium on Combinatorics, Keszthely, 1976), eds. Hajnal, A. and Sós, V. T., Colloquia Mathematica Societatis János Bolyai 18, Bolyai János matematikai társulat, Budapest — North-Holland, Amsterdam, 1978, pp. 697—703, MR **80g**: 05056. Cit. 7.19, 7.21, R 7.47.

Kotzig, A.: 'Problems', in *Proc. 10th Southeastern Conference on Combinatorics, Graph Theory and Computing* (Florida Atlantic Univ., Boca Raton, 1979), eds. Hoffman, F., McCarthy, D., Mullin, R. C. and Stanton, R. G., Congressus Numerantium 23—24, Utilitas Math., Winnipeg, 1979a, Vol. II, pp. 913—915. Cit. 9.9, 9.36.

Kotzig, A.: 'Selected Open Problems in Graph Theory', in *Graph Theory and Related Topics* (Proc. Conference, Univ. Waterloo, 1977), eds. Bondy, J. A. and Murty, U.S.R., Academic Press, New York—San Francisco—London, 1979b, pp. 358—367. Cit. 9.31, 9.35, 12.64, 14.15.

Kotzig, A.: '1-Factorizations of Cartesian Products of Regular Graphs', *J. Graph Theory* **3** (1979c) 23—34, MR **80c**: 05115. Cit. 11.25.

Kotzig, A. and Turgeon, J.: 'Sur l'existence de petites composantes dans tout système parfait d'ensembles de différences', in *Combinatorics '79*, Part I (Proc. Colloquium Univ. Montreal, 1979), eds. Deza, M. and Rosenberg, I. G., Annals of Discrete Mathematics 8—9, North-Holland, Amsterdam, 1980, pp. 71—75, MR **82c**: 05023. Cit. 7.19, R 7.47.

Kotzig, A.: 'Decomposition of Complete Graphs into Isomorphic Cubes', *J. Combin. Theory* **B31** (1981) 292—296, MR **83b**: 05103. Cit. 7.17, 7.19, 7.27, 9.30.

Kramer, E. S. and Mesner, D. M.: 'Intersections among Steiner Systems', *J. Combin. Theory* **A16** (1974) 273—285, MR 49: 78. Cit. 12.33.

Kropar, M. and Read, R. C.: 'On the Construction of the Self-Complementary Graphs on 12 Nodes', *J. Graph Theory* **3** (1979) 111—125, MR 80i: 05051. Cit. 14.6.

Kurek, Kh. E. and Petrenyuk, A. Ya. (1977): 'On Coverings of Graphs with Stars', in *Graph Theory*, eds. Khomenko, N. P., Inst. Mat. AN USSR, Kiev, 1977, pp. 145—156 (in Russian), MR 80f: 05037. Cit. 9.20.

Kuriki, S. and Jimbo, M.: On 1-Rotational $S_\lambda(2, 3, v)$-Designs', *Discrete Math.* **46** (1983) 33—40. Cit. 8.16.

Laskar, R. and Hare, W.: 'Chromatic Numbers for Certain Graphs', *J. London Math. Soc.* **4** (1971/72), No. 2, 489—492, MR 45: 6680. Cit. 12.28.

Laskar, R. and Auerbach, B.: 'On Decomposition of r-Partite Graphs into Edge Disjoint Hamilton Circuits', *Discrete Math.* **14** (1976) 265—268, MR 52: 10493. Cit. 11.16.

Laskar, R.: 'Decomposition of Some Composite Graphs into Hamilton Cycles', in *Combinatorics* (Proc. 5th Hungarian Colloquium on Combinatorics, Keszthely, 1976), eds. Hajnal, A. and Sós, V. T., Colloquia Mathematica Societatis János Bolyai 18, Bolyai János matematikai társulat, Budapest — North Holland, Amsterdam, 1978, pp. 705—716, MR 81g: 05078. Cit. 11.38.

Laskar, R. and Auerbach, B.: 'On Complementary Graphs with No Isolated Vertices', *Discrete Math.* **24** (1978) 113—118, MR 80d: 05045. Cit. 4.33, 4.34, 4.37.

Laufer, P. J.: 'Regular Perfect Systems of Difference Sets of Size 4 and Extremal Systems of Size 3', in *Theory and Practice of Combinatorics*, eds. Rosa, A., Sabidussi, G. and Turgeon, J., North-Holland Mathematics Studies 60, Annals of Discrete Mathematics 12, North-Holland, Amsterdam, 1982, pp. 193—201. Cit. 7.14.

Lawless, J. F.: 'Further Results Concerning the Existence of Handcuffed Designs' *Aeqs. Math.* **11** (1974a) 97—106, MR 50: 6884. Cit. 9.3, 9.11.

Lawless, J. F.: 'On the Construction of Handcuffed Designs', *J. Combin. Theory* **A16** (1974b) 76—86. Cit. 9.3, 9.11.

Lederberg, J.: 'Hamilton Circuits of Convex Trivalent Polyhedra (up to 18 Vertices)', *Amer. Math. Monthly* **74** (1967) 522—527, MR 35: 2770. Cit. 11.40.

Lenz, H.: 'A Few Remarks on Hanani's Design Constructions', *J. Geom.* **17** (1981) 161—170, MR 83b: 05021. Cit. 8.14, 8.26, 8.27.

Lenz, H.: 'A Few Simplified Proofs in Design Theory', *Exposition Math.* **1** (1983) 77—80, MR 84e: 05024. Cit. 8.20.

Lesniak-Foster, L. and Roberts, J.: 'On Ramsey Theory and Graphical Parameters', *Pac. J. Math.* **68** (1977) 105—114, MR 58: 364. Cit. 8.36.

Levi, F. W.: 'Remarks on Mr. V. Narasimba Murthi's Paper: On a Problem of Arrangements, I', *J. Ind. Math. Soc.* (N.S.) **4** (1940) 45—46, MR 2: 115. Cit. 11.11.

Li, Wei Xuan: 'The Connectivity of Complementary Graphs', *Kexue Tongbao* **22** (1977), No. 2, 78 (in Chinese), MR 58: 21796. Cit. 4.30.

Lick, D. R. and White, A. T.: 'k-Degenerate Graphs', *Can. J. Math.* **22** (1970) 1082—1096, MR 42: 1715. Cit. 4.19, 4.21.

Lick, D. R. and White, A. T.: 'Point Partition Numbers of Complementary Graphs', *Math. Japonicae* **19** (1974) 233—237, MR 54: 5051. Cit. 4.20.

Lie, Z.: 'A Short Disproof of Euler's Conjecture', *Ars Combin.* **12** (1982) 47—55. Cit. 12.30.

Lindner, C. C.: 'A Partial Steiner Triple System of Order n Can Be Embedded in a Steiner Triple System of Order $6n + 3$', *J. Combin. Theory* **A18** (1975) 349—351, MR 52: 129. Cit. 8.16.

Lindner, C. C., Mendelsohn, E. and Rosa, A.: 'On the Number of 1-Factorizations of the Complete Graph', *J. Combin. Theory* **B20** (1976) 265—282, MR 58: 10617. Cit. 13.4, 14.7.

Liouville, B.: 'Problèmes Hamiltoniens et somme des graphes', *Rev. Roumaine Math. Pures*

Appl. **26** (1981) 79—88, MR 83e: 05086. Cit. 11.25.

Lonc, Z. and Truszczyński, M.: 'Decompositions of Graphs into Graphs with Three Edges', Preprint Ia. Cit. 9.28.

Lonc, Z. and Truszczyński, M.: 'Decomposition of Graphs into Graphs with Bounded Maximum Degrees', Preprint Ib. Cit. 9.28.

Lorimer, P.: 'The Construction of Finite Projective Planes', in *Combinatorial Mathematics VIII* (Proc. 8th Australian Conference on Combinatorial Mathematics, Deakin Univ., Geelong, 1980), ed. McAvaney, K. L., Lecture Notes in Mathematics 884, Springer-Verlag, Berlin —Heidelberg—New York, 1981, pp. 64—76, MR 83a: 51001. Cit. 8.22.

Lovász, L.: *Combinatorial Problems and Exercises,* Akadémiai Kiadó, Budapest — North-Holland, Amsterdam, 1979a, MR 80m: 05001. Cit. 4.32.

Lovász, L.: 'On the Shannon Capacity of a Graph', *IEEE Trans. Inf. Theory* **25** (1979b) 1—7, MR 81g: 05095. Cit. 14.15, 14.17.

Lucas, E.: *Récréations mathématiques,* Vol. II, Gauthier-Villars, Paris, 1883. Cit. 11.5.

Mac Neish, H. F.: 'Euler Squares', *Ann. Math.* **23** (1922) 221—227. Cit. 12.30.

Maheo, M.: 'Strongly Graceful Graphs', *Discrete Math.* **29** (1980) 39—46, MR 82i: 05093. Cit. 7.17, 7.19, 7.52, 9.31, 9.35.

Maheo, M. and Thuillier, H.: 'On *d*-Graceful Graphs', *Ars Combin.* **13** (1982) 181—192, MR 84d: 05149. Cit. 7.8.

Majumdar, K. N.: 'On Some Theorems in Combinatorics Relating to Incomplete Block Designs', *Ann. Math. Statist.* **24** (1953) 377—389, MR 15: 93. Cit. 8.4.

Martin, P.: 'Cycles Hamiltoniens dans les graphes 4-réguliers 4-connexes', *Aeqs. Math.* (in press). Cit. 11.42.

Mathon, R .A., Phelps, K. T. and Rosa, A.: 'A Class of Steiner Triple Systems of Order 21 and Associated Kirkman Systems', *Math. Comput.* **37** (1981) 209—222, MR 82k: 05022. Cit. 8.35, 12.1, 12.8.

Mathon, R. A., Phelps, K. T. and Rosa, A.: 'Small Steiner Triple Systems and Their Properties', *Ars Combin.* **15** (1983) 3—110. Cit. 8.34, 8.35, 12.8.

Mathon, R. A. and Rosa, A.: 'A Census of 1-Factorizations of K_9^3: Solutions with Group of Order >4', *Ars Combin.* **16** (1983) 129—147. Cit. 12.33.

Mathon, R. A. and Rosa, A.: 'The 4-Rotational Steiner and Kirkman Triple Systems of Order 21', *Ars Combin.* **17A** (1984) 241—250. Cit. 8.35, 12.8.

Mathon, R. and Rosa, A.: 'Tables of Parameters of BIBDs with $r \leq 41$ Including Existence, Enumeration and Resolvability Results', *Ann. Discrete Math.* **26** (1985) 275—308. Cit. 8.20, 8.36, 8.49, 12.46, 12.49.

Mavron, V. C.: Construction for Resolvable and Related Designs', *Aeqs. Math.* **23** (1981) 131—145. Cit. 12.49.

McKay, B. D. and Stanton, R. G.: 'Isomorphism of Two Large Designs', *Ars Combin.* **6** (1978) 87—90, MR 80c: 05030. Cit. 8.27.

Meally, V.: 'Problem *164, the 13 Knights', *J. Recreat. Math.* **4** (1971) 136. Cit. 11.20.

Mendelsohn, E. and Rosa, A.: 'Embedding Maximal Packings of Triples', in *Proc. 14th Southeastern Conference on Combinatorics, Graph Theory and Computing* (Florida Atlantic Univ., Boca Raton, 1983), Congressus Numerantium 39—40, Utilitas Math., Winnipeg, 1983, Vol. II, pp. 235—247. Cit. 8.16.

Mendelsohn, E. and Rosa, A.: One-Factorizations of the Complete Graph — a Survey', *J. Graph Theory* **9** (1985) 43—65. Cit. 7.41, 14.7.

Mendelsohn, N. S.: 'Hamiltonian Decomposition of the Complete Directed *n*-Graphs', in *Theory of Graphs* (Proc. Colloquium, Tihany, 1966), eds. Erdős, P. and Katona, G., Akadémiai Kiadó, Budapest, 1968, pp. 237—241, MR 38: 4361. Cit. 11.8, 11.11, 11.61.

Mendelsohn, N. S.: 'A Natural Generalization of Steiner Triple Systems', in *Computers in*

Number Theory (Proc. Sci. Res. Council Atlas Symp., No. 2, Oxford, 1969), eds. Atkin, A. O. L. and Birch, B. J., Academic Press, New York, 1971, pp. 323—339, MR 48: 122. Cit. 8.35, 10.9, 10.12.

Meredith, G. H. J.: Regular *n*-Valent *n*-Connected Non-Hamiltonian Non-*n*-Edge Colorable Graphs', *J. Combin. Theory* **B14** (1973) 55—60, MR 47: 65. Cit. 11.41.

Merriel, D.: 'Partitioning the Directed Graph into 5-Cycles', Preprint. Cit. 10.12.

Mielants, W.: 'On the Nonexistence of a Class of Symmetric Group Divisible Partial Designs', *J. Geom.* **12** (1979) 89—98, MR 80e: 05030. Cit. 8.40, 8.42.

Miller, J. C. P.: 'Difference Bases, Three Problems in Additive Number Theory', in *Computers in Number Theory* (Proc. Sci. Res. Council Atlas Symp., No. 2, Oxford, 1969), eds. Atkin, A. O. L. and Birch, B. J., Academic Press, New York, 1971, pp. 299—322, MR 47: 4817. Cit. 7.14.

Mills, W. H.: 'A New Block Design', in *Proc. 4th Manitoba Conference on Numerical Mathematics and Computing* (Winnipeg, 1974), eds. Hartnell, B. L. and Williams, H. C., Congressus Numerantium 12, Utilitas Math., Winnipeg, 1975a, pp. 461—465, MR 56: 11820. Cit. 8.27.

Mills, W. H.: 'Two New Block Designs', *Utilitas Math.* **7** (1975b) 73—75, MR 51: 10123. Cit. 7.4, 8.27.

Mills, W. H.: 'The Construction of Balanced Incomplete Block Designs with $\lambda = 1$', in *Proc. 7th Manitoba Conference on Numerical Mathematics and Computing* (Univ. Manitoba, Winnipeg, 1977), eds. Hartnell, B. L. and Williams, H. C., Congressus Numerantium 20, Utilitas Math., Winnipeg, 1978a, pp. 131—148, MR 80g: 05013. Cit. 7.4, 8.27, 12.30.

Mills, W. H.: 'The Construction of BIBDs Using Nonabelian Groups', in *Proc. 9th Southeastern Conference on Combinatorics, Graph Theory and Computing* (Florida Atlantic Univ., Boca Raton, 1978), eds. Hoffman, F., McCarthy, D., Mullin, R. C. and Stanton, R. G., Congressus Numerantium 21, Utilitas Math., Winnipeg, 1978b, pp. 519—526, MR 80e: 05031. Cit. 8.27.

Mills, W. H.: 'The Construction of Balanced Incomplete Block Designs', in *Proc. 10th Southeastern Conference on Combinatorics, Graph Theory and Computing* (Florida Atlantic Univ., Boca Raton, 1979), eds. Hoffman, F., McCarthy, D., Mullin, R. C. and Stanton, R. G., Congressus Numerantium 23—24, Utilitas Math., Winnipeg, 1979, pp. 73—86, MR 81m: 05025. Cit. 8.27.

Mitchem, J.: 'On the Chromatic Number of Complementary Set Systems', *Tamkang J. Math.* 287—292, MR 43: 1875. Cit. 4.20.

Mitchem, J.: 'On The Chromatic Number of Complementary Set Systems', *Tamkang J. Math.* **5** (1974) 113—124, MR 50: 9671. Cit. 4.18.

Mohar, B.: 'On the Edge-Colorability of Products of Graphs', in *Preprint Series of the Department of Mathematics 19*, 1981, Društvo matematikov, fizikov in astronomov SR Slovenije, Yugoslavia, Ljubljana, 1982, pp. 125—134. Cit. 11.25.

Mohar, B. and Pisanski, T.: 'Edge Coloring of a Family of Regular Graphs', in *Preprint Series of the Department of Mathematics 19*, 1981, Društvo matematikov, fizikov in astronomov SR Slovenije, Yugoslavia, Ljubljana, 1982, pp. 5—18. Cit. 11.25.

Mohar, B., Pisanski, T. and Shawe-Taylor, J.: 'Edge Coloring of Composite Regular Graphs', in *Preprint Series of the Department of Mathematics 19*, 1981, Društvo matematikov, fizikov in astronomov SR Slovenije, Yugoslavia, Ljubljana, 1982, pp. 81—95. Cit. 11.25.

Moon, J. W.: *Topic on Tournaments*, Holt, Rinehart and Winston, New York, 1968, MR 41: 1574. Cit. 11.14, 14.19.

Moore, E. H.: 'Concerning Triple Systems', *Math. Ann.* **43** (1983) 271—285. Cit. 7.39.

Morris, P. A.: 'Self-Complementary Graphs and Digraphs', *Math. Comput.* **27** (1973) 216—217. Cit. 14.6.

Morris, P. A.: 'On Self-Complementary Graphs and Digraphs', in *Proc. 5th Southeastern*

226

Conference on Combinatorics, Graph Theory and Computing (Florida Atlantic Univ., Boca Raton, 1974), eds. Hoffman, F. et al., Congressus Numerantium 10, Utilitas Math., Winnipeg, 1974, pp. 583—590, MR 51: 245. Cit. 14.22.

Morris, P. A.: 'On Producing All Graceful Permutations', in *Proc. 3rd Caribbean Conference on Combinatorics and Computing* (Univ. West Indies, Bridgetown, 1981), eds. Cadogan, Ch. C., University of the West Indies, Cave Hill Campus, Bridgetown, 1981, pp. 144—152, MR 81f: 05003. Cit. 7.21.

Mulder, P.: 'Kirkman-Systemen', Academisch Proefschrift ter verkrijging van den graad van doctor in de Wis- en Natuurkunde aan de Rijksuniversiteit te Groningen, Leiden, 1917. Cit. 12.8.

Mullin, R. C., Schellenberg, P. J., Stinson, D. R. and Vanstone, S. A.: 'Some Results on the Existence of Squares', in *Combinatorial Mathematics, Optimal Designs and Their Applications* (Proc. Symposium on Combinatorial Mathematics and Optimal Design, Colorado State Univ., Fort Collins, 1978), ed. Srivastava, J., Annals of Discrete Mathematics 6, North-Holland, Amsterdam, 1980, pp. 257—274, MR 81m: 05036. Cit. 12.30.

Mullin, R. C. and Wallis, W. D.: 'An Exceptional Case in the Classification of Finite Linear Spaces', in *Proc. West Coast Conference on Combinatorics, Graph Theory and Computing* (Humboldt State Univ., Arcata, 1979), eds. Chinn, P. Z. and McCarthy, D., Congressus Numerantium 26, Utilitas Math., Winnipeg, 1980, pp. 229—279, MR 82f: 05020. Cit. 8.1, 14.7.

Myers, B. R.: 'Class of Hamiltonian-Partitionable Networks', *Electron. Lett.* **8** (1972a) 19—20, MR 49: 140. Cit. 11.27.

Myers, B. R.: 'Hamiltonian Factorization of the Product of a Complete Graph with Itself', *Networks* **2** (1972b) 1—9, MR 47: 78. Cit. 11.27, 11.33.

Nakamura, G., Kiyasu, Z. and Ikeno, N.: 'Solution of the Round Table Problem for the Case of $(p^k + 1)$ Persons', *Comment. Math. Univ. St. Paul* **29** (1980) 7—20, MR 81m: 05004. Cit. 11.20.

Narasimhamurti, V.: 'On a Problem of Arrangements, I', *J. Ind. Math. Soc.* **4** (1980) 39—43, MR 2: 115. Cit. 11.5.

Nash-Williams, C. St. J. A.: 'Decomposition of the *n*-Dimensional Lattice-Graph into Hamiltonian Lines', *Proc. Edinburgh Math. Soc.* **12** (1960/61), No. 2, 123—131, MR 24: A56. Cit. 11.25, 11.48.

Nash-Williams, C. St. J. A.: 'Decompositions of Graphs into Two-Way Infinite Paths', *Can. J. Math.* **15** (1963) 479—485, MR 27: 741. Cit. 10.20.

Nash-Williams, C. St. J. A.: 'Hamiltonian Lines in Products of Infinite Trees', *J. London Math. Soc.* **40** (1965) 37—40. Cit. 11.25.

Nash-Williams, C. St. J. A.: 'Hamiltonian Circuits in Graphs and Digraphs', in *The Many Facets of Graph Theory* (Proc. Conference Western Michigan Univ., Kalamazoo, 1968), eds. Chartrand, G. and Kapoor, S. F., Lecture Notes in Mathematics 110, Springer-Verlag, Berlin, 1969, pp. 237—243, MR 40: 5484. Cit. 11.54.

Nash-Williams, C. St. J. A.: 'Hamiltonian Lines in Graphs Whose Vertices Have Sufficiently Large Valencies', in *Combinatorial Theory and Its Applications* (Proc. Colloquium Balatonfüred, 1969), eds. Erdős, P., Rényi, A. and Sós, V. T., Colloquia Mathematica Societatis János Bolyai 4, North-Holland, Amsterdam — Bolyai János matematikai társulat, Budapest, 1970a, pp. 813—819, MR 45: 8565. Cit. 11.54.

Nash-Williams, C. St. J. A.: 'Problem', in *Combinatorial Theory and Its Application* (Proc. Colloquium, Balatonfüred, 1969), eds. Erdős, P., Rényi, A. and Sós, V. T., Colloquia Mathematica Societatis János Bolyai 4, North-Holland, Amsterdam — Bolyai János matematikai társulat, Budapest, 1970b, pp. 1179—1182. Cit. 8.16, 10.17.

Nash-Williams, C. St. J. A.: 'Edge-Disjoint Hamiltonian Circuits in Graphs with Vertices of

Large Valency', in *Studies in Pure Mathematics in Combinatorial Theory, Analysis, Geometry, Algebra and the Theory of Numbers*, ed. Mirsky, L., Academic Press, London—New York, 1971a, pp. 157—183, MR 44: 1594. Cit. 11.54.

Nash-Williams, C. St. J. A.: 'Possible Directions in Graph Theory', in *Combinatorial Mathematics and Its Applications* (Proc. Conference Math. Inst., Oxford, 1969), ed. Welsh, D. J. A., Academic Press, London—New York, 1971b, pp. 191—200, MR 44: 1593. Cit. 11.41.

Nebeský, L.: 'A Theorem on Hamiltonian Line Graphs', *Comment. Math. Univ. Carol.* **14** (1973) 107—112, MR 52: 2956. Cit. 5.6.

Nebeský, L.: 'On Pancyclic Line Graphs', *Czechoslov. Math. J.* **28** (1978) 650—655, MR 80a: 05137. Cit. 5.6.

Nebeský, L.: 'On the Existence of 1-Factors in Partial Squares of Graphs', *Czechoslov. Math. J.* **29** (1979) 349—352, MR 81b: 05098. Cit. 9.7, 10.17.

Nebeský, L.: 'On Upper Embedability of Complementary Graphs', *Čas. Pěstov. Mat.* **108** (1983) 214—217. Cit. 5.10.

Neumann-Lara, V.: 'k-Hamiltonian Graphs with Given Girth', in *Infinite and Finite Sets* (Proc. Colloquium, Keszthely, 1973), eds. Hajnal, A., Rado, R. and Sós, V. T., Colloquia Mathematica Societatis János Bolyai 10, Bolyai János matematikai társulat, Budapest — North-Holland, Amsterdam, 1975, pp. 1133—1142, MR 52: 188. Cit. 11.53.

Ninčák, J.: 'On a Conjecture by Nash-Williams', *Comment. Math. Univ. Carol.* **14** (1973) 135—138, MR 52: 2949. Cit. 11.54.

Ninčák, J.: 'Hamiltonian Circuits in Complete Bipartite Graphs', *Prikl. Mat. Programmirovanie* **11** (1974) 62—66 (in Russian), MR 49: 4856. Cit. 11.16.

Ninčák, J.: 'An Estimate of the Number of Hamiltonian Cycles in a Multigraph', in *Recent Advances in Graph Theory* (Proc. 2nd Czechoslovak Symposium, Prague, 1974), ed. Fielder, M., Academia, Prague, 1975, pp. 431—438, MR 52: 5477. Cit. 11.44.

Nirmala, K.: 'Complementary Graphs and Total Chromatic Numbers', in *Proc. Symposium on Graph Theory* (Indian Statist. Inst., Calcutta, 1976), ed. Rao, A. R., ISI Lecture Notes 4, Macmillan of India, New Delhi, 1979, pp. 227—245. Cit. 4.27.

Nordhaus, E. A. and Gaddum, J. W.: 'On Complementary Graphs', *Amer. Math. Monthly* **63** (1956) 175—177, MR 17: 1231. Cit. 4.1, 4.12.

Nordhaus, E. A.: 'On the Density and Chromatic Numbers of Graphs', in *The Many Facets of Graph Theory* (Proc. Conference Western Michigan Univ., Kalamazoo, 1968), eds. Chartrand, G. and Kapoor, S. F., Lecture Notes in Mathematics 110, Springer-Verlag, Berlin, 1969, pp. 245—249, MR 41: 1575. Cit. 5.10.

Novák, J.: 'The Use of Combinatorics in the Study of Plane Configurations $(12_4, 16_3)$', *Čas. Pěstov. Mat.* **84** (1959) 257—282 (in Czech), MR 24: A 447. Cit. 8.16.

Novák, J.: 'Contribution to the Theory of Combinations', *Čas. Pěstov. Mat.* **88** (1963) 129—141 (in Czech), MR 28: 2977. Cit. 8.16.

Novák, J.: 'Über gewisse minimale Systeme von k-Tupeln', *Comment. Math. Univ. Carol.* **11** (1970a) 397—402, MR 42: 4413. Cit. 8.16.

Novák, J.: 'Über gewisse Tripel-Systeme', in *Combinatorial Theory and Its Applications* (Proc. Colloquium, Balatonfüred, 1969), eds. Erdős, P., Rényi, A. and Sós, V. T., Colloquia Mathematica Societatis János Bolyai 4, North-Holland, Amsterdam — Bolyai János matematikai társulat, Budapest, 1970b, pp. 821—838, MR 45: 8543. Cit. 8.16.

Novák, J.: 'On Certain Minimal Systems of k-Tuples', in *Problèmes combinatoires et théorie des graphes* (Colloquium, Orsay, 1976), eds. Bermond, J.-C., Fournier, J.-C., Las Vergnas, M. and Sotteau, D., Colloq. Int. de C.N.R.S. 260, Centre national de la recherche scientifique, Paris, 1978, pp. 317—320. Cit. 8.16.

O'Keefe, E. S.: 'Verification of a Conjecture of Th. Skolem', *Math. Scand.* **9** (1961) 80—82. Cit. 7.35—7.37.

228

Olejník, F.: 'Theorems of the Nordhaus—Gaddum Type for k-Uniform Hypergraphs', *Math. Slov.* **31** (1981) 311—318, MR 83d: 05075. Cit. 4.18.

Olejník, F.: 'On Total Matching Numbers for k-Uniform Hypergraphs', *Math. Slov.* **34** (1984) 319—328. Cit. 4.37.

Ollerenshaw, K. and Bondi, H.: 'The Nine Prisoners Problem', *Bull. Inst. Math. Appl.* **14** (1978) 121—143, MR 58: 10529. Cit. 12.36.

Ore, O.: *Theory of Graphs,* Amer. Math. Soc., Providence, 1962, MR 27: 740. Cit. 1.1.

Ottaviani, M.: 'Alcune notevoli relazioni fra parametri relativi ai grafi semplici', *Boll. Un. Mat.Ital.* **5** (1972) 444—452, MR 46: 8914. Cit. 5.10.

Owens, P. J.: 'On Regular Graphs and Hamiltonian Circuits, Including Answers to Some Questions of Joseph Zaks', *J. Combin. Theory* **B28** (1980) 262—277, MR 81j: 05075. Cit. 11.54.

Palmer, E. M.: 'Asymptotic Formulas for the Number of Self-Complementary Graphs and Digraphs', *Mathematika* **17** (1970) 85—90, MR 45: 1803. Cit. 14.5, 14.7.

Palmer, E. M.: 'On the Number of n-Plexes', *Discrete Math.* **6** (1973) 377—390, MR 49: 127. Cit. 14.19.

Palmer, E. M.: Review, 1981, MR 81c: 05044. Cit. 14.5, 14.25.

Palumbíny, D. and Znám, Š.: 'On Decompositions of Complete Graphs into Factors with Given Radii', *Mat. Čas.* **23** (1973) 306—316, MR 49: 4869. Cit. 4.40.

Parker, E. T.: 'A Result in Balanced Incomplete Block Designs', *J. Combin. Theory* **3** (1967) 283—285, MR 35: 5342. Cit. 8.36.

Parthasarathy, K. R. and Sridharan, M. R.: 'Enumeration of Self-Complementary Graphs and Digraphs', *J. Math. Phys. Sci.* **3** (1969) 410—414, MR 41: 8308. Cit. 14.7, 14.8.

Peltesohn, R.: 'Eine Lösung der beiden Heffterschen Differenzenprobleme', *Compositio Math.* **6** (1939) 251—257. Cit. 7.36, 7.37.

Petrenyuk, A. Ya.: 'On a Problem of Ken'i Mano', *Kibernetika* (1970), No. 3, 86—88 (in Russian), MR 44: 5236. Cit. 12.57.

Petrenyuk, L. P. and Petrenyuk, A. Ya.: 'Intersection of Perfect 1-Factorizations of Complete Graphs', *Kibernetika* (1980), No. 1, 6—8 (Engl. transl. *Cybernetics* **16** (1980), No. 1, 6—9, MR 82c: 05076. Cit. 14.7.

Petrenyuk, A. Ya. and Shniter, V. Yu.: 'On the Construction of a Graph of H-Transformations of 1-Factorizations of Order 10', in *φ-Transformations and Combinatorial Properties of Graphs,* AN USSR Inst. Mat., Preprint, 1981, No. 1, pp. 34—48, MR 83c: 05101. Cit. 14.7.

Phelps, K. T.: 'On the Number of Commutative Latin Squares', *Ars Combin.* **10** (1980) 311—322, MR 82c: 05026. Cit. 8.34, 14.7.

Phelps, K. T. and Rosa, A.: 'Steiner Triple Systems with Rational Automorphisms', *Discrete Math.* **33** (1981) 57—66, MR 82a: 05018. Cit. 8.16, 8.35.

Pinto, G.: 'Sui grafi autocomplementari', *Riv. Mat. Univ. Parma* **7** (1981) 397—403, MR 83j: 05073. Cit. 14.17.

Pinto, G.: 'Sui grafi orientati autocomplementari', *Riv. Mat. Univ. Parma* **8** (1982) 107—113. Cit. 14.19.

Pisanski, T., Shawe-Taylor, J. and Mohar, B.: '1-Factorization of the Composition of Regular Graphs', in *Preprint Series of the Department of Mathematics 19*, 1981, Društvo matematikov, fizikov in astronomov SR Slovenije, Yugoslavia, Ljubljana, 1982, pp. 21—30. Cit. 11.25.

Plesník, J.: 'Bounds on Chromatic Numbers of Multiple Factors of a Complete Graph', *J. Graph Theory* **2** (1978) 9—17, MR 80a: 05096. Cit. 4.6, 4.9, 4.10, 4.16, 4.17.

Poljak, S. and Sůra, M.: 'An Algorithm for Graceful Labelling of a Class of Symmetrical Trees', *Ars Combin.* **14** (1982) 57—66, MR 84d: 05150. Cit. 7.21.

Poljak, S. and Sůra, M.: 'On a Construction of Graceful Trees', in *Graphs and Other Combinato-*

rial Topics (Proc. 3rd Czechoslovak Symp. on Graph Theory, Prague, 1982), ed. Fielder, M., Teubner-Texte zur Mathematik 59, Teubner, Leipzig, 1983, pp. 220—222. Cit. 7.21.

Pólya, G.: 'Kombinatorische Anzahlbestimmungen für Gruppen, Graphen und chemische Verbindungen', *Acta Math.* **68** (1937) 145—254. Cit. 14.5.

Porubský, Š.: 'Factorial Regular Representation of Groups in Complete Graphs', *Periodica Math. Hung.* **13** (1982) 9—14. Cit. 13.11, 13.16.

Pukanow, K. and Wilczyńska, K.: 'A Construction of Resolvable Quadruple Systems', *Colloq. Math.* **44** (1981) 359—364, MR 83e: 05025. Cit. 12.45, 12.55.

Quinn, S. J.: 'Factorization of Complete Bipartite Graphs into Two Isomorphic Subgraphs', in *Combinatorial Mathematics VI* (Proc. 6th Australian Conference, Univ. New England, Armidale, 1978), eds. Horadam, A. F. and Wallis, W. D., Lecture Notes in Mathematics 748, Springer-Verlag, Berlin, 1979, pp. 98—111, MR 81j: 05072. Cit. 14.7, 14.19, 14.23.

Quinn, S. J.: 'Isomorphic Factorizations of Complete Equipartite Graphs', *J. Graph Theory* **7** (1983) 285—310. Cit. 15.11,15.12.

Raghavarao, D.: *Construction and Combinatorial Problems in Design of Experiments*, J. Wiley, New York—London—Sydney—Toronto, 1971, MR 51: 2187. Cit. 8.4, 8.40, 8.42, 12.10, 12.49, R 12.66.

Rao, S. B.: 'Graph and Its Complement', *Ind. Nat. Sci. Acad.* **A41** (1975), No. 3, 297—304, MR 58: 27657. Cit. 4.30.

Rao, S. B.: 'Characterization of Self-Complementary Graphs with 2-Factors', *Discrete Math.* **17** (1977a) 225—233, MR 55: 12577. Cit. 14.17.

Rao, S. B.: 'Cycles in Self-Complementary Graphs', *J. Combin. Theory* **B22** (1977b) 1—9, MR 56: 5357. Cit. 14.16.

Rao, S. B.: 'Explored, Semi-Explored and Unexplored Territories in the Structure Theory of Self-Complementary Graphs and Digraphs', in *Proc. Symposium on Graph Theory* (Indian Statist. Inst., Calcutta, 1976), ed. Rao, A. R., ISI Lecture Notes 4, Macmillan of India, New Delhi, 1979a, pp. 10—35, MR 81a: 05036. Cit. 14.16, 14.17.

Rao, S. B.: 'Solution of the Hamiltonian Problem for Self-Complementary Graphs', *J. Combin. Theory* **B** (1979b) 13—41, MR 81c: 05058. Cit. 14.16.

Rao, S. B.: 'The Number of Open Chains of Length Three and the Parity of the Number of Open Chains of Length *k* in Self-Complementary Graphs', *Discrete Math.* **28** (1979c) 291—301, MR 80j: 05081. Cit. 14.16.

Rao, S. B.: 'The Range of the Number of Triangles in Self-Complementary Graphs of Given Order', in *Proc. Symposium on Graph Theory* (Indian Statist. Inst., Calcutta, 1976), ed. Rao, A. R., ISI Lecture Notes 4, Macmillan of India, New Delhi, 1979d, pp. 74—76, MR 81a: 05072. Cit. 14.16.

Rao, S. B.: 'Characterization of Forcibly Self-Complementary Degree Sequences', Preprint. Cit. 14.14.

Ray-Chaudhuri, D. K. and Wilson, R. M.: 'Solution of Kirkman's Schoolgirl Problem', in *Combinatorics* (Proc. Symposium University California, Los Angeles, 1968), ed. Motzkin, T. S., Proc. of Symposia in Pure Mathematics 19, American Math. Soc., Providence, 1971, pp. 187—203, MR 47: 3195. Cit. 12.6, 12.44.

Ray-Chaudhuri, D. K. and Wilson, R. M.: 'The Existence of Resolvable Block Designs', in *A Survey of Combinatorial Theory* (Proc. Int. Symposium on Combinatorial Mathematics and Its Applications, Univ. Colorado, Fort Collins, 1971), eds. Srivastava, J. N., Harary, F., Rao, C. R., Rota G.-C. and Shrikhande, S. S., North-Holland, Amsterdam, 1973, pp. 361—375, MR 50: 12761. Cit. 12.47.

Read, R. C.: 'On the Number of Self-Complementary Graphs and Digraphs', *J. London Math. Soc.* **38** (1963) 99—104, MR 26: 4339. Cit. 14.5, 14.7, 14.19, 14.21, 14.22.

Read, R. C.: 'Some Applications of a Theorem of de Bruijn', in *Graph Theory and Theoretical*

Physics, ed. Harary, F., Academic Press, London—New York, 1967, pp. 273—280, MR 38: 64. Cit. 14.5, 14.7.

Read, R. C.: 'A Survey of Graph Generation Techniques', in *Combinatorial Mathematics VIII* (Proc. 8th Australian Conference on Combinatorial Mathematics, Deakin Univ., Geelong, 1980), ed. McAvaney, K. L., Lecture Notes in Mathematics 884, Springer-Verlag, Berlin —Heidelberg—New York, 1981, pp. 77—89, MR 83c: 68081. Cit. 14.7.

Redfield, J. H.: 'The Theory of Group-Reduced Distributions', *Amer. J. Math.* **49** (1927) 433—455. Cit. 14.5.

Rees, D. H.: 'Some Designs for Use in Serology', *Biometrics* **23** (1967) 779—791. Cit. 11.3.

Reid, K. B. and Thomassen, C.: 'Strongly Self-Complementary and Hereditarily Isomorphic Tournaments', *Monatsh. Math.* **81** (1976) 291—304, MR 55: 2640. Cit. 14.9.

Reid, K. B. and Beineke, L. W.: 'Tournaments', in *Selected Topics in Graph Theory*, eds. Beineke, L. W. and Wilson, R. J., Academic Press, London—New York—San Francisco, 1978, pp. 169—204, MR 81e: 05059. Cit. 14.19.

Reiss, M.: 'Über eine Steinersche combinatorische Aufgabe, welche im 45sten Bande dieses Journals, Seite 181, gestellt worden ist', *J. Reine Angew. Math.* **56** (1859) 326—344. Cit. 7.39.

Ringel, G.: 'Über drei kombinatorische Probleme am *n*-dimensionalen Würfel und Würfel-gitter', *Abh. Math. Sem. Univ. Hamburg* **20** (1955) 10—19, MR 17: 772. Cit. 11.25, 11.31.

Ringel, G.: 'Selbstkomplementäre Graphen', *Arch. Math.* **14** (1963) 354—358, MR 25: 22. Cit. 13.9, 13.20, 14.2—14.4, 14.11, 14.27, 14.28.

Ringel, G.: 'Problem 25', *in Theory of Graphs and Its Applications* (Proc. Symposium, Smolenice, 1963), ed. Fielder, M., Nakladatelstvi ČSAV, Prague, 1964, p. 162. Cit. 7.19.

Robinson, R. W.: 'Asymptotic Number of Self-Converse Oriented Graphs', in *Combinatorial Mathematics* (Proc. Int. Conference on Combinatorial Theory, Australian Nat. Univ., Canberra, 1977), eds. Holton, D. A. and Seberry, J., Lecture Notes in Mathematics 686, Springer-Verlag, Berlin—New York — Australian Academy of Science, Canberra, 1978, pp. 255—266, MR 80m: 05069. Cit. 14.5, 14.25.

Robinson, R. W.: 'Isomorphic Factorizations, VI, Automorphisms', in *Combinatorial Mathematics VI* (Proc. 6th Australian Conference, Univ. New England, Armidale, 1978), eds. Horadam, A. F. and Wallis, W. D., Lecture Notes in Mathematics 748, Springer-Verlag, Berlin, 1979, pp. 127—136. Cit. 13.4, 13.5, 13.12, 13.13, 13.18—13.20 13.22, 13.23, 15.4.

Roditty, Y.: 'Packing and Covering of the Complete Graph with a Graph *G* of Four Vertices or Less', *J. Combin. Theory* **A34** (1983) 231—243, MR 84e: 05087. Cit. 8.16, 9.4.

Rogers, D. G.: 'A Graceful Algorithm', *Southeast Asian Bull. Math.* **2** (1978) 42—44, MR 81i: 05119. Cit. 7.21.

Rokowska, B.: 'On Resolvable Quadruple Systems', *Discuss. Math.* **4** (in press). Cit. 12.45. Wrocław Technical University, No. 19, *Studies and Research* **14** (Prace Nauk Inst. Mat. Politech. Wrocław, Ser. Stud. Materialy No. 14, Analiza Dyskretna), 1980, pp. 3—9, MR 81k: 05017. Cit. 12.49.

Rokowska, B.: 'On Resolvable Quadruple Systems', *Discuss. Math.* **4** (*in press*). Cit. 12.45.

Rosa, A.: 'The Use of Graphs for the Solution of Kirkman's Problem', *Mat.-Fyz. Čas.* **13** (1963a) 105—113 (in Russian), MR 28: 1615. Cit. 12.1, 12.7, 12.8.

Rosa, A.: 'A Problem in the Theory of Graphs', *Mat.-Fyz. Čas.* **13** (1963b) 238—240 (in Russian), MR 32: 2349. Cit. 12.8.

Rosa, A.: 'On the Cyclic Decompositions of Complete Graphs', PhD. Thesis, Bratislava, 1965 (in Slovak). Cit. 7.9, 7.21.

Rosa, A.: 'On the Cyclic Decompositions of the Complete Graph into Polygons with Odd Number of Edges', *Čas. Pěstov. Mat.* **91** (1966a) 53—63 (in Slovak), MR 33: 1250. Cit. 7.1, 10.6.

Rosa, A.: 'On Cyclic Decompositions of the Complete Graph into $(4m + 2)$-gons', *Mat.-Fyz. Čas.* **16** (1966b) 349—353, MR 35: 4120. Cit. 7.1, 10.5, 10.6.

Rosa, A.: 'A Note on Cyclic Steiner Triple Systems', *Mat.-Fyz. Čas.* **16** (1966c) 285—290 (in Slovak), MR 35: 2759. Cit. 7.32, 7.35—7.37.

Rosa, A.: 'On Certain Valuations of the Vertices of a Graph', in *Théorie des graphes — Theory of Graphs* (Journées int. d'étude, Rome, 1966), ed. Rosenstiehl, P., Dunod, Paris — Gordon and Breach, New York, 1967a, pp. 349—355, MR 36: 6319. Cit. 7.8—7.11, 7.15—7.17, 7.19—7.21, 7.23, 7.26, 7.51, 8.15, 9.20.

Rosa, A.: 'On Decompositions of the Complete Graph into $4k$-gons', *Mat. Čas.* **17** (1967b) 242—246, MR 37: 1271. Cit. 10.6.

Rosa, A. and Huang, C.: 'Complete Classification of Solutions to the Problem of 9 Prisoners', in *Proc. 25th Summer Meeting of the Canadian Mathematical Congress* (Lakehead Univ., Thunder Bay, 1971), eds. Eames, W R., Stanton, R. G. and Thomas, R. S. D., Lakehead University, Thunder Bay, 1971, pp. 553—562, MR 49: 92. Cit. 12.36.

Rosa, A. and Huang, C.: 'Another Class of Balanced Graph Designs: Balanced Circuit Designs', *Discrete Math.* **12** (1975) 269—293, MR 54: 7292. Cit. 9.2, 10.3.

Rosa, A.: 'Labeling Snakes', *Ars Combin.* **3** (1977) 67—73, MR 57: 12301. Cit. 7.21.

Rosa, A.: 'On Generalized Room Squares', in *Problèmes combinatoires et théorie des graphes* (Colloquium, Orsay, 1976), eds. Bermond, J.-C., Fournier, J.-C., Las Vergnas, M. and Sotteau, D., Colloq. Int. de C.N.R.S. 260, Centre national de la recherche scientifique, Paris, 1978, pp. 353—358, MR 81h: 05028. Cit. 12.34.

Rosa, A.: 'Intersection Properties of Steiner Systems' in *Topics on Steiner Systems*, eds. Lindner, C. C. and Rosa, A., Annals of Discrete Mathematics 7, North-Holland, Amsterdam, 1980a, pp. 115—128. Cit. 12.8.

Rosa, A.: 'Room Squares Generalized', in *Combinatorics '79*, Part I (Proc. Colloquium Univ. Montreal, Montreal, 1979), eds. Deza, M. and Rosenberg, I. G., Annals of Discrete Mathematics 8—9, North-Holland, Amsterdam, 1980b, pp. 43—57, MR 82b: 05036. Cit. 12.34.

Rosenberg, I. G.: 'Regular and Strongly Regular Self-Complementary Graphs', in *Theory and Practice of Combinatorics,* eds. Rosa, A., Sabidussi, G. and Turgeon, J., North-Holland Mathematics Studies 60, Annals of Discrete Mathematics 12, North-Holland, Amsterdam, 1982, pp. 223—238. Cit. 14.15.

Ruiz, S.: 'On a Family of Self-Complementary Graphs', in *Combinatorics '79*, Part II (Proc. Colloquium Univ. Montreal, Montreal, 1979), eds. Deza, M. and Rosenberg, I. G., Annals of Discrete Mathematics 8—9, North-Holland, Amsterdam, 1980, pp. 267—268, MR 82c: 05054. Cit. 14.17.

Ruiz, S.: 'On Strongly Regular Self-Complementary Graphs', *J. Graph Theory* **5** (1981) 213—215, MR 82i: 05044. Cit. 14.15.

Ruiz, S.: 'Solution to Problem 2', *Graph Theory Newslett.* **13** (1984), No. 3, 10—11. Cit. 14.15, 14.31.

Rumov, B. T.: 'On the Existence of Some Block Designs', *Mat. Zametki* **32** (1982) 869—887 (in Russian), MR 84d: 05037. Cit. 12.49.

Rybnikov, K. A.: *Introduction to Combinatorial Analysis,* Izd. Moskovskogo Universiteta, Moscow, 1972 (in Russian), MR 50: 118. Cit. 8.4, 8.45.

Ryser, H. J.: *Combinatorial Mathematics,* Carus Math. Monograph 14, Mathematical Association of America, 1963, MR 27: 51. Cit. 8.5.

Ryser, H. J.: 'An Extension of a Theorem of de Bruijn and Erdős on Combinatorial Designs', *J. Algebra* **10** (1968) 246—261, MR 37: 5103. Cit. 8.5, 8.57.

Ryser, H. J.: 'Symmetric Designs and Related Configurations', *J. Combin. Theory* **A24** (1972) 98—111, MR 44: 2635. Cit. 8.5.

232

Ryser, H. J.: 'Combinatorial Matrix Theory', in *Studies in Combinatorics*, ed. Rota, G.-C., MAA Studies in Mathematics 17, Mathematical Association of America, 1978, pp. 1—21, MR 58: 21695. Cit. 8.5.

Sachs, H.: 'Über selbstkomplementäre Graphen', *Publ. Math., Debrecen* **9** (1962) 270—288, MR 27: 1934. Cit. 13.9, 13.20, 14.2—14.4, 14.15, 14.26.

Sachs, H.: 'Bemerkung zur Konstruktion zyklischer selbstkomplementärer gerichteter Graphen', *Wiss. Z. Tech. Hochschule Ilmenau* **11** (1965) 161—162, MR 33: 5522. Cit. 14.19.

Sachs, H.: 'Construction of Non-Hamiltonian Planar Regular Graphs of Degrees 3, 4 and 5 with Highest Possible Connectivity', in *Théorie des graphes — Theory of Graphs* (Journées int. d'étude, Rome, 1966), ed. Rosenstiehl, P., Dunod, Paris — Gordon and Breach, New York, 1967, pp. 373—382, MR 36: 3671. Cit. 11.40.

Sachs, H. and Schäuble, M.: 'Konstruktion von Graphen mit gewissen Färbungseigenschaften', in *Beiträge zur Graphentheorie* (Kolloquium, Manebach, 1967), eds. Sachs, H., Voss, H.-J. and Walther, H., Teubner, Leipzig, 1968, pp. 131—135, MR 40: 1304. Cit. 4.23.

Sade, A.: 'Produit direct-singulier de quasigroupes orthogonaux et anti-abelians', *Ann. Soc. Sci. Bruxelles* **74** (1960) 91—99, MR 25: 4017. Cit. 12.30.

Salvi Zagaglia, N.: 'Self-Complementary and Self-Converse Graphs', *Ist. Lombardo Acad. Sci. Lett. Rend.* **A111** (1977) 163—172 (in Italian), MR 58: 21775. Cit. 14.19, 14.25.

Salvi Zagaglia, N.: 'Self-Complementary and Self-Converse Graphs', *Ist. Lombardo Acad. Sci. Lett. Rend.* **A111** (1977) 163—172 (in Italian), MR 58: 21775. Cit. 14.19, 14.25.

Salvi Zagaglia, N.: 'Construction of Self-Converse Oriented Graphs', *Ist. Lombardo Acad. Sci. Sci. Lett. Rend.* **A113** (1979) 426—435 (in Italian), MR 82m: 05051. Cit. 14.19.

Salvi Zagaglia, N.: 'On the Circuits of a Self-Complementary Tournament', *Ist. Lombardo Acad. Converse Tournament', Ist. Lombardo Acad. Sci. Lett. Rend.* **A114** (1980) 169—178 (in Italian), MR 84h: 05059. Cit. 14.19.

Sampathkumar, E. and Walikar, H. B.: 'The Connected Domination Number of a Graph', *J. Math. Phys. Sci.* **13** (1979) 607—613. Cit. 4.37.

Sauer, N.: 'On the Maximal Number of Edges in Graphs with a Given Number of Edge-Disjoint Triangles', *J. London Math. Soc.* **4** (1971), No. 2, 153—156, MR 45: 8567. Cit. 8.39.

Schönheim, J.: 'On Maximal Systems of k-Tuples', *Studia Sci. Math. Hung.* **1** (1966) 363—368, MR 34: 2485, Cit. 8.16.

Schönheim, J.: 'Partition of the Edges of the Directed Complete Graph Into 4-Cycles', *Discrete Math.* **11** (1975) 67—70, MR 51: 251. Cit. 10.12.

Schönheim, J. and Bialostocki, A.: 'Packing and Covering of the Complete Graph with 4-Cycles', *Can. Math. Bull.* **18** (1975) 703—708, MR 53: 10661. Cit. 8.16.

Schönheim, J. and Bialostocki, A.: 'Decomposition of K_n into Subgraphs of Prescribed Type', *Arch. Math.* **31** (1978) 105—112, MR 58: 2746c. Cit. 15.3.

Schrijver, A.: 'Bounds on Permanents, and the Number of 1-Factors and 1-Factorizations of Bipartite Graphs', in *Surveys in Combinatorics*, ed. Lloyd, K., London Math. Soc. Lecture Note Series 82, Cambridge University Press, Cambridge, 1983, pp. 107—134. Cit. 14.7.

Schürger, K.: 'Inequalities for the Chromatic Numbers of Graphs', *J. Combin. Theory* **B16** (1974) 77—85, MR 48: 10875. Cit. 4.18.

Schwenk, A. J.: 'The Number of Color Cyclic Factorizations of a Graph', in *Enumeration and Design*, Academic Press, Toronto, 1984, pp. 297—311, MR 86e: 81071. Cit. 13.4, 13.20, 14.5, 14.7.

Seberry, J.: 'A Class of Group Divisible Designs', *Ars Combin.* **6** (1978) 151—152. Cit. 8.40, 8.42.

Seberry, J. and Skillicorn, D. B.: 'All Directed BIBDs with $k = 3$ Exist', *J. Combin. Theory* **A29** (1980) 244—248, MR 81h: 05026. Cit. 8.37, 10.11.

Sedláček, J.: 'On the Spanning Trees of Finite Graphs', *Čas. Pěstov. Mat.* **91** (1966) 221—227, MR 33: 3961. Cit. 14.17.

233

Seiden, E.: A Method of Construction of Resolvable BIBD', *Sankhyā* **A25** (1963) 393—394, MR 30: 2632. Cit. 8.30.

Severn, E. A.: 'Maximal Partial Steiner Triple Systems', Thesis, University of Toronto, 1984. Cit. 8.16.

Sheehan, J.: 'Unsolved Problem', in *Proc. 5th British Combinatorial Conference* (Aberdeen, 1975), eds. Nash-Williams, C. St. J. A. and Sheehan, J., Congressus Numerantium 15, Utilitas Math., Winnipeg, 1976, pp. 691—692, MR 53: 141, Cit. 11.43.

Sheehan, J.: 'Graphs with Exactly One Hamiltonian Circuit', *J. Graph Theory* (in press). Cit. 11.43.

Sheppard, D. A.: 'The Factorial Representation of Balanced Labelled Graphs', *Discrete Math.* **15** (1976) 379—388, MR 56: 5369. Cit. 7.9, 7.19.

Shrikhande, S. S.: 'The Impossibility of Certain Symmetrical Balanced Incomplete Block Designs', *Ann. Math. Statist.* **212** (1950) 106—111, MR 11: 106. Cit. 8.20.

Shrikhande, S. S.: 'Relations Between Certain Incomplete Block Designs', in *Contributions to Probability and Statistics*, Stanford University Press, Stanford, 1960, pp. 388—395, MR 22: 11465. Cit. 8.51.

Sierpiński, W.: *Elementary Theory of Numbers*, Państwowe wyd. naukowe, Warsaw, 1964, MR 31: 116. Cit. 8.21.

Simonovits, M. and Sós, V. T.: 'Intersection Theorems on Structures', in *Combinatorial Mathematics, Optimal Designs and Their Applications* (Proc. Symposium on Combinatorial Mathematics and Optimal Design, Colorado State Univ., Fort Collins, 1978), ed. Srivastava, J., Annals of Discrete Mathematics 6, North-Holland, Amsterdam, 1980, pp. 301—313, MR 83c: 05072. Cit. 8.5.

Singer, J.: 'A Theorem in Finite Projective Geometry and Some Applications to Number Theory', *Trans. Amer. Math. Soc.* **43** (1938) 377—385. Cit. R 7.56.

Skillicorn, D. B.: 'A Note in Directed Coverings of Pairs by Triples', *Ars Combin.* **11** (1981) 231—234, MR 82k: 05043. Cit. 10.11.

Skillicorn, D. B.: 'A Note on Directed Packings of Pairs into Triples', *Ars Combin.* **13** (1982a) 227—229, MR 83j: 05026. Cit. 10.11.

Skillicorn, D. B.: 'Directed Coverings and Packings of Pair and Quadruples', in *Combinatorial Mathematics IX* (Proc. 9th Australian Conference on Combinatorial Mathematics, Univ. Queensland, Brisbane, 1981), eds. Billington, E. J., Oates-Williams, S. and Street, A. P., Lecture Notes in Mathematics 952, Springer-Verlag, Berlin, 1982b, pp. 387—391, MR 83j: 05025. Cit. 8.38.

Skolem, T.: 'On Certain Distribution of Integers in Pairs with Given Differences', *Math. Scand.* **5** (1957) 57—68, MR 19: 1159. Cit. 7.32, 7.35—7.37.

Skolem, T.: 'Some Remarks on the Triple Systems of Steiner', *Math. Scand.* **6** (1958) 273—280, MR 21: 5582. Cit. 7.32, 7.33, 7.35, 7.36.

Skupień, Z.: 'On Homogeneously Traceable Non-Hamiltonian Digraphs and Oriented Graphs', in *The Theory and Applications of Graphs* (Proc. 4th Int. Conference Western Michigan Univ., Kalamazoo, 1980), eds. Chartrand, G., Alavi, Y., Goldsmith, D. L., Lesniak-Foster, L. and Lick, D. R., J. Wiley, New York, 1981, pp. 517—527, MR 82m: 05052. Cit. 11.12.

Slater, P. J.: 'On k-Sequential and Other Numbered Graphs', *Discrete Math.* **34** (1981) 185—193, MR 82d: 05092. Cit. 7.8, 7.19, 7.21.

Sloane, N. J. A.: 'Hamiltonian Cycles in a Graph of Degree 4', *J. Combin. Theory* **6** (1969) 311—312, MR 38: 5668. Cit. 11.44.

Smith, P.: 'A Doubly Divisible Nearly Kirkman System', *Discrete Math.* **18** (1977) 93—96, MR 57: 5783. Cit. 12.8, 12.34, 12.57.

Sotteau, D.: 'Decomposition of K_n^* into Circuits of Odd Length', *Discrete Math.* **15** (1976) 185—191, MR 55: 7832. Cit. 10.16.

Sotteau, D.: Thesis, Université Paris-Sud, Paris, 1980. Cit. 10.16.

Sotteau, D.: 'Decomposition of $K_{m,n}$ ($K^*_{m,n}$) into Cycles (Circuits) of Length $2k$', *J. Combin. Theory* **B30** (1981) 75—81, MR 82e: 05088. Cit. 10.18, 10.19, 11.18.

Spencer, J.: 'Maximal Consistent Families of Triples', *J. Combin. Theory* **5** (1968) 1—8, MR 37: 2616. Cit. 8.16.

Sprott, D. A.: 'A Series of Symmetrical Group Divisible Designs', *Ann. Math. Statist.* **30** (1959) 249—251, MR 21: 1671. Cit. 8.42.

Sridharan, M. R.: 'Self-Complemetary and Self-Converse Oriented Graphs', Nederl. Akad. Wetensch. Proc. A73, *Indagationes Math.* **32** (1970) 441—447, MR 43: 6138. Cit. 14.19, 14.25.

Sridharan, M. R. and Parthasarathy, K. R.: 'Isographs and Oriented Isographs', *J. Combin. Theory* **B13** (1972) 99—111, MR 46: 7088. Cit. 14.7, 14.19.

Sridharan, M. R.: 'Mixed Self-Complementary and Self-Converse Graphs', *Discrete Math.* **14** (1976) 373—376, MR 54: 2520. Cit. 14.19, 14.25.

Sridharan, M. R.: 'Note on an Asymptotic Formula for a Class of Digraphs', *Can. Math. Bull.* **21** (1978) 377—381, MR 80b: 05037. Cit. 14.5, 14.17, 14.19.

Stanton, R. G. and Gryte, D. G.: 'A Family of BIBDs', in *Combinatorial Structures and Their Applications* (Proc. Conference Univ. of Calgary, Calgary, 1969), eds. Guy, R., Hanani, H., Sauer, N. and Schönheim, J., Gordon and Breach, New York—London—Paris, 1970, pp. 411—412. Cit. 8.27.

Stanton, R. G. and Zarnke, C. R.: 'Labelling of Balanced Trees, in *Proc. 4th Southeastern Conference on Combinatorics, Graph Theory and Computing* (Florida Atlantic Univ., Boca Raton, 1973), eds. Hoffman, F., Levow, R. B. and Thomas, R. S. D., Congressus Numerantium 8, Utilitas Math., Winnipeg, 1973, pp. 479—495, MR 51: 2959. Cit. 7.21.

Stanton, R. G. and Goulden, I. P.: 'Graph Factorization, General Triple Systems, and Cyclic Triple Systems'. *Aeqs. Math.* **22** (1981) 1—28, MR 83h: 05016. Cit. 8.16, 8.35.

Stanton, R. G. and Mullin, R. C.: 'Some Properties of H-Designs', in *Combinatorial Mathematics VIII* (Proc. 8th Australian Conference on Combinatorial Mathematics, Deakin Univ., Geelong, 1980), ed. McAvaney, K. L., Lecture Notes in Mathematics 884, Springer-Verlag, Berlin—Heidelberg—New York, 1981, pp. 1—7, MR 83d: 05026. Cit. 8.43.

Stanton, R. G. and Rogers, M. J.: 'Packings and Coverings by Triples', *Ars Combin.* **13** (1982) 61—69, MR 84i: 05042. Cit. 8.16.

Stanton, R. G., Rogers, M. J., Quinn, R. F. and Cowan, D. D.: 'Bipackings of Pairs into Triples, and Isomorphism Classes of Small Bipackings', *J. Aust. Math. Soc.* A **34** (1983) 214—228. Cit. 8.16.

Steiner, J.: 'Kombinatorische Aufgabe', *J. Reine Angew. Math.* **45** (1853) 181—182. Cit. 7.29.

Stewart, B. M.: 'On a Theorem of Nordhaus and Gaddum', *J. Combin. Theory* **6** (1969) 217—218, MR 43: 104. Cit. 4.13.

Stinson, D. R.: 'Applications and Generalizations of the Variance Method in Combinatorial Designs', *Utilitas Math.* **22** (1982) 323—333, MR 83m: 05027. Cit. 8.1, 8.5.

Stinson, D. R.: 'The Nonexistence of Certain Finite Linear Spaces', *Geom. Dedicata* **13** (1983) 429—434, MR 84g: 05039. Cit. 8.1.

Stinson, D. R.: 'A Short Proof of the Nonexistence of a Pair of Orthogonal Latin Squares of Order Six', *J. Combin. Theory* **A36** (1984) 373—376. Cit. 12.30.

Street, A. P. and Willis, W. D.: *Combinatorial Theory: An Introduction*, Charles Babbage Research Centre, Winnipeg, 1977, MR 56: 2831. Cit. 8.23.

Street, D. J. and Seberry, J.: 'All BIBDs with Block Size Four Exist', *Utilitas Math.* **18** (1980) 27—34, MR 82a: 05019. Cit. 8.37.

Street, D. J. and Wilson, W. H.: 'On Directed Balanced Incomplete Block Designs with Block Size Five', *Utilitas Math.* **18** (1980) 161—174, MR 84h: 05021. Cit. 8.37.

Sumner, D. P.: 'Graphs with 1-Factors', *Proc. Amer. Math. Soc.* **42** (1974) 8—12, MR 48: 2004. Cit. 9.7.

Surányi, L.: 'On a Problem of P. Erdős and A. Hajnal', in *Combinatorial Theory and Its Applications* (Proc. Colloquium, Balatonfüred, 1969), eds. Erdős, P., Rényi, A. and Sós, V. T., Colloquia Mathematica Societatis János Bolyai 4, North-Holland, Amsterdam — Bolyai János matematikai társulat, Budapest, 1970, pp. 1029—1041, MR 45: 6669. Cit. 8.39.

Sylvester, J. J.: 'Note on a Nine Schoolgirls Problem', *Messenger Math.* **22** (1892/93) 159—160, Letters', *London, Edinburgh and Dublin Philos. Mag. and J. Sci.* **21** (1861) 369—377. Cit. 12.33.

Sylvester, J. J.: 'Note on a Nine Schoolgirls Problem', *Messenger Math.* **22** (1892—93) 159—160, Correction: 192. Cit. 12.33.

Szemerédi, E.: 'On a Problem of R. Erdős', in *Combinatorial Theory and Its Applications* (Proc. Colloquium, Balatonfüred, 1969), eds. Erdős, P., Rényi, A. and Sós, V. T., Colloquia Mathematica Societatis János Bolyai 4, North-Holland, Amsterdam — Bolyai János matematikai társulat, Budapest, 1970, pp. 1051—1053, MR 45: 6670. Cit. 8.39.

Tarry, G.: 'Le problème des 36 officiers', *C. R. Ass. Franc. Acad. Sci.* **1** (1900) 122—123; **2**, (1901) 170—203. Cit. 12.30.

Tarsi, M.: 'Decomposition of Complete Multigraphs into Stars', *Discrete Math.* **26** (1979) 273—278, MR 81b: 05028. Cit. 9.21, 9.23.

Tarsi, M.: 'On the Decomposition of a Graph into Stars', *Discrete Math.* **36** (1981) 299—304, MR 84c: 05036. Cit. 9.20, 9.23.

Tarsi, M.: 'Decomposition of a Complete Multigraph into Simple Paths: Nonbalanced Handcuffed Designs', *J. Combin. Theory* **A34** (1983) 60—70, MR 84i: 05037. Cit. 9.10.

Tazawa, S., Ushio, K. and Yamamoto, S.: 'Partite-Claw-Decomposition of a Complete Mutlipartite Graph', *Hiroshima Math. J.* **8** (1978) 195—206, MR 80i: 05033a. Cit. 9.22.

Tazawa, S.: 'Claw-Decompositions and Evenly-Partite-Claw-Decomposition of Complete Multipartite Graphs', *Hiroshima Math. J.* **9** (1979) 503—531, MR 81b: 05100. Cit. 9.22.

Teichert H.-M.: 'Über einige Eigenschaften spezialer *B*-Produkte von Graphen', in *Das 27. Internationale Wissenschaftliche Kolloquium* (Ilmenau, 1982), Heft 6, Vortragsr. B1, B2, Ilmenau, 1982, pp. 37—39. Cit. 11.25.

Teichert, H.-M.: 'Hamiltonian Properties of the Lexicographic Product of Undirected Graphs', *Elektron. Informationsverarb. Kybernet.* **19** (1983a) 67—77. Cit. 11.38.

Teichert, H.-M.: 'On the Pancyclicity of Some Product Graphs', *Elektron. Informationsverarb. Kybernet.* **19** (1983b) 345—356. Cit. 11.25.

Thomason, A. G.: 'Hamiltonian Cycles and Uniquely Edge Colourable Graphs', in *Advances in Graph Theory* (Cambridge Combinatorial Conference, Trinity College, Cambridge, 1977), ed. Bollobás, B., Annals of Discrete Mathematics 3, North-Holland, Amsterdam, 1978, pp. 259—268, MR 80e: 05077. Cit. 11.44, 11.46, 11.47, 11.49, 11.50.

Thomassen, C.: 'Long Cycles in Digraphs with Constraints on the Degrees', in *Surveys in Combinatorics* (Proc. 7th British Combinatorial Conference, Cambridge, 1979), ed. Bollobás, B., London Math. Soc. Lecture Note Series 38, Cambridge University Press, Cambridge, 1979, pp. 211—228, MR 81c: 05045. Cit. 11.13, R 7.62.

Thomassen, C.: 'Hamiltonian-Connected Tournaments', *J. Combin. Theory* **B28** (1980) 142—163, MR 82d: 05065. Cit. 11.14.

Thomassen, C.: 'Long Cycles in Digraphs', *Proc. London Math. Soc.* **42** (1981) 231—251, MR 83f: 05043. Cit. 11.13, 11.62.

Thomassen, C.: 'Edge-Disjoint Hamiltonian Paths and Cycles in Tournaments', *Proc. London Math. Soc.* **45** (1982) 151—168. Cit. 11.14, 11.54.

Tillson, T. W.: 'A Hamiltonian Decomposition of K^*_{2m}, $2m \geq 8$', *J. Combin. Theory* **B29** (1980) 68—74, MR 82e: 05075. Cit. 10.16, 11.5, 11.7, 11.10.

236

Todorov, D. T. and Tonchev, V. D.: 'On Some Coverings of Triples' *Dokl. Bolg. AN* **35** (1982) 1209—1211, MR 84c: 05033. Cit. 8.33.

Tomasta, P.: 'Decompositions of Graphs and Hypergraphs into Isomorphic Factors with a Given Diameter', *Czechoslov. Math. J.* **27** (1977) 598—608, MR 57: 12304. Cit. 13.20.

Tomasta, P. and Zelinka, B.: 'Decomposition of a Complete Equipartite Graph into Isomorphic Subgraphs', *Math. Slov.* **31** (1981) 165—169, MR 82e: 05110. Cit. 8.39, 8.42, 12.27, 15.11.

Tran van Trung: 'The Existence of Symmetric Block Designs with Parameters (41, 16, 6) and (66, 26, 10)', *J. Combin. Theory* **A33** (1982) 201—204, MR 83m: 05029. Cit. 8.36.

Treash, C. A: 'The Completion of Finite Incomplete Steiner Triple Systems with Applications to Loop Theory', *J. Combin. Theory* **A10** (1971) 259—265, MR 43: 397. Cit. 8.16.

Truszczyński, M.: 'Decompositions of Complete Multipartite Graphs into Isomorphic Stars', in *Graphs and Other Combinatorial Topics* (Proc. 3rd Czechoslovak Symposium on Graph Theory, Prague, 1982), ed. Fielder, M., Teubner-Texte zur Mathematik 59, Teubner, Leipzig, 1983, pp. 333—343. Cit. 9.22.

Turner, J. M.: 'Point-Symmetric Graphs with a Prime Number of Points', *J. Combin. Theory* **3** (1967) 136—145, MR 35: 2783. Cit. 11.52.

Tutte, W. T.: 'On Hamiltonian Circuits', *J. London Math. Soc.* **21** (1946) 98—101, MR 8, 397. Cit. 11.43.

Tutte, W.: 'A Theorem on Planar Graphs', *Trans. Amer. Math. Soc.* **82** (1956) 99—116, MR 18, 408. Cit. 11.42.

Ushio, K., Tazawa, S. and Yamamoto, S.: 'On Claw-Decomposition of a Complete Multipartite Graph', *Hiroshima Math. J.* **8** (1978) 207—210, MR 80i: 05033b. Cit. 9.22.

Ushio, K.: 'On Balanced Claw-Decomposition of a Complete Multipartite Graph', *Memoires of Niihama Technical College* **16** (1980) 29—33. Cit. 9.25.

Ushio, K.: 'Bipartite Decomposition of Complete Multipartite Graphs', *Hiroshima Math. J.* **11** (1981) 321—345, MR 82m: 05076. Cit. 9.20, 9.22.

Ushio, K.: 'On Balanced Claw Designs of Complete Multipartite Graphs', *Discrete Math.* **38** (1982) 117—119, MR 84e: 05041. Cit. 9.25, 9.26.

Vanstone, S. A. and Schellenberg, P. J.: 'A Construction for BIBDs Based on an Intersection Property', *Utilitas Math.* **11** (1977) 313—324, MR 55: 10289. Cit. 12.44.

Vanstone, S. A.: 'Resolvable r, λ-Designs and the Fisher Inequality', *J. Aust. Math. Soc.* **A28** (1979) 471—478, MR 82c: 05022. Cit. 8.5, 12.49, R 12.66.

Vanstone, S. A. 'Doubly Resolvable Designs', *Discrete Math.* **29** (1980) 77—86, MR 81f: 05030. Cit. 8.40, 12.2, 12.8, 12.34, 12.57.

Vasiliev, Yu. L.: 'Vector Fields on the Vertex-Set of the n-Dimensional Unit Cube', *Diskret. Analiz* **11** (1967) 21—59 (in Russian), MR 37: 3553. Cit. 10.17.

Venkatachalam, C. V.: 'Self-Complementary Graphs on Eight Points', *Math. Educat.* **10** (1976), No. 2, A43—A44, MR 54: 10070. Cit. 14.6.

Vizing, V. G.: 'The Cartesian Product of Graphs', in *Computing Systems (Novosibirsk)* **9** (1963) 30—43 (in Russian), MR 35: 81. Cit. 11.25.

Vizing, V. G.: 'The Chromatic Class of a Multigraph', *Kibernetika* (1965), No. 3, 29—39 (in Russian); *Cybernetics* **1** (1965), No. 3, 32—41 (English transl.), MR 32: 7333. Cit. 4.26.

Walikar, H. B.: 'Isomorphic Factorization: Tripartite Conjecture', *Graph Theory Newslett.* **9** (1980), No. 4, 3. Cit. 15.8.

Wallis, W. D., Street, A. P. and Wallis, J. S.: *Combinatorics: Room Squares, Sum-Free Sets, Hadamard Matrices*, Lecture Notes in Mathematics 292, Springer-Verlag, Berlin, 1972, MR 52: 13397. Cit. 13.4, 14.7.

Wallis, W. D.: 'Which Isomorphic Factorizations of Regular Graphs Are Block Designs?', *J. Comb. Inf. Syst. Sci.* **2** (1977) 104—106, MR 56: 15503. Cit. 8.39.

Wallis, W. D.: 'One-Factorizations of Wreath Products', in *Combinatorial Mathematics VIII*

(Proc. 8th Australian Conference on Combinatorial Mathematics, Deakin Univ., Geelong, 1980), ed. McAvaney, K. L., Lecture Notes in Mathematics 884, Springer-Verlag, Berlin—Heidelberg—New York, 1981, pp. 337—345, MR 83d: 05078. Cit. 11.25.

Wang-Jiangfang: 'Isomorphic Factorizations of Complete Equipartite Graphs — the Proof of the Harary—Robinson—Wormald Conjecture', *Sci. Sinica* **A25** (1982) 1152—1164. Cit. 15.11.

Wang-Jiangfang: 'Isomorphic Factorizations of Multipartite Directed Graphs — the Proof of Conjecture of Harary, Robinson and Wormald in Isomorphic Factorizations of Directed Graphs', *Sci. Sinica* **A26** (1983) 1167—1177. Cit. 15.12.

Wang, L. L.: 'A Test for Sequencing of a Class of Finite Groups with Two Generators', *Notices Amer. Math. Soc.* **20** (1973) A632. Cit. 11.11.

White, A. T.: 'A Note on Conservative Graphs', *J. Graph Theory* **4** (1980) 423—425, MR 83d: 05040. Cit. 7.19.

White H. S., Cole, F. N. and Cummings, L. D.: 'Complete Classification of the Triad Systems on Fifteen Elements', *Mem. Nat. Acad. Sci. U.S.A.* **14**, (1919) 1—89. Cit. 8.34.

Wilczyńska, K.: On Nonisomorphic Resolvable Quadruple Systems', *Discussiones Math.* **5** (1982) 3—8. Cit. 12.45, 12.49.

Wille, D.: 'On the Enumeration of Self-Complementary *m*-Placed Relations', *Discrete Math.* **10** (1974) 189—192, MR 50: 1903. Cit. 14.7, 14.19, 14.21.

Wille, D.: 'Enumeration of Self-Complementary Structures', *J. Combin. Theory* **B25** (1978) 143—150, MR 80e: 05070. Cit. 14.7, 14.19, 14.21.

Wilson, R. J.: *Introduction to Graph Theory*, Oliver and Boyd, Edinburgh, 1972, MR 50: 9643. Cit. 1.1.

Wilson, R. M.: 'An Existence Theory For Pairwise Balanced Designs, I—II' (I. Composition Theorems and Morphisms, II. The Structure of PBD-Closed Sets and the Existence Conjectures), *J. Combin. Theory* **A13** (1972a) 220—245, 246—273, MR 46: 3338. Cit. 8.12, 8.40, 8.43, 12.28.

Wilson, R. M.: 'Cyclotomy and Difference Families in Elementary Abelian Groups', *J. Number Theory* **4** (1972b) 17—47, MR 46: 8860. Cit. 8.27.

Wilson, R. M.: 'Construction and Uses of Pairwise Balanced Designs' in *Combinatorics, Part 1, Theory of Designs, Finite Geometry and Coding Theory* (Proc. NATO Advanced Study Inst. Comb., Breukelen, 1974), eds. Hall, M. and van Lint, J. M., Math. Centre Tracts 56, Mathematisch Centrum, Amsterdam — Reidel, Dordrecht, 1974a, pp. 18—41, MR 51: 194. Cit. 6.11.

Wilson, R. M.: 'Nonisomorphic Steiner Triple Systems', *Math. Z.* **135** (1974b) 303—313, MR 49: 4803. Cit. 8.34.

Wilson, R. M.: 'An Existence Theory of Pairwise Balanced Designs, III, Proof of the Existence Conjecture', *J. Combin. Theory* **A18** (1975) 71—79, MR 51: 2942. Cit. 6.11, 8.9, 8.12, 8.36, 8.40, 8.43.

Wilson R. M.: 'Decomposition of Complete Graphs into Subgraphs Isomorphic to a Given Graph', in *Proc. 5th British Combinatorial Conference* (Aberdeen, 1975), eds. Nash-Williams, C. St. J. A. and Sheehan, J., Congressus Numerantium 15, Utilitas Math., Winnipeg, 1976, pp. 647—659, MR 53: 214. Cit. 6.5, 6.8, 6.9, 6.11, 6.16.

Wilson, R. M.: 'Edge Decompositions of Colored Graphs', in *Higher Combinatorics* (Proc. NATO Advanced Study Inst. Comb., Berlin-West, 1976), ed. Aigner, M., NATO Advanced Study Inst. Ser. C 31, Reidel, Dordrecht, 1977, pp. 201—202. Cit. 6.18, 8.43.

Witte, D., Letzter, G. and Gallian, J. A.: 'On Hamiltonian Circuits in Cartesian Products of Cayley Digraphs' *Dicrete Math.* **43** (1983) 297—307, MR 84b: 05054. Cit. 11.25.

Woodall, D. R.: 'Square *λ*-Linked Designs', *Proc. London Math. Soc.* **20** (1970) 669—687, MR 41: 8264. Cit. 8.1, 8.4, 8.5, 8.57.

238

Woodall, D. R.: 'The Inequality $b \geq v$', in *Proc. 5th British Combinatorial Conference* (Aberdeen, 1975), eds. Nash-Williams, C. St. J. A. and Sheehan, J., Congressus Numerantium 15, Utilitas Math., Winnipeg, 1976, pp. 661—664, MR 52: 13444. Cit. 8.4, 8.5.

Yamamoto, S., Ikeda, H., Shige-eda, S., Ushio, K. and Hamada, N.: 'Design of a New Balanced File Organization Scheme with the Least Redundancy', *Inf. Control* **28** (1975a) 156—175, MR 51: 7462. Cit. 9.20.

Yamamoto, S., Ikeda, H., Shige-eda, S., Ushio, K. and Hamada, N.: 'On Claw-Decomposition of Complete Graphs and Complete Bigraphs', *Hiroshima Math. J.* **5** (1975b) 33—42, MR 52: 205. Cit. 9.22—9.24.

Yamamoto, S. and Tazawa, S.: 'Hyperclaw Decomposition of Complete Hypergraphs', in *Combinatorial Mathematics, Optimal Designs and Their Applications* (Proc. Symposium on Combinatorial Mathematics and Optimal Design, Colorado State Univ., Fort Collins, 1978), ed. Srivastava, J., Annals of Discrete Mathematics 6, North-Holland, Amsterdam, 1980, pp. 385—391, MR 82b: 05101. Cit. 9.20.

Yang Shi-hui: 'Some Sufficient Conditions for Complete Bipartite Graph $K(A, B, C)$ to Have Isomorphic Factorization', *Acta Math. Appl. Sin.* **6** (1983) 393—405 (in Chinese). Cit. 15.9.

Yavorskii, E. B.: 'Representations of Directed Graphs and φ-Transformations', in *Theoretical and Applied Questions of Differential Equations and Algebra*, ed. Sharkovskii, A. N., Naukova dumka, Kiev, 1978, pp. 247—250 (in Russian), MR 81i: 05060. Cit. 9.6, 9.7.

Zaks, J.: 'Pairs of Hamiltonian Circuits in 5-Connected Planar Graphs', *J. Combin. Theory* **B21** (1976) 116—131, MR 55: 180. Cit. 11.54.

Zaks, J.: 'Non-Hamiltonian Simple Planar Graphs', in *Theory and Practice of Combinatorics*, eds. Rosa, A., Sabidussi, G. and Turgeon, J., North-Holland Mathematics Studies 60, Annals of Discrete Mathematics 12, North-Holland, Amsterdam, 1982, pp. 255—263. Cit. 11.42.

Zaretskii, K. A.: 'On Hamiltonian Cycles and Hamiltonian Paths in the Cartesian Product of Two Graphs', *Kibernetika* (1966), No. 5, 4—11 (in Russian), MR 35: 2784. Cit. 11.23, 11.25.

Zelinka, B.: 'Decomposition of a Graph into Isomorphic Subgraphs', *Čas. Pěstov. Mat.* **90** (1965) 147—152 (in Czech), MR 33: 1253. Cit. 13.20, 14.24, 15.3.

Zelinka, B.: 'Decomposition of the Complete Graphs According to a Given Group', *Mat. Čas.* **17** (1967) 234—239 (in Czech), MR 36: 5030. Cit. 13.4, 13.11, 13.14, 13.17.

Zelinka, B.: 'The Decomposition of a Generalized Graph into Isomorphic Subgraphs', *Čas. Pěstov. Mat.* **93** (1968) 278—283, MR 43: 94. Cit. 12.35, 13.20, 14.29.

Zelinka, B.: 'Decomposition of a Digraph into Isomorphic Subgraphs', *Mat. Čas.* **20** (1970a) 92—100, MR 46: 5163. Cit. 13.20, 14.19, 14.20, 14.24.

Zelinka, B.: 'On (H, K)-Decompositions of a Complete Graph', *Mat. Čas.* **20** (1970b) 116—121, MR 46: 5183. Cit. 13.4, 13.20.

Zelinka, B.: 'Decompositions of Simplex-Like Graphs and Generalized Bipartite Graphs', *Čas. Pěstov. Mat.* **95** (1970c) 1—6 (in Czech), MR 42: 1708. Cit. 13.20, 14.19, 14.24.

Zelinka, B.: 'The Decomposition of a General Graph According to a Given Abelian Group', *Mat. Čas.* **20** (1970d) 281—292, MR 46: 5166. Cit. 13.11, 13.20.

Zelinka, B.: 'The Decomposition of a Digraph into Isotopic Subgraphs', *Mat. Čas.* **21** (1971) 221—226, MR 46: 3365. Cit. 13.20.

Zelinka, B.: 'Quasigroups and Factorization of Complete Digraphs', *Mat. Čas.* **23** (1973) 333—341, MR 50: 12799. Cit. 13.20.

Zelinka, B.: 'Zweiseitig unendliche Züge in lokalendlichen Graphen', *Čas. Pěstov. Mat.* **99** (1974) 386—393, MR 58: 21824. Cit. 10.20.

Zelinka, B.: 'Self-Complementary Vertex-Transitive Undirected Graphs', *Math. Slov.* **29** (1979) 91—95, MR 81h: 05116. Cit. 14.15.

Zykov, A. A.: 'On Some Properties of Linear Complexes', *Mat. Sbor.* **24** (1949) 163—188 (in Russian), MR 11: 73. Cit. 4.6.

SUBJECT INDEX

244

248

— of *n*-matrices, orthogonal 12.23
— of triangles, cyclic Steiner 7.29
— —, Kirkman 12.2
— —, Mendelsohn 8.35
— —, Steiner 7.29, 8.32
—, nearly Kirkman 12.57
—, Steiner 8.32, 8.35

T

t-uniform hypergraph 1.51, 8.33
tensor product of graphs 11.23
total achromatic number 4.4, 4,5
— chromatic number 4.2, 4.3, 4.5, 4.27, 4.46
— colouring 2.3
— graph 1.19
— multicolouring 3.4
— pseudoachromatic number 4.4, 4.5
tournament 1.21
—, acyclic 1.21
—, trasitive 1.21
trail 1.44
transitive graph 1.45
— orientation 1.45
— tournament 1.21
tree 1.49
triangle 1.47
tripartite conjecture 15.8
trivial graph 1.13
t-(*v*, *k*, λ)-design 8.32
two-way infinite path 1.44, 10.20, 11.25
— — semiwalk 1.37

U

undirected edge 1.5
— graph 1.5
— semiwalk 1.35
uniform hypergraph 1.51, 8.33

union of subgraphs 1.32
— —, disjoint 1.32
— —, edge-disjoint 1.32

V

(*v k*, λ) difference set 8.20
v-gon 1.47
valency 1.9, 1.11
value of edge 7.8
vertex 1.2, 1.51
— arboricity 4.19
— automorphism group of graph 1.30
— — of graph 1.29
— isomorphism of graphs 1.29
— *n*-connected graph 4.28
— *n*-partition number 4.19—4.22
—, central 5.3, 5.4
—, initial 1.5, 1.35
—, isolated 1.5
—, null 1.5, 1.9
—, pendant 1.9
—, peripheral 5.3, 5.4
—, terminal 1.5, 1.35
vertex-colouring 2.3
vertex-connectivity 4.28
vertex-covering number 4.31
vertex-cut 1.40
vertex-independence number 4.31
vertex-labelling 7.8
vertex-multicolouring 3.4
vertex-set 1.2
vertex-transitive graph 11.52, 14.15, 15.13
vertices, adjacent 1.5
—, independent 4.31
(*v*, *k*, λ)-configuration 8.2, 8.5
(*v*, *k*, λ)-design 8.1, 8.2, 8.40

W

wheel 7.18

DECOMPOSITIONS
OF GRAPHS

RNDr. Juraj Bosák, DrSc.

Zo slovenského originálu Rozklady grafov, ktorý vyšiel vo Vede v Bratislave roku 1986, preložili doc. RNDr. Jozef Širán, CSc. a RNDr. Martin Škoviera, CSc.

Obálku navrhol Peter Šilhan
Zodpovedné redaktorky Zuzana Malíková a Oľga Silnická
Výtvarná redaktorka Viera Miková
Technická redaktorka Marcela Janálová

Rukopis zadaný do tlačiarne 30. 6. 1989

Prvé vydanie v angličtine. Vydala Veda, vydavateľstvo Slovenskej akadémie vied, v koedícii s vydavateľstvom Kluwer Academic Publishers, Dordrecht, Holandsko, v Bratislave roku 1990 ako svoju 2898. publikáciu. Počet strán XVIII + 248. AH 17,92 (text 17,49, ilustr. 0,43), VH 18,49. Náklad 680 výtlačkov.
Vytlačila Slovenská polygrafia, š. p., Západoslovenské tlačiarne, z. p., závod Svornosť, Bratislava.

ISBN 80-224-0083-1
03   Kčs 37,—